Aquaculture Management Techniques

Aquaculture Management Techniques

Nasreen Akhter

RANDOM

RANDOM PUBLICATIONS
NEW DELHI (INDIA)

Aquaculture Management Techniques

ISBN 978-93-5111-586-1

Published in 2015 in India by

RANDOM PUBLICATIONS

4376-A/4B, Gali Murari Lal, Ansari Road
New Delhi-110 002
Phone : +9111-43580356, 011-23289044, 011-43142548
e-mail: sales@randompublications.com,
info@randompublications.com, randomexports@gmail.com
Reprinted 2019

Type Setting by : Friends Media, Delhi-110089
Digitally Printed at Replika Press Pvt. Ltd.

Preface

Aquaculture is the farming of aquatic organisms such as fish, crustaceans, molluscs and aquatic plants. Aquaculture involves cultivating freshwater and saltwater populations under controlled conditions, and can be contrasted with commercial fishing, which is the harvesting of wild fish. Broadly speaking, finfish and shellfish fisheries can be conceptualized as akin to hunting and gathering while aquaculture is akin to agriculture. Mariculture refers to aquaculture practiced in marine environments and in underwater habitats. According to the FAO, aquaculture "is understood to mean the farming of aquatic organisms including fish, molluscs, crustaceans and aquatic plants. Farming implies some form of intervention in the rearing process to enhance production, such as regular stocking, feeding, protection from predators, etc.

Farming also implies individual or corporate ownership of the stock being cultivated." The reported output from global aquaculture operations would supply one half of the fish and shellfish that is directly consumed by humans; however, there are issues about the reliability of the reported figures. Further, in current aquaculture practice, products from several pounds of wild fish are used to produce one pound of a piscivorous fish like salmon. Particular kinds of aquaculture include fish farming, shrimp farming, oyster farming, mariculture, algaculture, and the cultivation of ornamental fish. Particular methods include aquaponics and integrated multi-trophic aquaculture, both of which integrate fish farming and plant farming.

I would like to thank my team for standing beside me throughout my career and writing this book. My special thanks go to "Random Publications" who have published the book.

– Nasreen Akhter

Contents

1

Key Issues in the Safe use of Wastewater and Excreta in Aquaculture

WASTEWATER IN AQUACULTURE

SOIL-AQUIFER TREATMENT (SAT)

Where soil and groundwater conditions are favourable for artificial recharge of groundwater through infiltration basins, a high degree of upgrading can be achieved by allowing partially-treated sewage effluent to infiltrate into the soil and move down to the groundwater. The unsaturated or "vadose" zone then acts as a natural filter and can remove essentially all suspended solids, biodegradable materials, bacteria, viruses, and other microorganisms. Significant reductions in nitrogen, phosphorus, and heavy metals concentrations can also be achieved.After the sewage, treated in passage through the vadose zone, has reached the groundwater it is usually allowed to flow some distance through the aquifer before it is collected.

This additional movement through the aquifer can produce further purification (removal of microorganisms, precipitation of phosphates, adsorption of synthetic organics, etc.) of the sewage. Since the soil and aquifer are used as natural treatment, systems are called soil-aquifer treatment systems or SAT systems. Soil-aquifer treatment is, essentially, a low-technology, advanced wastewater treatment system. It also has an aesthetic advantage over conventionally treated sewage in that water recovered from an SAT system is not only clear and odour-free but it comes from a well, drain, or via natural drainage to a stream or low area, rather than from a sewer or sewage treatment plant. Thus, the water has lost its connotation of sewage and the public see it water more as coming out of the ground (groundwater) than as sewage effluent. This could be an important factor in the public acceptance of sewage reuse schemes.

SAT SYSTEM LAYOUTS

Various types of SAT system, the simplest being where the sewage effluent is applied to infiltration basins on high ground from where it moves down to the groundwater and eventually drains naturally through an aquifer into a lower area. This lower area can be a natural depression or seepage area, a stream or lake, or a surface drain. SAT systems as in Figure also serve to reduce the pollution of surface waters. Instead of discharging wastewater directly into streams or lakes, it is applied to infiltration basins at a higher elevation so that it receives soil-aquifer treatment before entering the stream or lake.

Where the groundwater is too deep to collect the renovated sewage water by gravity, pumped wells must be used and there are two basic layouts. In one, the infiltration basins are arranged in two parallel strips and the wells are located on the line midway between the two strips. In the other, the infiltration basins are located close together in a cluster and the wells are on a circle around this cluster. The wells pump essentially all renovated sewage water and no native groundwater from the aquifer outside the SAT system. A mixture of renovated sewage water and native groundwater. Systems 11C and 11D can be used both for seasonal underground storage of sewage water, allowing the groundwater mound to rise during periods of low irrigation water demand (winter), and for pumping the groundwater mound down in periods of high irrigation water demands (summer). The type of SAT system would be suitable for small systems where there are only a few basins around a centrally located well.

WASTEWATER REUSE FOR AQUACULTURE

Wastewater reuse for aquaculture has been practised in many countries for a considerable period of time. It has the potential of wider application in the tropics. There is great diversity of systems involving cultivation of aquatic species, (mainly fish) and plants (mainly aquatic vegetables such as water spinach).Farmers and local communities have developed most reuse systems; the primary motivating factor has been reuse of nutrients for food production rather than wastewater treatment, and with scant attention to either waste treatment or to public health. In most aquaculture systems, wastewater is not reused directly in aquaculture and the nutrients contained in the wastewater are used as fertilizer to produce natural food such as plankton for fish. These nutrients, mainly nitrogen and phosphorus, are also taken up directly by large aquatic plants such as duckweed which is cultivated for animal feed, and aquatic vegetables such as water spinach and water mimosa cultivated for human food.As wastewater provides a source of nutrients for aquaculture, it is technically feasible to link it up with most sanitation technologies, providing that land is available at reasonable cost. Farmers have learned by experience how to culture fish, first in static-water nightsoil-fed ponds and more recently in conventional wastewater-fed fishponds. Research has provided a scientific basis for the key parameters in wastewater-fed aquaculture practice developed earlier by farmers and these can be found in the Source Book, published by IWA and IETC.

BIOTA IN AQUACULTURE PONDS

FOOD CHAINS

The objective in fertilizing an aquaculture pond with excreta, nightsoil or wastewater is to produce natural food for fish. Since several species of fish feed directly on faecal solids, use of raw sewage or fresh nightsoil as influent to fish ponds should be prohibited for health reasons.

Edwards has represented the complex food chains in an excreta-fed fish pond, involving ultimate decomposers or bacteria, phytoplankton, zooplankton and invertebrate detritivores. Inorganic nutrients released in the bacterial degredation of organic solids in sewage, nightsoil or excreta are taken up by phytoplankton. Zooplankton graze phytoplankton and small detritus particles coated with bacteria, the latter also serving as food for benthic invertebrate detritivores. Plankton, particularly phytoplankaton, are the major sources of natural food in a fish pond but benthic invertebrates, mainly chironomids, also serve as fish food, although they are quantitatively less important. To optimise fish production in a human waste fed pond, the majority of the fish should be filter feeders, to exploit the plankton growth.

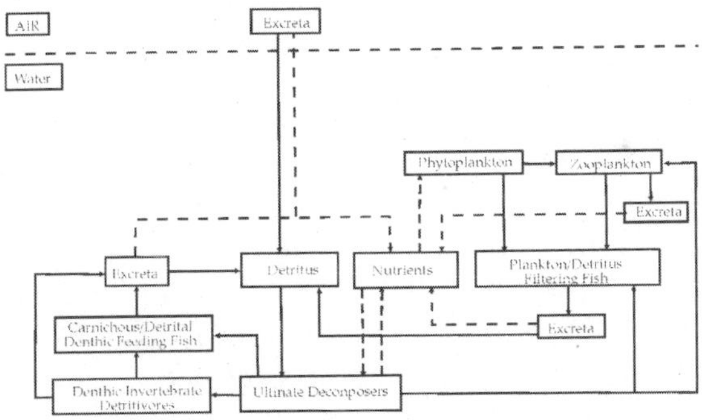

Fig. Food chains in an excreta-fed aquaculture system

A wide range of fish species has been cultivated in aquaculture ponds receiving human waste, including common carp *(Cyprinùs carpio),* Indian major carps (*Catla catlax, Cirrhina mrigala and Labeo rohita),* Chinese silver carp *(Hypophthalmichthys molitrix),* bighead carp *(Aristichthys nobilis),* grass carp *(Ctenopharyngodon idella),* crucian carp (*Carassius auratus),* Nile carp *(Osteochilus hasseltii),* tilapia *(Oreochromis spp.),* milkfish *(Chanos chanos),* catfish *(Pangasius spp.),* kissing gouramy *(Helostoma temmincki),* giant gourami *(Osphronemus goramy),* silver barb *(Puntius gonionotus)* and freshwater prawn *(Macrobrachium lanchesterii).*

FISH SPECIES

The selection reflects local culture rather than fish optimally-suited to such environments. For example, Chinese carps and Indian major carps are the major species in excreta-fed systems in China and India, respectively. In some countries, a polyculture of several fish species is used.

Tilapia are generally cultured to a lesser extent than carps in excreta-fed systems although, technically, they are more suitable for this environment because they are better able to tolerate adverse environmental conditions than carp species. Milkfish have been found to have poorer growth and survival statistics compared with Indian major carps and Chinese carps in ponds fed with stabilisation pond effluent in India.

Edwards gives a thorough review of current knowledge on the various fish species which can be cultured in ponds fed with human waste.

It would appear that considerable confusion still exists with regard to fish feeding on natural food. Although fish are generally divided into types according to their natural nutritional habits - those that feed on phytoplankton, or zooplankton or benthic animals - several species are known to feed on whatever particles are suspended in the water.

There is also uncertainty about the types of phytoplankton fed upon by filter-feeding fish. For example, although blue-green algae are thought to be indigestible to fish, Tilapia have been shown to readily digest these algae and there is evidence that silver carp can do the same.

AQUATIC PLANTS

Aquatic macrophytes grow readily in ponds fed with human waste and their use in wastewater treatment has been discussed. Some creeping aquatic macrophytes are cultivated as vegetables for human consumption in aquaculture ponds and duckweeds are also cultivated, mainly for fish feed.

Among the aquatic plants grown for use as vegetables are water spinach *(Ipomoea aquatica),* water mimosa (*Neptunia oleracea),* water cress *(Rorippa nasturtium-aquaticum)* and Chinese water chestnut *(Eleocharis dulcis)*. The duckweeds *Lemna, Spirodela* and *Wolffia* are cultivated in some parts of Asia in shallow ponds fertilized with excreta, mainly as feed for Chinese carps but also for chickens, ducks and edible snails.

TECHNICAL ASPECTS OF FISH CULTURE

ENVIRONMENTAL FACTORS

In a successful aquaculture system there must be both an organismic balance, to produce an optimal supply of natural food at all levels, and a chemical balance, to ensure sufficient oxygen supply for the growth of fish and their natural food organisms and to minimize the build-up of toxic metabolic products.

Chemical balance is usually achieved through organismic balance in waste-fed ponds because the most important chemical transformations are biologically mediated. It is now recognised that depletion of dissolved oxygen in fertilized fish ponds is due primarily to the high rates of respiration at night of dense concentrations of phytoplankton. Romaire *et al.* introduced Equation to cover the factors influencing waste-fed fish pond dissolved oxygen (DO) at dawn:

$$DO_{dn} = DO_{dk} \pm DO_{df} - DO_{m} - DO_{f} - DO_{p}$$

where:

DO_{dn} = DO concentration at dawn
DO_{dk} = DO concentration at dusk
DO_{df} = DO gain or loss due to diffusion
DO_{m} = DO consumed by mud
DO_{f} = DO consumed by fish
DO_{p} = DO consumed by plankton

Bacterial respiration is not specifically mentioned in this equation but is included in the mud consumption of DO and in the planktonic DO consumption. In a well-managed waste-fed fish pond the DO in the morning should be only a few mg/l whereas in late afternoon the pond should be supersaturated with DO.

Mud respiration probably lowers DO by less than 1 mg/l overnight and a fish population weighing 3000 kg/ha would also lower DO by only about 1 mg/l overnight. Phytoplankton photosynthesis is the major source of oxygen during daylight hours and, during the night, the major cause of oxygen depletion is respiration. It has been estimated that respiration of plankton (bacterioplankton, phytoplankton and zooplankton) can lower pond DO by 8-10 mg/l overnight. By far the greatest proportion of the DO depletion overnight is caused by the respiration of the phytoplankton that develops as a result of the nutrients contained in the waste.

Phytoplanktons provide feed for the largest percentage of fish farmed in Asia. They also exhibit a positive net primary productivity on a 24-hour basis and are net oxygen contributors to a fish pond. The objective in a waste-fed fish pond should be to maintain an algal standing crop at an optimum level for net primary productivity by balancing the production of phytoplankton biomass, in response to waste fertilization, with the grasing of phytoplankton biomass by filter-feeding fish.

Fish mortality in a waste-fed pond can result from at least three possible causes. First, the depletion of oxygen due to bacterial oxygen demand caused by an increase in organic load. Second, the depletion of oxygen overnight due to the respiratory demand of too large a concentration of phytoplankton, having grown in response to an increase in inorganic nutrients, caused by an organismic imbalance. The third possible cause is high ammonia concentration in the waste feed. All three causes of fish mortality have been reported in respect of sewage-fertilized fish ponds. The sensitivity of fish to low levels of DO varies with species, life stage (eggs, larvae, adults) and life process (feeding, growth, and reproduction). A minimum constant DO concentration of 5 mg/l is considered satisfactory, although an absolute minimum consistent with the presence of fish is probably less than 1 mg/l.Fish cultured in waste-fed ponds appear to be able to tolerate very low DO concentrations, for at least short periods of time, with air-breathing fish (such as walking catfish *(Clarias batrachus)* being the most tolerant, followed in decreasing order of tolerance by tilapia, carps, channel catfish and trout. Reducing phytoplankton biomass to maintain a reasonable DO in the early morning hours might well depress fish growth more than exposure to a few hours of low DO. A wastewater fertilized aquaculture system might occasionally require a stand-by mechanical oxygenation system for use during periods when DO would otherwise be very low. However, if the system is well managed to avoid overloading, this expense can be avoided.

Unionised ammonia (NH_3) is toxic to fish in the concentration range 0.2 - 2.0 mg/l. However, the tolerance of different species of fish varies, with tilapa species being least affected by high ammonia levels. Bartone *et al.* found that satisfactory growth and survival of tilapia was possible in fish ponds fed with tertiary effluent in Lima, Peru when the average total ammonia concentration was less than 2 mg N/l and the average unionised ammonia concentration was less than 0.5 mg N/l, with the latter only exceeding 2 mg N/l for short periods. In ponds receiving large quantities of organic matter, sediments tend to accumulate and release anaerobic breakdown products, such as methane and sulphides, which can inhibit fish growth. Bottom feeding fish, such as the common carp (*Cyprinus carpio),* are most affected by such conditions, especially if the macrozoobenthos disappear.

FISH YIELDS AND POPULATION MANAGEMENT

A wide range of yields has been reported from waste-fed aquaculture systems, for example: 2-6 tons/ha yr in Indonesia, 2.7 - 9.3 tons/ha yr in China and 3.5 - 7.8 tons/ha yr in Taiwan. Although the majority of waste-fed fish ponds stocks carps, research in Peru and Thailand has demonstrated the potential of tilapia for such systems. Management of fish ponds can have a significant effect on fish yields but the maximum attainable yield in practice is of the order of 10 - 12 tons/ha yr.Increase in weight of small fingerlings stocked in a pond follows a sigmoidal curve. The first phase of growth is slow, so a high stocking density can be adopted to better utilise the spatial and nutritional resources of the pond. Alternatively, this can be achieved by stocking with larger fish having a higher initial weight, following growth in nursery ponds.

Fish yield is positively correlated with the size of the stocked fish at a given stocking density. In South China, tilapia are stocked once a year at rates of either 30g fish and 0.15/m^2 or 1.3g fish at 2.3 - 3.0/m^2 stocking density. An increase in weight of fish in a pond leads initially to an increase in yield or production but there is subsequently a reduction in the growth rate of individual fish because of the limitation of natural food production in the system.The third phase of slow growth in Figure is because the total weight of fish in the pond is approaching the carrying capacity. Intermediate harvesting when the rapid growth ceases, at the end of phase 2, should lead to significant increases in total yield. The high yields of tilapia reported in South China sewage-fed ponds are due to high stocking density and frequent harvesting.

Clearly, the key to achieving high yields in a waste-fed pond is to determine the carrying capacity of the pond, the maximum standing stock of fish.

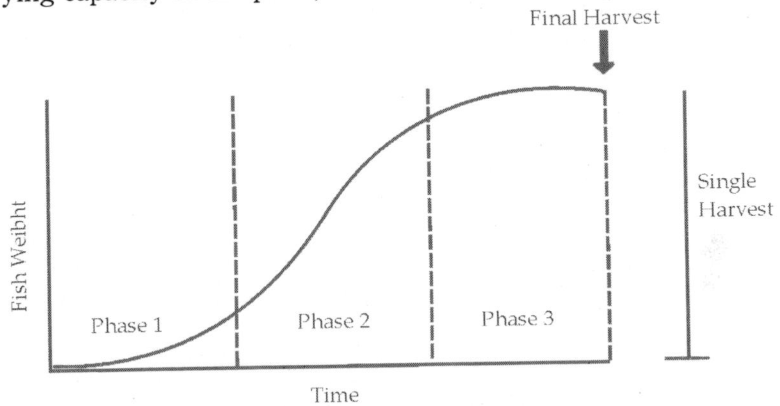

Fig. Fish growth cycle

This can be assessed by varying the waste load and determining the maximum production of natural food consistent with satisfactory water quality, sustainable through a fish culture cycle. Fish stocking density is related to carrying capacity according to the desired weight of individual fish at harvest, as follows:

$$\text{Fish stocking density, (Number/ha)} = \frac{\text{Carrying capacity (kg/ha)}}{\text{Harvesable weight of individual fish (kg)}}$$

Experience has shown that there is a limit to the fish yield attainable from a waste-fed fish pond. Higher yields can be achieved by addition of energy-rich supplementary feed, such as cereals, cereal brans or pelleted-feed. The highest yields are only achieved with a sufficiently high fish stocking density to benefit from the improvement in pond nutrition. There appears to be increased efficiency of utilisation of supplementary feed by fish in ponds receiving sewage effluent. Marketable weights of fish vary with species and local market preferences but, in general, desirable sizes of the following fish range from 0.25-0.6 kg for tilapia, 0.5-1.5 kg for Indian major carps (mrigal 0.5, rohu 1.0, catla 1.5 kg) and perhaps 1-2 kg for Chinese carps. Thus, for a particular carrying capacity, Chinese carps should be stocked at an intermediate density. The length of culture cycle, or frequency of harvesting, depends on the time it takes stocked fish to reach marketable size. It should be recognised that the size of individual fish is only significant if the product is to be consumed by humans. When fish are raised as high-protein feed for carnivorous fish or livestock, size is relatively unimportant. Nevertheless, it is now appreciated that sustainable yields of even high densities of small-size fish with a high specific growth rate are not significantly different from the yield of table fish for human consumption (6.2-7.8 tons/ha yr).

HEALTH RELATED ASPECTS OF FISH CULTURE

Although it is good practice to limit the discharge of toxic materials to sewerage systems, inevitably some of these materials gain access and heavy metals and pesticides are frequently present in municipal sewage. This gives rise to concern about bioaccumulation when sewage effluent is used in aquaculture. Algae are known to accumulate various heavy metals but, with the possible exception of mercury, fish raised in sewage-fed ponds have not been observed to accumulate high concentrations of these toxic substances.

It would appear that the concentrations of heavy metals in the pond water may be accumulated at slower rates than new tissues develop in rapidly growing fish, such as tilapia. In the case of mercury, the position of fish in the food chain seems to be important in determining their mercury uptake, with carnivorous fish accumulating more than herbivores.

Fish, apparently, have the ability to regulate the heavy metal content of their tissues, except for mercury, and tend to accumulate metals in parts other than muscle tissue. There is little information on the uptake of toxics other than heavy metals but a high phenol content in the sewage fed to fish ponds in Wuhon, China caused the fish flesh to become unpalatable due to the odour of phenol. Weis *et al.* have reported on the effects of treated municipal wastewater on the early life stages of three species of fish and indicated that moderately toxic effluent (organic fractions) caused cardiovascular and skeletal defects, depression of heart rate and poor hatching, larval and juvenile growth rates. The health effects of aquacultural use of human wastes in respect of pathogenic organisms have been discussed. Depuration was mentioned as a means to decontaminate fish grown in waste-fed aquaculture. It is generally believed that holding fish in clean-water ponds for several weeks at the end of the growing cycle will remove residual objectionable odours and pathogens and provide fish acceptable for market.

FRESHWATER FISH

Freshwater fish are fish that spend some or all of their lives in fresh water, such as rivers and lakes, with a salinity of less than 0.05 per cent. These environments differ from marine conditions in many ways, the most obvious being the difference in levels of salinity. To survive fresh water, the fish need a range of physiological adaptations.41.24 per cent of all known species of fish are found in fresh water. This is primarily due to the rapid speciation that the scattered habitats make possible. When dealing with ponds and lakes, one might use the same basic models of speciation as when studying island biogeography.

PHYSIOLOGY OF FRESHWATER FISH

Freshwater fish differ physiologically from salt water fish in several respects. Their gills must be able to diffuse dissolved gasses while keeping the salts in the body fluids inside. Their scales reduce water diffusion through the skin: freshwater fish that have lost too many scales will die. They also have well developed kidneys to reclaim salts from body fluids before excretion.

MIGRATING FISH

Many species of fish do reproduce in freshwater, but spend most of their adult lives in the sea. These are known as anadromous fish, and include, for instance, salmon, trout and three-spined stickleback. Some other kinds of fish are, on the contrary, born in salt water, but live most of or parts of their adult lives in fresh water; for instance the eels. These are known as catadromous fish. Species migrating between marine and fresh waters need adaptations for both environments; when in salt water they need to keep the bodily salt concentration on a level lower than the surroundings, and vice versa. Many species solve this problem by associating different habitats with different stages of life. Both eels, anadromous salmoniform fish and the sea lamprey have different tolerances in salinity in different stages of their lives.

Status in North America

About four in ten North American freshwater fish are endangered, according to a pan-North American study, the main cause being human pollution. The number of fish species and subspecies to become endangered has risen from 40 to 61, since 1989.

MAJORITY OF FRESHWATER FISH PONDS

RENOVATION OF EXISTING PONDS

The majority of freshwater fish ponds in the Indian subcontinent are the dugout ponds of an undrainable nature which at times lack proper embankments. During the course of culture operations, such ponds receive huge amounts of feed, fertilizers and manures as critical inputs and sediment particles carried down by rain water from the catchment area. A portion of the organic production in the pond also undergoes death and decay and gradually adds to the pond bottom sediment. Thus, with the advancement of time, a thick sediment layer is formed reducing the depth of the pond. They are quite rich in organic and inorganic nutrients, but due to slow bacterial action under prevailing anaerobic conditions the nutrients are almost locked up in the sediment and are not available for primary production. Further, the anaerobic decomposition of the organic matter accumulated in the sediment releases harmful gases and depletes dissolved oxygen level in the water. Thus, it becomes necessary to renovate the existing ponds periodically every 4–6 years by removing sediment from the pond bottom, redressing and repairing the dykes, etc., in order to make the ponds more suitable and to regain their fertility. For this, the following practical measures are recommended.

DEWEEDING

It has been observed that most of the rural ponds are not properly managed and become weed-infested in course of time. Before dewatering the pond, large floating weeds such as water hyacinth should be eradicated by pulling them out manually or mechanically. Otherwise, collection and removal of such weeds will require more labour and time. Other rooted emergent or submerged weeds can be taken care of only after draining the pond.

WHEN TO TAKE UP THE RENOVATION WORK

As soon as the water table of the area surrounding the pond goes down the renovation work can be initiated. Summer is the most suitable period for this Purpose as complete drying of the water body is possible. In this period pond renovation can be carried out efficiently and economically. Removal of slushy silt from the partially dried pond bottom is difficult, laborious and expensive.

DEWATERING AND DRYING

Dewatering of the existing pond is possible either by draining the water after cutting a portion of the embankment or by pumping out. If the water table in the surrounding area is high, there is considerable inflow of water from the pond bottom. This phenomenon of sub-surface secretion is called percolation. In case the rate of percolation is high, several furrows or ditches may be made towards the lowest contour point where a pit may be dug out to drain all the percolated wall. Periodical pumping of water from the pit facilitates keeping the bed dry.

DESILTING

After complete dewatering, the pond bed is allowed to dry and develop cracks in the silt mass. The texture of silt is different from that of the bottom hard soil and cracks quickly. At this stage dried silt is cut and removed manually or mechanically and heaped at a suitable place for its utilisation in agricultural fields. Where complete drying is not possible due to high rate of percolation, walking platforms made up of bamboos or wooden planks may be put on the slushy bed to facilitate desilting work.In some larger ponds it becomes difficult to dry the central portion of the pond bottom as it is nearer to underground water table. In such cases the slushy and loose silt should be scrapped and spread to the sides with the help of wooden planks tied with ropes for pulling. This helps in drying the silt and easy removal thereafter.

CONTOURING

Where the bed is found to be uneven, contouring is necessary to estimate the amount of silt to be removed. Itis done by taking the level measurements at certain spots on the pond bed. It will also help in redesigning the pond taking into consideration the highest flood level and maximum rain water level.

MAINTENANCE OF DYKES

In general, rural ponds lack proper embankments. During high rainfall or peak irrigation periods in canal-irrigated areas such ponds get inundated with water from the neighbouring agricultural fields causing stocked fish to escape, predators and unwanted species to enter and at times results in mass fish kills due to pesticide pollution. Hence, provision of proper dykes is a must. Existing pond dykes should be repaired every year after the monsoon.

Rats and crabs cause great harm to pond dykes by making holes. Such holes allow serious leakage and if not checked immediately, may endanger the stability of the dykes. Periodically and especially at the time of renovation, such spots should be properly repaired by stuffing binding clay, claylime mixture or any other locally cheap cementing material.Due to poor consolidation, erosion from the top of the dyke during heavy rains usually results in grooving out of small channels. These areas should be covered with earth, levelled, thoroughly rammed and grass turfed. In relatively larger ponds, wave action due to wind also causes large-scale dyke erosion.

By putting large floating aquatic plants such as water hyacinth along the sides of the dykes exposed to wave action during the windy season such erosion can be checked. Frequent erosion in steep dykes during heavy rain or wind can be avoided by strengthening the inner sides of the dykes with poles or bamboos or corrugated cement planks.

Most of the traditional pond dykes are below the required height; as a result, overflow of water occurs during heavy rains or flood. These dykes should be properly raised and the height may be kept at a minimum of one meter above the maximum water level recorded in that area. While raising the dykes, the top width may be kept at a minimum of 1.5 m with 2:1 slope (horizontal: vertical).

Cutting the dyke to allow water into the pond from the surrounding area without any secured screening is a normal practice, which however creates many management problems. The silt mass is very rich in organic and inorganic nutrients making it most suitable for application in agriculture and horticulture. Being non-cohesive and unstable, it is unsuitable for making dykes as it may be washed back in the pond.

RECLAMATION OF DERELICT WATER BODIES

Derelict waters in millions of hectares, lying unutilised, are common sights in most of the South Asian countries. Such untapped water bodies with potential for aquacultural production may be reclaimed and made suitable for fish culture by adopting more or less similar procedures.In case of larger water areas, it would be better if they are connected temporarily to nearby natural or man-made drainage systems having relatively lower bed level for complete dewatering by gravity and making the entire area completely dry. However, if such topographic facilities are not available, heavy duty water pumps may be put into use for quicker dewatering.In extensively large areas dewatering by draining or by pumping is not feasible. Moreover, the dry period of the year also may not last long enough to permit the work to be completed. It has been experienced that such areas can also be successfully reclaimed and renovated by partitioning into smaller units by raising cross bundhs, farm roads, etc. Each newly formed unit then can be dewatered, dried and desilted.

MICROORGANISMS

Microorganisms are of major important in industrial wastewater treatment, agricultural and aquaculture. They reside in the sediment and other substrates, and in the water of aquaculture facilities, as well as in and on the cultured species.Microorganisms may have positive or negative effects on the outcome of aquaculture operations. Positive microbial activities include elimination of toxic materials such as ammonia, nitrite, and hydrogen sulfide, degradation of uneaten feed, and nutrition of aquatic animals such as shrimp, fish; production of aqua-farmer. These and other functions make microorganisms key players in the health and sustainability of aquaculture. Yet, microorganisms are among the least known and understood elements in aquaculture. Like other areas in aquaculture, microorganisms require management and manipulation.

MAJOR MICROBIAL GROUPS

The world of microorganisms is made of bacteria, fungi, algae, protozoa, and viruses. They are group together only because of their small size, and not by their function. If, for example, the same taxonomical rules were applied to larger animals, some fish, shrimp, green plants, birds and mammals would be grouped together. Some microorganisms such as viruses, bacteria, and protozoa are notoriously small, under one mm. Others, like algae and fungi, have large size relatives (such as the brown algae that is among the largest living organisms). Unlike larger organisms, the morphology of microorganisms is relatively poor and is confined to few shapes and colours. However, their poor morphology is compensated by great physiological versatility.

VIRUSES

Viruses are very small, ranging between 0.01 and 0.03 microns, and only visible by using an electron microscope. They cannot live independently, and only multiply inside the cells of other organisms. However, their demand for a host is fairly specific. For example, it is unlikely that a crustacean virus will attack humans or fish. Viruses are also the simplest of all organisms and are made of nucleic acid (either DNA or RNA), frequently coated with a protein layer.

ALGAE

Algae are photosynthetic organisms (contain chlorophyll) and obtain their energy from the sun and their carbon from carbon dioxide. Their size ranges from one micron to many meters. All organisms that use carbon dioxide for their carbon requirement are called autotrophs. Algae are generally beneficial in aquaculture by supplying oxygen and a natural food base for the cultured animals, such as dinoflagellates that cause the red ties.

FUNGI

Fungi are similar to algae, but they do not contain chlorophyll and require pre-formed organic matter as energy and carbon sources (*e.g.,* sugars, fat, protein, and other carbohydrates). Such organisms are called heterotrophs. Fungi, ranging in size from a few microns to several centimeters, grow either independently by feeding on decaying matter, or in association with plants and animals.

PROTOZOA

Protozoa are heterotrophs, mostly free-living, feeding mainly by devouring smaller microorganisms. Their size ranges between two and 200 micron meters. A large group of

protozoa, the Sporozoa, are parasites. Small numbers of protozoa contain chlorophyll and can switch between autotrophic and heterotrophic modes of feeding, based on light conditions.

BACTERIA

Bacteria range in size from 0.1 to 15 micron, with some giants that may reach half a millimeter. They make up the most metabolically diverse group of living organisms. Although some are parasitic to animals and plants, the majority of bacteria are free-living, having either a neutral or beneficial relationship with humans and other animals and plants. Their metabolic versatility is incredible: while most are heterotrophs, using either light or chemical energy. One of their most remarkable characteristics is their ability to multiply rapidly, with generation times usually ranging between minutes to hours.

MICROBIAL PROCESS

Bacteria and other microorganisms, most notably fungi, are able to metabolise and transform numerous organic and inorganic compounds.

Therefore, man has used them for thousands of years for making yogurt, pickles, bread, cheese, wine, and more recently for waste purification and wastewater purification. Process controlled by microorganisms can occur aerobically (in the presence of oxygen) or anaerobically (with no oxygen present). The starting materials and the end products of such processes vary based on the microorganisms' capabilities (as reflected in their genetic makeup), and the environment in which these processes occur (*e.g.,* availability of oxygen, temperature, salinity, pH, etc.

AEROBIC MICROBIAL PROCESS IN AQUACULTUTRE

Generally, aerobic microbial processes yield compounds which can be beneficial, and are either not toxic or have lox toxicity levels in aquaculture ponds or tanks. Oxidation of organic matter to carbon dioxide, a process which is the main consumer of oxygen in aquaculture ponds or tanks. Oxidation of ammonia to nitrate via nitrite, which also consumes large quantity of oxygen. xidation of reduced sulfur compounds (such as hydrogen sulfide and elemental sulfur) to sulfate, a process that generally has low oxygen demand in aquaculture. Conversion of carbon dioxide to biomass by autotrophic bacteria (such as the nitrifying bacteria) with a relatively mall amount of biomass produced in aquaculture facilities, when compared to the conversion of carbon dioxide to biomass by algae.

Conversion of carbon dioxide to biomass by algae depending on the availability of light. Excluding feeding, the photosynthetic process in aquaculture is the main input of carbon source and natural food for aquatic animals.

ANAEROBIC MICROBIAL PROCESSES IN AQUACULTURE

Microbial anaerobic processes, if not controlled, can produce compounds that are highly toxic to cultured animals.

These processes include:

Consumption of organic matter, without the utilisation of free oxygen, resulting in products which are usually not fully oxidised (such as alcohols, organic acids). Reduction of nitrate and nitrite, which can yield either nitrogen gas or ammonia. In aquaculture, due to the toxicity of ammonia and nitrite, ammonia production is not welcomed, while nitrogen gas production is beneficial. However, in agriculture, the opposite is true - the conversion of nitrate and nitrite to nitrogen das result in a loss of fertilizers. Reduction of sulfur compounds to hydrogen sulfide as a final product, a compound, which is toxic to most animals at even very low concentrations.

FUNCTION OF ENZYMES

Enzymes are "biological catalysts." "Biological" means the substance in question is produced or is derived from some living organism. "Catalyst" denotes a substance that has the ability to increase the rate of a chemical reaction, and is not changed or destroyed by the chemical reaction that it accelerates.

Generally speaking, catalysts are specific in nature as to the type of reaction they can catalyse. Enzymes, as a subclass of catalysts, are very specific in nature. Each enzyme can act to catalyse only very select chemical reactions and only with very select substances. An enzyme has been described as a "key" which can "unlock" complex compounds. An enzyme, as the key, must have a certain structure or multi-dimensional shape that matches a specific section of the "substrate" (a substrate is the compound or substance which undergoes the change). Once these two components come together, certain chemical bonds within the substrate molecule change much as a lock is released, and just like the key in this illustration, the enzyme is free to execute its duty once again.Many chemical reactions do proceed but at such a slow rate that their progress would seem to be imperceptible at normally encountered environmental temperature. Consider for example, the oxidation of glucose or other sugars to useable energy by animals and plants. For a living organism to derive heat and other energy from sugar, the sugar must be oxidised (combined with oxygen) or metabolically "burned"

However, in a living system, the oxidation of sugar must meet an additional condition; that oxidation of sugar must proceed essentially at normal body temperature. Obviously, sugar surrounded by sufficient oxygen would not oxidise very rapidly at this temperature. In conjunction with a series of enzymes created by the living organism, however, this reaction does proceed quite rapidly at temperatures up to 100°F (38 °C). Therefore, enzymes allow the living organism to make use of the potential energy contained in sugar and other food substances.

Enzymes or biological catalysts allow reactions that are necessary to sustain life proceed relatively quickly at the normal environmental temperatures. Enzymes often increase the rate of a chemical reaction between 10 and 20 million times what the speed of reaction would be when left uncatalysed (at a given temperature). Nutrients locked in certain organics are complex macromolecules, or in hard-to-digest matrices may be released or predigested by a high degree of heat or concentrated acid treatment. In an alternative manner, specific enzymes can promote the pre-digestion of certain complex nutrients and facilitate the release of highly digestible nutrients in organics during processing without the need of excessive heat or rigourous chemical treatment.

Naming of Enzymes

One researcher reports treating grain, sorghum or barley with the enzyme "gumase" while another reports the same with the enzyme "beta-glucanase" When methodologies are examined, it is discovered that both of these preparations are the same product. Unfortunately, this apparent contradiction in terms happens often. Enzymes have been named by several methods and this fact has been known to cause confusion in their classification. For example, common or "trivial" names of enzymes, generally contain a prefix representing the name of the substance or substrate upon which they act or affect, followed by the suffix "ase". The "ase" simply denotes or identifies that the substance is an enzyme. Examples of this system of nomenclature includes the enzyme that catalyses the conversion of proteins into their component amino acids, the name of this enzyme is "protease" or "proteinase".

Another example is the enzyme that accelerates the breakdown of the two components of starch into sugars. The components of starch are known as "amylose" and "amylo-pectin", thus, the enzyme helping to break them down is called "amylase". Confusion may exist, however, when older names of enzymes are used. Included in these older terms are ficin, pepsin, bromelin and trypsin, which are older trivial names of individual types of protease preparations, the enzymes that accelerate digestion of proteins. There are also many sub- classes of enzymes. Amylases are a prime example; subclasses of amylase include: alpha-amylase, beta-amylase, and gluco-amylase, to name a few. All these enzymes do is accelerate the digestion of starch and are broadly classified as amylases, but their actions are all slightly different in nature.

To help sort this out, the International Union of Biochemistry in 1961 proposed a system for enzymes' classification and naming which is finding acceptance mainly in this discussion. One example of this system, however, is the term: "alpha 1, 4-glucan glucanohydrolase" which is a name for alpha-amylase.All these systems of nomenclature may become confusing to someone who has use for only a few types of enzymes or uses them for industrial or agricultural purposes. Therefore, the use of the more widely known terms such as "amylase" and "protease" are more or less universally in these fields. It should be remembered, however, that there are many types of enzymes that fit into these broad categories that may be more or less suitable for specific agriculturally related application. The final selection for a specific application should be made only after consulting a knowledgeable individual well versed in the technical aspects of the particular enzyme requirements and applicable characteristics.

ISOLATED FORM OF ENZYMES

Enzymes have been isolated from every type of living organism. Many of these biological catalysts are significant only from an academic or medical standpoint, but some of the available enzyme from this vast repertoire have been utilised for agricultural and industrial purposes for years. The table below lists several of the industrially on sequential enzymes and their sources in nature. It is significant to note that animals plants and microorganisms all yield industrially important enzymes. Some enzymes of animal or plant origin have been used in agricultural applications; however, those enzymes most broadly used are of microbial origin.

Source of Enzyme

Plant

- Malted grains or tubers Amylase
- Pineapple Bromelin (Protease)
- Fig Tree Ficin (Protease)
- Papaya Papain (Protease)

Animal

- Liver Catalase (Peroxide Breakdown)
- Calf Stomach Rennet/Chymosin (Milk Clotting)
- Hog Stomach Pepsin (Protease)
- Hog Pancreas Pancreatic Enzymes (Several)
- Digestive Tract Trypain (Protease)

Microbial

- Fungi (Molds and Yeast) amylase, beta glucanase, hemicellulase, protease, cellulase, pectinase, lipase, (many types of each), lactase

Bacteria

- Amylase, protease, isomerase, lactase (many types of each), rennet, oxidase, catalase, beta-glucanase, hemicellulase.

One encounters many digestive or hydrolysing enzymes in the digestive tract of human and other animals. These biological catalysts are necessary for the full utilisation of foods ingested.

Microorganisms, many being as small as 1/10,000 th of an inch in length, are much too minute to have complicated digestive systems as animals do. Therefore, these microbes must predigest their potential foods outside of their cell boundaries so that they can absorb the very small nutrient compounds of predigested foods.In order to predigest the potential food sources outside their cell boundaries, many microbes excrete enzymes out through their enveloping membrane with its supportive cell wall and into the surrounding environment. Hence these "extra-cellular enzymes" must function in the environment outside the protection of the cell's wall and membrane, they must be reasonably stable and have relatively high resistance to chemicals and must function over a relatively broad temperature range. To realise the effects of the enzymes they produce, microorganisms also must produce relatively large quantities of these catalysts. All of these factors contribute to the industrial significance and durability of extra-cellular microbial enzymes. It should be noted that most of the agriculturally and industrial important enzymes, are those that catalyse the digestion or "hydrolysis" of certain large organic molecules like starch, cellulose, and protein. The enzymes actually attack these complex molecules, accelerating their digestion and yielding simpler substances.Hence this process of digestion is referred to as hydrolysis, the enzymes that catalyse the process are considered to be "hydrolysing enzymes" or "hydrolases".

The hydrolysing enzymes include:

- Amylases, which catalyse the digestion of starch into small segments of multiple sugars and into individual soluble sugars.
- Proteases, (or proteinase), which split up proteins into their component amino acid building blocks.
- Lipase, which split up animal and vegetable fats and oils into their component part: glycerol and fatty acids.
- Cellulase (of various types) which breaks down the complex molecule of cellulose into more digestible components of single and multiple sugars.
- Beta-glucanase, (or gumase) which digest one type of vegetable gum into sugars and/or dextrins.
- Pectinase which digests pectin and similar carbohydrates of plant origin.

ACTIVITY AND STABILITY OF ENZYMES

Wastes processing levels usually dictate some variation in physical conditions under which the enzyme products must function. In order to utilise enzymes to their optimal potential in catalytic ability, we must be familiar with the basic principles that can affect the activity and stability of these enzymes. Enzymes, being biological compounds and being comprised of a high percentage of protein, are subject to many environmental effects. Although the following principles hold true for most biological enzymes produced for commercial agricultural use: The pH of the environment has a profound affect on enzyme activity and stability.Activity optimal for pH's of various enzymes vary; however, the optimal pH's for the biological catalysts produced by most commercial strains of microorganisms lies between pH 4.0 and 7.5. This range is from moderately acidic to mildly alkaline in nature. These are the pH levels normally encountered.

Another major affect of enzyme activity and stability is temperature. Hence enzymes are biochemical catalysts, made up at least partially of protein, they are sensitive in varying degrees to heat. Raising temperatures of the environment generally multiplies the degree of activity by the enzyme. Once an optimum temperature has been reached, however, even higher temperatures cause rapid degradation of the enzyme with concurrent and irreversible loss in activity. Optimal temperatures generally range from 98 °F to 140 °F (37 °C to 60 °C) for most hydrolytic enzymes. High temperatures (over 150 °F, 66 °C) generally have detrimental effects on the enzymes.

However, there is broad variation in resistance and sensitivity to heat among the enzymes' types. Bacterial enzymes such as those from Bacillus subtilis are less sensitive to heat than are the fungal enzymes of *A oryzae.* Some amylase preparations prepared by the fermentation of Bacillus species can withstand even boiling for short periods and have optimal activities in the 158 °F - 176 °F (70 °C - 80 °C) range.

Our laboratory has determined that approximately 85 per cent of the activity from Bacillus subtilis/licheniformis alpha-amylase survives high heat. A. oryzae amylases, however, showed a greater than 90 per cent loss activity in high heat. When the enzyme-bearing, dried fermentation products of these two microorganisms are kept dry, they are much more resistant to environmental temperature stress than if they are moistened. In fact, very few stability problems are encountered with most enzymes in typical situations.

Working System of Bacteria and Enzymes to Digest Organic Wastes

The following discussion outlines the biological process. This process is responsible for the digestion of organic waste, no matter where it occurs.

With minor variations, this same process digest waste in:

- Aquaculture,
- Agriculture,
- Composting,
- Livestock,
- Municipal,
- Industrial Wastewater Treatment

BACTERIAL/ENZYME DIGESTION

Bacterial digestion is the process of bacteria, consuming organic matter. Enzymes act to break the organic matter into water soluble nutrients, which the bacteria digest. Using complex chemical reactions, the organic waste is metabolised down to water and carbon dioxide (the final metabolic waste products), providing the bacteria with energy for growth and reproduction.

It may be simply shown by the following equation:

Aerobic Digestion:

$$\text{Organic waste + water} \xrightarrow{\text{Enzyme}} \text{water soluble nutrients + oxygen} \xrightarrow{\text{Bacteria}} \text{water + carbon dioxide}$$

Anaerobic Digestion:

$$\text{Organic waste + Water} \xrightarrow{\text{Enzyme}} \text{Water Soluble Nutrients} \xrightarrow{\text{Bacteria}} \text{Water + Carbon Dioxide}$$

Organic waste is consumed by the bacteria, used as nutrients by the bacteria, and is no longer present to produce odours, sludge, pollution, or unsightly mess.

SELECTED BACTERIA CULTURES

Thousand of different types of bacteria exist everywhere in our world, and most of them carry on bacterial organic digestion in some way. However, some of them are found only in a particular environment, require specialised types of food, and/or have very unique biological roles (niches).A bacteria is a single cell life of form - each individual cell is a separate, unique organism. Bacteria often grow into colonies that appear as jelly-like masses, but each cell remains as independent.Bacteria reproduce by a process called cell division. A mature bacteria reproduces by dividing into two cells, each identical to each other and the parent bacteria. Under ideal conditions, bacteria can reproduce very rapidly, producing a new generation every 20 to 30 minutes.Following this reproduction process, we see that the number of individual bacteria doubles with each generation. The population explodes as the number of organisms increases logarithmically. This population boom begins soon after the bacteria is introduced into a favourable environment, after a short lag time when the bacteria becomes acclimated to the new conditions.

Obviously, this population cannot increase forever. At some point, the food source will be depleted, waste products will accumulate, or some other change in the environment. Will cause the population to level off or decrease (such as a change in pH, temperature, or oxygen content of the environment). Also, introduction of any poisons into the environment may have negative effects on the population, as well as competition from other types of bacteria.

Bacteria can be classified into different types:

- Aerobic types (which require oxygen to live)
- Anaerobic (which can live without oxygen)
- Facultative types can thrive under both aerobic and anaerobic conditions.

For waste digestion, we can identify several beneficial characteristics that we want our chosen bacteria to have.

The "good" bacteria that we will choose must:

- Consume (digest) a wide variety of organic material that are present in wastes.
- Digest waste quickly and completely, without producing significant odours of noxious gas.
- Not cause any disease in man or animals - they must be *non-pathogenic.*
- Grow and reproduce quickly and readily in the environmental conditions found in waste disposal systems.

Certain bacteria belonging to the Bacillus species have these desirable characteristics. They consume organic waste thousands of times faster than the bacteria that are naturally present in the waste. They grow and reproduce easily, are non-pathogenic, and do not produce foul odours or gas as they digest waste.

These "good" bacteria are cultured (grown by artificial means) on liquid or dry nutrient medium. These cultured bacteria are then freeze dried to put them in a state of suspension.

They remain alive, ready to swim, eat, and reproduce as soon as they are activated (rehydrated) and put into the proper environment.

The proper environment needed for rapid growth and reproduction of these good bacteria must have these characteristics:

- A water medium containing food (organic waste).
- Dissolved oxygen (for the aerobic types that require it) in sufficient quantities.
- Proper pH—not too acid nor too alkaline—between 6 and 9 on the pH scale.
- Moderate temperature, between 50 °F and 110 °F.

ENZYMES AID IN DIGESTION

An enzyme is a chemical catalyst that breaks up long, complex waste molecules (Hydrolytic Reaction) into smaller pieces, which can then be digested directly by the bacteria.

Enzymes are simply chemicals - they are not living things, and they cannot grow or reproduce themselves. Enzymes are manufactured by bacteria, and used by the bacteria in order to digest waste. The enzymes that are mixed into the products are actually produced by special bacteria, extracted from them in dry form, and blended into the mixture.

Enzyme are added to help them go to work faster. When added to the organic waste, the enzymes immediately go to work breaking down the waste into water - soluble nutrients for the bacteria to digest. The enzymes break the large, complex molecules of starches, proteins, carbohydrates, and cellulose into smaller, simpler pieces. These enzymes act like chemical "knives", chopping the large molecules of waste into smaller pieces of water - soluble nutrients for the bacteria. The growing bacteria will then start to produce more enzymes on their own, creating a continuing cycle of enzyme production.

The Role of Bacteria in the Sewage Treatment Process

One area of sewage treatment that is not well understood is the bacterial decomposition process. Bacteria may be aerobic, anaerobic or facultative. Aerobic bacteria require oxygen for life support whereas anaerobes can sustain life without oxygen. Facultative bacteria have the capability of living either in the presence or in the absent of oxygen. In the typical sewage treatment plant, oxygen is added to improve the functioning of aerobic bacteria and to assist them in maintaining superiority over the anaerobes. Agitation, settling, pH and other controllable are carefully considered and employed as a means of maximizing the potential of bacterial reduction of organic in the wastewater.

Single-celled organisms grow and when they have attained a certain size, divide, becoming two. Assuming an adequate food supply, they then grow and divide again like the original cell. Every time a cell splits, approximately every 20 to 30 minutes, a new generation occurs. This is known as the exponential or logarithmic growth phase. At the exponential growth rate, the largest number of cells are produced in the shortest period of time.

In nature and in the laboratory, this growth cannot be maintained indefinitely, simply because the optimum environment of growth cannot be maintained. The amount of growth is the function of two variables: - environment and food. The pattern which actually results is known as the bacterial growth rate curve. Initially dehydrated products (dry) must first re-hydrate and acclimate in a linear growth phase before the exponential rate is reached.Microorganisms and their enzyme systems are responsible for many different chemical reactions produced in the degradation of organic matter. As the bacteria metabolise, grow and divide they produce enzymes. These enzymes are high molecular weight proteins.

It is important to recognise the fact that colonies of bacteria are literally factories for the production of enzymes. The enzymes which are manufactured by the bacteria will be appropriate to the substrate in which the enzyme will be working and so you have automatic production of the right enzyme for the biological reduction of any waste material, provided you have the right bacteria to start with. Enzymes do not reproduce whereas as bacteria do.

Enzymes in biochemical reactions act as organic catalysts. The enzymes actually become a part of the action, but after having caused it, split off from it and are themselves unchanged. After the biochemical reactions are complete and products formed, the enzyme is released to catalyse another reaction. The rate of reaction may be increase by increasing the quantity of the substrate or temperature up to a certain point, but beyond this, the rate of reaction ceases to increase because the enzyme concentration limits it.

All treatment plants should be designed to take advantage of the decomposition of organic materials by bacterial activity. This is something you can equate to lower costs, increased capacity, and an improved quality of effluent; even freedom from bad odours which may typically result when anaerobe bacteria become dominant and in their decomposition process, produce hydrogen sulfide gas and similar by-products. Consider the fact that the total organic load of wastewater or sewage is composed of constantly changing constituent, it would be quite difficult to degrade all of these organics by the addition of one enzyme, or even several enzymes. Enzymes are specific catalysts and do not reproduce. What is needed is the addition of an enzyme manufacturing system right in the sewage that can be pre - determined as to its activity and performance and which has the initial or continuing capacity to reduce waste.

At the present time, the addition of specifically cultured bacteria seems to be the least expensive and most generally reliable way to accomplish desirable results. When you add the right bacteria in proper proportions to the environment, you have established entirely new parameters of potential for the treatment situation.

WASTEWATER TREATMENT

Wastewater is water that has been used and must be treated before it is released into another body of water, so that it does not cause further pollution of water sources. Wastewater comes from a variety of sources. Everything that you flush down your toilet or rinse down the drain is wastewater.

Rainwater and run-off, along with various pollutants, go down street gutters and eventually end up at a wastewater treatment facility. Wastewater can also come from agricultural and industrial sources. Some wastewaters are more difficult to treat than others; for example, industrial wastewater can be difficult to treat, whereas domestic wastewater is relatively easy to treat (though it is increasingly difficult to treat domestic waste, due to increased amounts of pharmaceuticals and personal care products that are found in domestic wastewater.

RESPONSIBILITY FOR MAKING SURE THAT WASTEWATER IS TREATED PROPERLY

Similar to drinking water provisions, the federal government has delegated responsibility for wastewater treatment to the provinces and territories.There are two federal acts, however, that may apply to wastewater. The Fisheries Act prohibits the release of harmful substances into waters that fish live in. The Canadian Environmental Protection Act governs the release of toxic substances into the environment and allows the federal government to develop regulations for the use of toxic substances.

Most provincial and territorial governments have legislation regarding wastewater treatment standards and requirements. Operators of wastewater treatment facilities must obtain permits or licenses from the provincial or territorial government, and these permits may also require additional treatments or limits on effluent discharges. For example, in British Columbia, all municipalities are required to have a Liquid Waste Management Plan; without an approved plan, discharges are illegal. Provincial and territorial governments generally assist municipal governments with funds to build and maintain infrastructure.

Municipal governments directly oversee the wastewater treatment process, and are able to pass additional by-laws. For example, the Regional Municipality of Ottawa-Carleton has developed a programme to eliminate toxic substances from the wastewater treatment system, requiring all industrial, institutional and commercial facilities to limit the amount of certain pollutants that are allowed into sewers.

Treatment of Wastewater, to Make it Safe for Discharge

There are several levels of wastewater treatment; these are primary, secondary and tertiary levels of treatment. Most municipal wastewater treatment facilities use primary and secondary levels of treatment, and some also use tertiary treatments. The type and order of treatment may vary from one treatment plant to another, but this diagram of the Ottawa-Carleton wastewater treatment plant illustrates the basic components.

The primary level of treatment uses screens and settling tanks to remove the majority of solids. This step is extremely important, because solids make up approximately 35 per cent of the pollutants that must be removed. The screens usually have openings of about 10 millimetres, which is small enough to remove sticks, garbage and other large materials from the wastewater. This material is removed and disposed of at the landfill.The water is then put into settling tanks (or clarifiers), where it sits for several hours, allowing the sludge to settle and a scum to form on the top. The scum is then skimmed off the top, the sludge is removed from the bottom, and the partially treated wastewater moves on to the secondary treatment level.

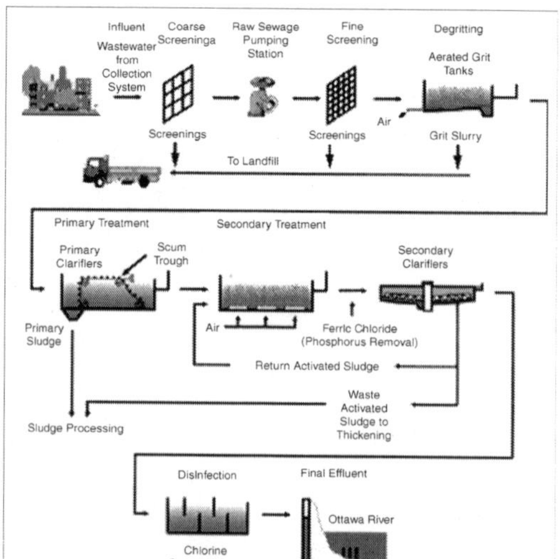

The primary treatment generally removes up to 50 per cent of the Biological Oxygen Demand (BOD; these are substances that use up the oxygen in the water), around 90 per cent of suspended solids, and up to 55 per cent of fecal coliforms. While primary treatment removes a significant amount of harmful substances from wastewater, it is not enough to ensure that all harmful pollutants have been removed.Secondary treatment of wastewater uses bacteria to digest the remaining pollutants.

This is accomplished by forcefully mixing the wastewater with bacteria and oxygen. The oxygen helps the bacteria to digest the pollutants faster. The water is then taken to settling tanks where the sludge again settles, leaving the water 90 to 95 per cent free of pollutants. The picture below shows the settling tanks in the Winnipeg Wastewater Treatment Plant. Secondary treatment removes about 85 to 90 per cent of BOD and suspended solid, and about 90 to 99 per cent of coliform bacteria.

Some treatment plants follow this with a sand filter, to remove additional pollutants. The water is then disinfected with chlorine, ozone, or ultraviolet light, and then discharged.

The sludge that is removed from the settling tanks and the scum that is skimmed off the top during the primary steps are treated separately from the water. Anaerobic bacteria (anaerobic bacteria do not require oxygen) feed off of the sludge for 10 to 20 days at temperatures around 38 degrees Celsius.This process decreases the odour and organic matter of the sludge, and creates a highly combustible gas of methane and carbon dioxide, which can be used as fuel to heat the treatment plant. Finally, the sludge is sent to a centrifuge, like the one shown in the picture below. A centrifuge is a machine that spins very quickly, forcing the liquid to separate from the solid. The liquid can then be processed with the wastewater and the solid is used as fertilizer on fields.

Tertiary (or advanced) treatment removes dissolved substances, such as colour, metals, organic chemicals and nutrients like phosphorus and nitrogen. There are a number of physical, chemical and biological treatment processes that are used for tertiary treatment. One of the biological treatment processes is called Biological Nutrient Removal (BNR). This diagram shows the treatment steps that Saskatoon wastewater goes through.

In this treatment plant, wastewater first undergoes primary and secondary treatment. For the tertiary treatment, the BNR process occurs in the bioreactors. The BNR process uses bacteria in different conditions in several tanks, to digest the contaminants in the water. The three tanks have unique environments, with different amounts of oxygen. As the water has passes through the three tanks, the phosphorus is removed and the ammonia is broken down into nitrate and nitrogen gas, which other bacterial processes can not do. The BNR process can remove over 90 per cent of phosphates, while traditional processes remove much less than 90 per cent. The water spends approximately nine hours in the bioreactors, before entering the secondary clarifier, which is a settling tank, where the bacteria-laden sludge settles to the bottom of the tank.

Small Communities Treat Waste Ewater, to Make it Safe for Sischarge

In small communities, wastewater treatment facilities may consist of individual septic systems, simple collection systems that directly discharge effluent to surface waters, or municipal lagoons that are emptied annually. These facilities usually treat and disperse the waste as close as possible to its source, thus minimizing operational costs and maintenance requirements. The longer the waste can sit in a lagoon before being discharged, the less likely it will be to contaminate drinking water sources. Some communities store the waste in lagoons, but others release the waste directly into water sources.

Lagoons are reservoirs in the ground that store waste for a time until it is discharged, either to the soil or a water body. Shallow lagoons, that are less than 1.5 metres deep, are used for primary treatment, which allows the solid waste to settle to the bottom of the lagoon over a period of 6 to 20 days. Shallow lagoons, however, cannot effectively remove the majority of contaminants that pose problems for ground and surface waters. Deep lagoons, which are more than three metres deep, can provide long-term storage and treatment for six months to one year.

Many lagoons in small communities are emptied once per year. Rural communities often make use of surrounding land to dispose of wastewater. When the soil is adequate, and there are no water sources nearby, the bacteria in the soil can remove and break down the contaminants in wastewater. Due to the availability of land in many rural areas, this can be an effective method to treat wastewater. However, there are other communities that dispose of waste in a way that threatens the quality of the lake, river or groundwater source that provides drinking water.

The Environmental Protection Agency estimates that between 10 and 20 per cent of small community wastewater treatment facilities in the United States are not operating properly; state water quality agencies have identified malfunctioning wastewater treatment systems as the second greatest threat to water quality (after underground storage tanks). When the inadequate wastewater treatments are combined with ineffective drinking water treatment, the result is a serious contamination issue for a great number of rural communities.Rural communities typically find it difficult to install and maintain wastewater treatment operations. And while many communities have inadequate methods of treating wastewater, there are some communities that are leading the way with innovative methods of treatment and water conservation measures. In several Arctic communities, including Iqaluit, Nunavut, the high cost of water has led to wastewater treatments that allow water to be reused. Wastewater is passed through a septic tank, filtered, and disinfected with ozone treatment; it is then reused for non-consumptive uses, such as toilets and laundry.These conservation measures allow them to reuse up to 55 per cent of wastewater, while decreasing pressure on wastewater treatment and storage processes. For more information about water conservation, including the ways in which rural and First Nations communities are leading efforts to reduce water use.

WORKING PROCESS OF SEPTIC SYSTEMS

There are many people living in rural areas that are not served by wastewater treatment plants. In fact, according to Environment Canada, as of 2000, only 57 per cent of Canadians were served by wastewater treatment plants, compared with 74 per cent of Americans, 86.5 per cent of Germans, and 99 per cent of Swedes. Many people in rural areas use septic systems to safely store waste. Wastewater travels through pipes, from the house to a buried septic tank. The diagram below illustrates the basic components of a septic system.

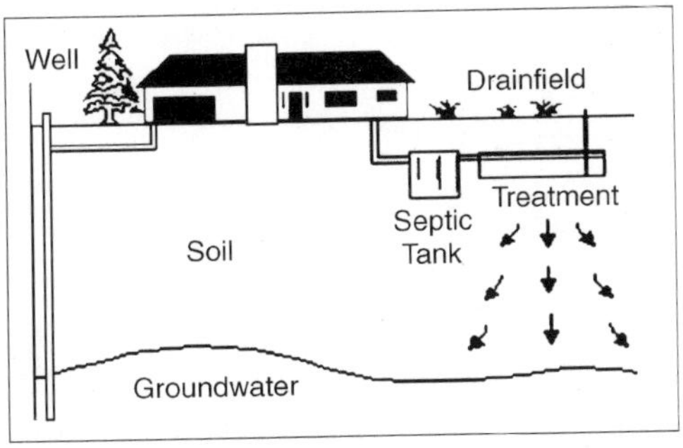

Fig. Septic Tank

The diagram below illustrates the tank in more detail. In the septic tank, the solids settle to the bottom and a scum forms on the top, similar to the process that occurs in settling tanks in municipal wastewater treatment plants. Once separated, the water flows out to the drainfield, and screens and compartments keep the sludge and scum inside the tank, where bacteria begin to partially digest the sludge. When the partially treated wastewater enters the drainfield, it begins to infiltrate the soil and percolate downwards. So long as the soil is appropriate, the microbes in the soil digest the pollutants, removing the bacteria, viruses and excess nutrients by the time the water reaches the source.

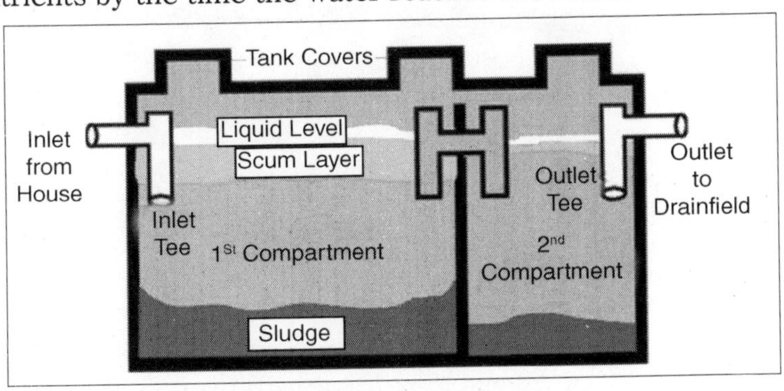

Fig. Inside of Septic System.

When the soil is not suitable, or there are nearby water sources, an alternative system may be used, so that water sources do not become contaminated. The alternative system may use sand, peat, or plastic instead of soil. Constructed wetlands, lagoons, aerators, or disinfection devices are also effective in treating the wastewater.

It is estimated that around half of all rural wells are contaminated, many from septic tanks. The United States Centers for Disease Control and Prevention suggests that wells should be at least 15 metres from septic tanks, but this distance also varies according to the type of soil. In a recent study of groundwater sources in southeastern Michigan by the United States Geological Survey, 38 wells were tested between 1999 and 2001. Of the wells that were near sewerlines, viruses were detected in only two of the 18 wells. Of the 20 wells that were near septic systems, viruses were found in seven wells. This suggests that septic systems are a major cause of groundwater contamination. Previous studies of contaminated wells that were cited by the USGS study found coliforms in up to 80 per cent of the wells, as well as some with significant numbers of the E. coli bacterium.

If you are using a septic system, it is important to maintain it properly, as failure to do so could result in the leakage of pollutants into the soil or water sources. The United States Environmental Protection Agency (EPA) recommends that you have your septic system inspected at least once every three years, so that it can be checked for leaks and malfunctions, and also be pumped out when it gets full (typically once every three to five years).

Following some water conservation practices can greatly reduce pressure on your septic system.

Here are a few things that you can do to care for your septic system:

- Do not use your drain or toilet as a garbage disposal; avoid putting dental floss, diapers, coffee grounds and paper towel down the drain, as they can clog up your septic system.
- Spread your loads of laundry out over the week. When too much water is added to the septic tank, it does not have time to treat wastes, and you could be flooding your drainfield with wastewater.
- Plant grass on your drainfield, but keep trees and shrubs away from it, because roots can clog the system and cause damage.
- Do not drive on your drainfield, because this can compact the soil and damage the septic system components.

NATURAL WAYS TO TREAT WASTEWATER

If nature itself can cleanse water, then imitating nature's processes may be the most effective and sustainable ways of treating wastewater. A great deal of water renewal occurs naturally in wetlands. Constructed wetlands consist of a lined cell, which the water flows into. Plants are planted in the cell and the roots filter the contaminants out of the water. Below is a diagram of a constructed wetland. Notice that many of the processes in a wetland are similar to the Biological Nutrient Removal process that was described above.

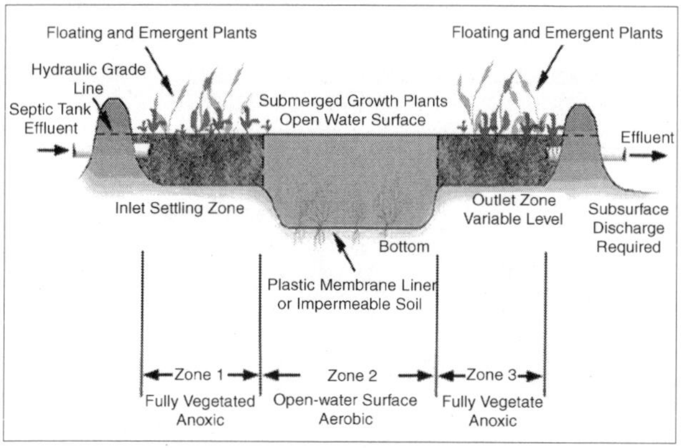

Fig. Constructed Wetland

Another natural method is called rapid infiltration, which is a process where a basin is filled with wastewater, which has already gone through a pre-treatment. The ground acts as a filter and removes the pollutants from the water. This method is similar to what happens in a septic system. A third "natural" process is overland flow, which is used in regions of nearly impermeable ground. The water flows down a sloped surface that is planted with thick grasses. Because the soil is highly impermeable, the water is forced through the vegetation, which effectively removes the pollutants.

Slow rate irrigation is a process that uses a portion of land, and allows the water to flow slowly enough that the land's capacity to infiltrate the water and remove the impurities is not overburdened. Silviculture is similar to slow rate irrigation, in that it uses a large amount of land to treat wastewater, by planting crops or trees that will flourish during the treatment process. Aquiculture uses aquatic plant and animal species to treat wastewater, similar to the constructed wetland process.

There are also alternative separation systems that can conserve water. One such system separates blackwater (from toilets) from greywater (from showers and dishwashers), so that greywater can be minimally treated and used for watering the lawn. As well, there are incinerating, chemical, or composting toilets that release the waste when it is safe.

CONVENTIONAL WASTEWATER TREATMENT PROCESSES

Conventional wastewater treatment consists of a combination of physical, chemical, and biological processes and operations to remove solids, organic matter and, sometimes, nutrients from wastewater. General terms used to describe different degrees of treatment, in order of increasing treatment level, are preliminary, primary, secondary, and tertiary and/or advanced wastewater treatment. In some countries, disinfection to remove pathogens sometimes follows the last treatment step. A generalised wastewater treatment diagram is shown in Figure.

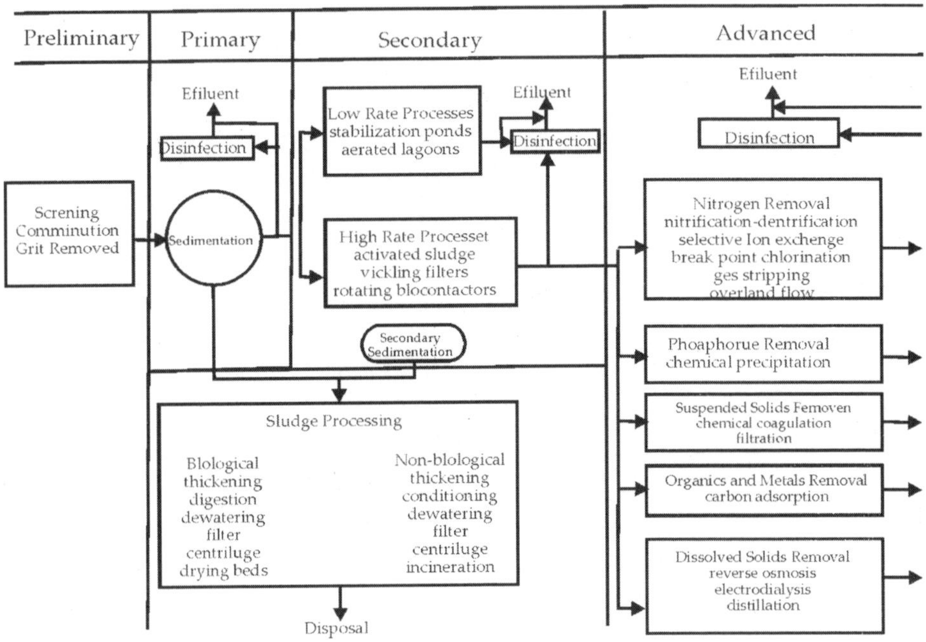

Fig. Generalised flow diagram for municipal wastewater treatment

PRELIMINARY TREATMENT

The objective of preliminary treatment is the removal of coarse solids and other large materials often found in raw wastewater. Removal of these materials is necessary to enhance the operation and maintenance of subsequent treatment units. Preliminary treatment operations typically include coarse screening, grit removal and, in some cases, comminution of large objects. In grit chambers, the velocity of the water through the chamber is maintained sufficiently high, or air is used, so as to prevent the settling of most organic solids. Grit removal is not included as a preliminary treatment step in most small wastewater treatment plants. Comminutors are sometimes adopted to supplement coarse screening and serve to reduce the size of large particles so that they will be removed in the form of a sludge in subsequent treatment processes.

Flow measurement devices, often standing-wave flumes, are always included at the preliminary treatment stage.The objective of primary treatment is the removal of settleable organic and inorganic solids by sedimentation, and the removal of materials that will float (scum) by skimming. Approximately 25 to 50 per cent of the incoming biochemical oxygen demand (BOD_5), 50 to 70 per cent of the total suspended solids (SS), and 65 per cent of the oil and grease are removed during primary treatment. Some organic nitrogen, organic phosphorus, and heavy metals associated with solids are also removed during primary sedimentation but colloidal and dissolved constituents are not affected. The effluent from primary sedimentation units is referred to as primary effluent.

In many industrialised countries, primary treatment is the minimum level of preapplication treatment required for wastewater irrigation. It may be considered sufficient treatment if the wastewater is used to irrigate crops that are not consumed by humans or to irrigate orchards, vineyards, and some processed food crops. However, to prevent potential nuisance conditions in storage or flow-equalising reservoirs, some form of secondary treatment is normally required in these countries, even in the case of non-food crop irrigation. It may be possible to use at least a portion of primary effluent for irrigation if off-line storage is provided.

Primary sedimentation tanks or clarifiers may be round or rectangular basins, typically 3 to 5 m deep, with hydraulic retention time between 2 and 3 hours. Settled solids (primary sludge) are normally removed from the bottom of tanks by sludge rakes that scrape the sludge to a central well from which it is pumped to sludge processing units. Scum is swept across the tank surface by water jets or mechanical means from which it is also pumped to sludge processing units.

In large sewage treatment plants, primary sludge is most commonly processed biologically by anaerobic digestion. In the digestion process, anaerobic and facultative bacteria metabolise the organic material in sludge, thereby reducing the volume requiring ultimate disposal, making the sludge stable (non-putrescible) and improving its dewatering characteristics. Digestion is carried out in covered tanks (anaerobic digesters), typically 7 to 14 m deep. The residence time in a digester may vary from a minimum of about 10 days for high-rate digesters (well-mixed and heated) to 60 days or more in standard-rate digesters.

Gas containing about 60 to 65 per cent methane is produced during digestion and can be recovered as an energy source. In small sewage treatment plants, sludge is processed in a variety of ways including: aerobic digestion, storage in sludge lagoons, direct application to sludge drying beds, in-process storage (as in stabilisation ponds), and land application. ***Example:*** Biological treatment biochemistry

SECONDARY TREATMENT

The objective of secondary treatment is the further treatment of the effluent from primary treatment to remove the residual organics and suspended solids.

In most cases, secondary treatment follows primary treatment and involves the removal of biodegradable dissolved and colloidal organic matter using aerobic biological treatment processes.

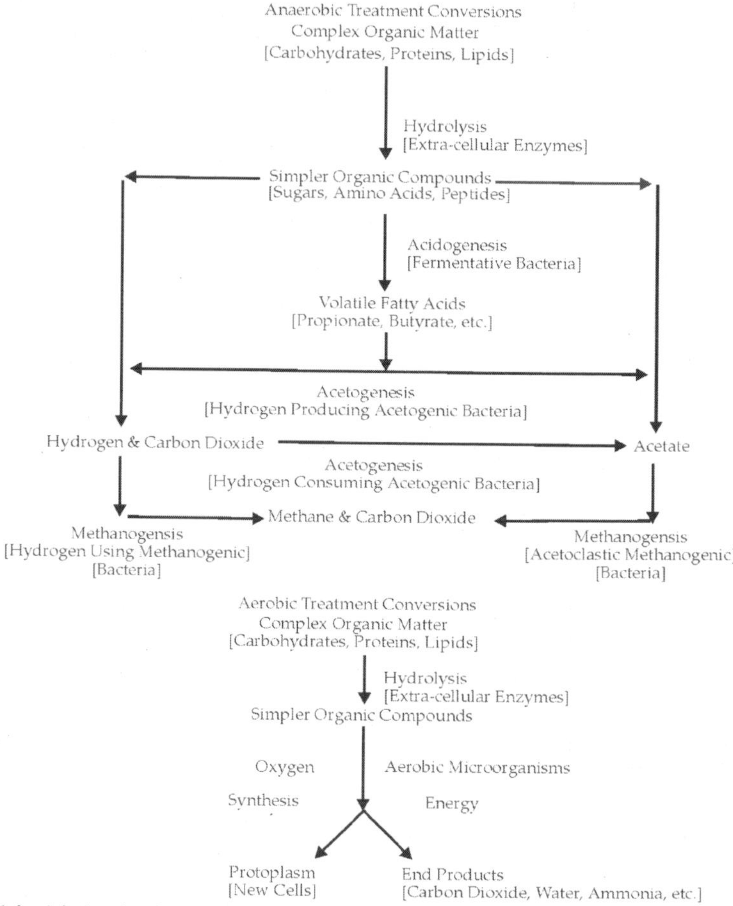

Aerobic biological treatment is performed in the presence of oxygen by aerobic microorganisms (principally bacteria) that metabolise the organic matter in the wastewater, thereby producing more microorganisms and inorganic end-products (principally CO_2, NH_3, and H_2O). Several aerobic biological processes are used for secondary treatment differing primarily in the manner in which oxygen is supplied to the microorganisms and in the rate at which organisms metabolise the organic matter.

High-rate biological processes are characterised by relatively small reactor volumes and high concentrations of microorganisms compared with low rate processes. Consequently, the growth rate of new organisms is much greater in high-rate systems because of the well controlled environment. The microorganisms must be separated from the treated wastewater by sedimentation to produce clarified secondary effluent. The sedimentation tanks used in secondary treatment, often referred to as secondary clarifiers, operate in the same basic manner as the primary clarifiers described previously. The biological solids removed during secondary sedimentation, called secondary or biological sludge, are normally combined with primary sludge for sludge processing.

Common high-rate processes include the activated sludge processes, trickling filters or biofilters, oxidation ditches, and rotating biological contactors (RBC). A combination of two of these processes in series (*e.g.*, biofilter followed by activated sludge) is sometimes used to treat municipal wastewater containing a high concentration of organic material from industrial sources.

Activated Sludge

In the activated sludge process, the dispersed-growth reactor is an aeration tank or basin containing a suspension of the wastewater and microorganisms, the mixed liquor. The contents of the aeration tank are mixed vigourously by aeration devices which also supply oxygen to the biological suspension.

Aeration devices commonly used include submerged diffusers that release compressed air and mechanical surface aerators that introduce air by agitating the liquid surface. Hydraulic retention time in the aeration tanks usually ranges from 3 to 8 hours but can be higher with high BOD_5 wastewaters. Following the aeration step, the microorganisms are separated from the liquid by sedimentation and the clarified liquid is secondary effluent.

A portion of the biological sludge is recycled to the aeration basin to maintain a high mixed-liquor suspended solids (MLSS) level. The remainder is removed from the process and sent to sludge processing to maintain a relatively constant concentration of microorganisms in the system. Several variations of the basic activated sludge process, such as extended aeration and oxidation ditches, are in common use, but the principles are similar.

Trickling Filters

A trickling filter or biofilter consists of a basin or tower filled with support media such as stones, plastic shapes, or wooden slats. Wastewater is applied intermittently, or sometimes continuously, over the media. Microorganisms become attached to the media and form a biological layer or fixed film. Organic matter in the wastewater diffuses into the film, where it is metabolised.

Oxygen is normally supplied to the film by the natural flow of air either up or down through the media, depending on the relative temperatures of the wastewater and ambient air. Forced air can also be supplied by blowers but this is rarely necessary. The thickness of the biofilm increases as new organisms grow. Periodically, portions of the film 'slough off the media. The sloughed material is separated from the liquid in a secondary clarifier and discharged to sludge processing. Clarified liquid from the secondary clarifier is the secondary effluent and a portion is often recycled to the biofilter to improve hydraulic distribution of the wastewater over the filter.

Rotating Biological Contactors

Rotating biological contactors (RBCs) are fixed-film reactors similar to biofilters in that organisms are attached to support media. In the case of the RBC, the support media are slowly rotating discs that are partially submerged in flowing wastewater in the reactor. Oxygen is supplied to the attached biofilm from the air when the film is out of the water and from the liquid when submerged, since oxygen is transferred to the wastewater by surface turbulence created by the discs' rotation. Sloughed pieces of biofilm are removed in the same manner described for biofilters.High-rate biological treatment processes, in combination with primary sedimentation, typically remove 85 per cent of the BOD_5 and SS originally present in the raw wastewater and some of the heavy metals. Activated sludge generally produces an effluent of slightly higher quality, in terms of these constituents, than biofilters or RBCs. When coupled with a disinfection step, these processes can provide substantial but not complete removal of bacteria and virus. However, they remove very little phosphorus, nitrogen, non-biodegradable organics, or dissolved minerals.

TERTIARY AND/OR ADVANCED TREATMENT

Tertiary and/or advanced wastewater treatment is employed when specific wastewater constituents which cannot be removed by secondary treatment must be removed. Individual treatment processes are necessary to remove nitrogen, phosphorus, additional suspended solids, refractory organics, heavy metals and dissolved solids. Because advanced treatment usually follows high-rate secondary treatment, it is sometimes referred to as tertiary treatment. However, advanced treatment processes are sometimes combined with primary or secondary treatment (*e.g.*, chemical addition to primary clarifiers or aeration basins to remove phosphorus) or used in place of secondary treatment (*e.g.*, overland flow treatment of primary effluent).

The Bardenpho Process adopted is shown in simplified form in Figure. Effluent from primary clarifiers flows to the biological reactor, which is physically divided into five zones by baffles and weirs. In sequence these zones are:

- Anaerobic fermentation zone (characterised by very low dissolved oxygen levels and the absence of nitrates);
- Anoxic zone (low dissolved oxygen levels but nitrates present);
- Aerobic zone (aerated);
- Secondary anoxic zone;
- Final aeration zone.

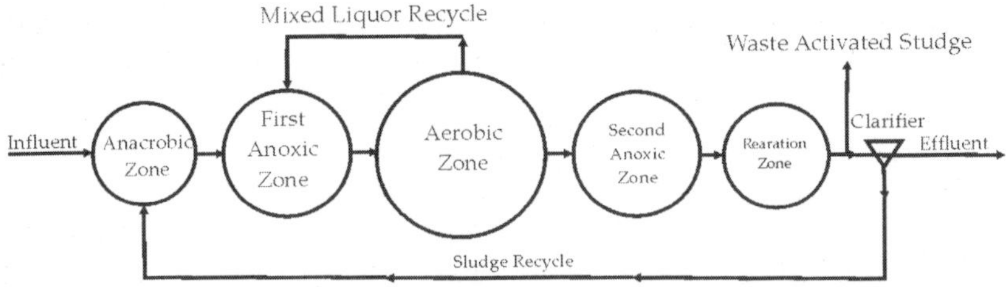

Above: Figure 1-Bandenpbo process train

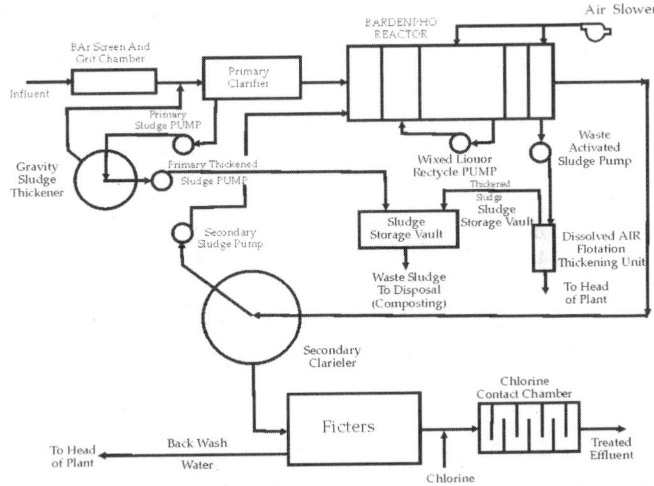

Fig. Simplified flow diagram of Bardenpho-plant

The function of the first zone is to condition the group of bacteria responsible for phosphorus removal by stressing them under low oxidation-reduction conditions, which results in a release of phosphorus equilibrium in the cells of the bacteria. On subsequent exposure to an adequate supply of oxygen and phosphorus in the aerated zones, these cells rapidly accumulate phosphorus considerably in excess of their normal metabolic requirements. Phosphorus is removed from the system with the waste activated sludge.

Most of the nitrogen in the influent is in the ammonia form, and this passes through the first two zones virtually unaltered. In the third aerobic zone, the sludge age is such that almost complete nitrification takes place, and the ammonia nitrogen is converted to nitrites and then to nitrates. The nitrate-rich mixed liquor is then recycled from the aerobic zone back to the first anoxic zone. Here denitrification occurs, where the recycled nitrates, in the absence of dissolved oxygen, are reduced by facultative bacteria to nitrogen gas, using the influent organic carbon compounds as hydrogen donors. The nitrogen gas merely escapes to atmosphere. In the second anoxic zone, those nitrates which were not recycled are reduced by the endogenous respiration of bacteria. In the final re-aeration zone, dissolved oxygen levels are again raised to prevent further denitrification, which would impair settling in the secondary clarifiers to which the mixed liquor then flows.

An experimentation programme on this plant demonstrated the importance of the addition of volatile fatty acids to the anaerobic fermentation zone to achieve good phosphorus removal. These essential short-chain organics (mainly acetates) are produced by the controlled fermentation of primary sludge in a gravity thickener and are released into the thickener supernatent, which can be fed to the head of the biological reactor.

Without this supernatent return flow, overall phosphorus removal quickly dropped to levels found in conventional activated sludge plants. Performance data over three years have proved that, with thickener supernatent recycle, effluent quality median values of 0.5-1.38 mg/l Ortho-P, 1.4-1.6 mg/l Total nitrogen and 1.4-2.0 mg/l nitrate-N are achievable.

This advanced biological wastewater treatment plant cost only marginally more than a conventional activated sludge plant but nevertheless involved considerable investment. Furthermore, the complexity of the process and the skilled operation required to achieve consistent results make this approach unsuitable for developing countries. In many situations, where the risk of public exposure to the reclaimed water or residual constituents is high, the intent of the treatment is to minimize the probability of human exposure to enteric viruses and other pathogens. Effective disinfection of viruses is believed to be inhibited by suspended and colloidal solids in the water, therefore these solids must be removed by advanced treatment before the disinfection step. The sequence of treatment often specified in the United States is: secondary treatment followed by chemical coagulation, sedimentation, filtration, and disinfection.

This level of treatment is assumed to produce an effluent free from detectable viruses. In Near East countries adopting tertiary treatment, the tendency has been to introduce pre-chlorination before rapid-gravity sand filtration and post-chlorination afterwards. A final ozonation treatment after this sequence has been considered in at least one country.

DISINFECTION

Disinfection normally involves the injection of a chlorine solution at the head end of a chlorine contact basin. The chlorine dosage depends upon the strength of the wastewater and other factors, but dosages of 5 to 15 mg/l are common. Ozone and ultra violet (uv) irradiation can also be used for disinfection but these methods of disinfection are not in common use. Chlorine contact basins are usually rectangular channels, with baffles to prevent short-circuiting, designed to provide a contact time of about 30 minutes. However, to meet advanced wastewater treatment requirements, a chlorine contact time of as long as 120 minutes is sometimes required for specific irrigation uses of reclaimed wastewater. The bactericidal effects of chlorine and other disinfectants are dependent upon pH, contact time, organic content, and effluent temperature.

EFFLUENT STORAGE

Although not considered a step in the treatment process, a storage facility is, in most cases, a critical link between the wastewater treatment plant and the irrigation system. Storage is needed for the following reasons:

- To equalise daily variations in flow from the treatment plant and to store excess when average wastwater flow exceeds irrigation demands; includes winter storage.
- To meet peak irrigation demands in excess of the average wastewater flow.
- To minimize the effects of disruptions in the operations of the treatment plant and irrigation system.

Storage is used to provide insurance against the possibility of unsuitable reclaimed wastewater entering the irrigation system and to provide additional time to resolve temporary water quality problems.

RELIABILITY OF CONVENTIONAL AND ADVANCED WASTEWATER TREATMENT

Wastewater reclamation and reuse systems should contain both design and operational requirements necessary to ensure reliability of treatment. Reliability features such as alarm systems, standby power supplies, treatment process duplications, emergency storage or disposal of inadequately treated wastewater, monitoring devices, and automatic controllers are important.

From a public health standpoint, provisions for adequate and reliabile disinfection are the most essential features of the advanced wastewater treatment process. Where disinfection is required, several reliability features must be incorporated into the system to ensure uninterrupted chlorine feed.

WASTEWATER TREATMENT PROCESSES

Relatively simple wastewater treatment technologies can be designed to provide low cost sanitation and environmental protection while providing additional benefits from the reuse of water. These technologies use natural aquatic and terrestrial systems. They are in use in a number of locations throughout Latin America and the Caribbean.

These systems may be classified into three principal types, as shown in Figure. Mechanical treatment systems, which use natural processes within a constructed environment, tend to be used when suitable lands are unavailable for the implementation of natural system technologies. Aquatic systems are represented by lagoons; facultative, aerated, and hydrograph controlled release (HCR) lagoons are variations of this technology. Further, the lagoon-based treatment systems can be supplemented by additional pre- or post-treatments using constructed wetlands, aquacultural production systems, and/or sand filtration. They are used to treat a variety of wastewaters and function under a wide range of weather conditions. Terrestrial systems make use of the nutrients contained in wastewaters; plant growth and soil adsorption convert biologically available nutrients into less-available forms of biomass, which is then harvested for a variety of uses, including methane gas production, alcohol production, or cattle feed supplements.

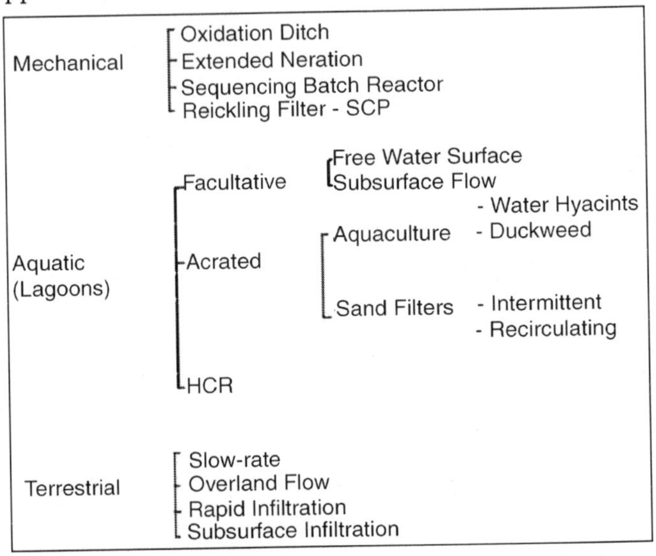

Fig. Summary of Wastewater Treatment Technologies.

TECHNICAL DESCRIPTION

Mechanical Treatment Technologies

Mechanical systems utilise a combination of physical, biological, and chemical processes to achieve the treatment objectives. Using essentially natural processes within an artificial environment, mechanical treatment technologies use a series of tanks, along with pumps, blowers, screens, grinders, and other mechanical components, to treat wastewaters. Flow of wastewater in the system is controlled by various types of instrumentation. Sequencing batch reactors (SBR), oxidation ditches, and extended aeration systems are all variations of the activated-sludge process, which is a suspended-growth system. The trickling filter solids contact process (TF-SCP), in contrast, is an attached-growth system. These treatment systems are effective where land is at a premium.

Aquatic Treatment Technologies

Facultative lagoons are the most common form of aquatic treatment-lagoon technology currently in use. The water layer near the surface is aerobic while the bottom layer, which includes sludge deposits, is anaerobic. The intermediate layer is aerobic near the top and anaerobic near the bottom, and constitutes the facultative zone. Aerated lagoons are smaller and deeper than facultative lagoons. These systems evolved from stabilisation ponds when aeration devices were added to counteract odours arising from septic conditions. The aeration devices can be mechanical or diffused air systems. The chief disadvantage of lagoons is high effluent solids content, which can exceed 100 mg/l. To counteract this, hydrograph controlled release (HCR) lagoons are a recent innovation. In this system, wastewater is discharged only during periods when the stream flow is adequate to prevent water quality degradation. When stream conditions prohibit discharge, wastewater is accumulated in a storage lagoon. Typical design parameters are summarised in Table.

Constructed wetlands, aquacultural operations, and sand filters are generally the most successful methods of polishing the treated wastewater effluent from the lagoons. These systems have also been used with more traditional, engineered primary treatment technologies such as Imhoff tanks, septic tanks, and primary clarifiers. Their main advantage is to provide additional treatment beyond secondary treatment where required. In recent years, constructed wetlands have been utilised in two designs: systems using surface water flows and systems using subsurface flows. Both systems utilise the roots of plants to provide substrate for the growth of attached bacteria which utilise the nutrients present in the effluents and for the transfer of oxygen.

Bacteria do the bulk of the work in these systems, although there is some nitrogen uptake by the plants. The surface water system most closely approximates a natural wetland. Typically, these systems are long, narrow basins, with depths of less than 2 feet, that are planted with aquatic vegetation such as bulrush *(Scirpus* spp.) or cattails *(Typha*spp.). The shallow groundwater systems use a gravel or sand medium, approximately eighteen inches deep, which provides a rooting medium for the aquatic plants and through which the wastewater flows.

Table. Typical Design Features Aquatic Treatment Units

Technology Loading (Feet) (lb/ac/day)	**Treatment**	**Detention**	**Depth Goal**	**Organic Time (Days)**
Oxidation pond	Secondary	10-40	3-4.5	36-110
Facultative pond	Secondary	25-180	4.5-7.5	20-60
Aerated pond	Secondary, polishing	7-20	6-18	45-180
Storage pond, HCR pond	Secondary, storage, polishing	100-200	9-15	20-60
Root zone Treatment, Hyacinth pond	Secondary	30-50	<4.5	<45

Aquaculture systems are distinguished by the type of plants grown in the wastewater holding basins. These plants are commonly water hyacinth *(Eichhornia crassipes)* or duckweed *(Lemna* spp.). These systems are basically shallow ponds covered with floating plants that detain wastewater at least one week. The main purpose of the plants in these systems is to provide a suitable habitat for bacteria which remove the vast majority of dissolved nutrients. The design features of such systems are summarised in Table.

Table. Typical Design Features for Constructed Wetlands

Design Factor	Surface Water Flow	Subsurface Water Flow
Minimum surface area	23-115 ac/mgd	2.3-46 ac/mgd
Maximum water depth	Relatively shallow	Water level below ground surface
Bed depth	Not applicable	12.30m
Minimum hydraulic residence time	7 days	7 days
Maximum hydraulic loading rate	0.2-1.0 gpd/sq ft	0.5-10 gpd/sq ft
Minimum pretreatment	Primary (secondary optional)	Primary
Range of organic loading as BOD	9-18 lb/ac/d	1.8-140 lb/ac/d

Sand filters have been used for wastewater treatment purposes for at least a century in Latin America and the Caribbean. Two types of sand filters are commonly used: intermittent and recirculating. They differ mainly in the method of application of the wastewater. Intermittent filters are flooded with wastewater and then allowed to drain completely before the next application of wastewater. In contrast, recirculating filters use a pump to recirculate the effluent to the filter in a ratio of 3 to 5 parts filter effluent to 1 part raw wastewater. Both types of filters use a sand layer, 2 to 3 feet thick, underlain by a collection system of perforated or open joint pipes enclosed within graded gravel. Water is treated biologically by the epiphytic flora associated with the sand and gravel particles, although some physical filtration of suspended solids by the sand grains and some chemical adsorption onto the surface of the sand grains play a role in the treatment process.

Terrestrial Treatment Technologies

Terrestrial treatment systems include slow-rate overland flow, slow-rate subsurface infiltration, and rapid infiltration methods. In addition to wastewater treatment and low maintenance costs, these systems may yield additional benefits by providing water for groundwater recharge, reforestation, agriculture, and/or livestock pasturage. They depend upon physical, chemical, and biological reactions on and within the soil. Slow-rate overland flow systems require vegetation, both to take up nutrients and other contaminants and to slow the passage of the effluent across the land surface to ensure maximum contact times between the effluents and the plants/soils. Slow-rate subsurface infiltration systems and rapid infiltration systems are "zero discharge" systems that rarely discharge effluents directly to streams or other surface waters. Each system has different constraints regarding soil permeability.

Although slow-rate overland flow systems are the most costly of the natural systems to implement, their advantage is their positive impact on sustainable development practices. In addition to treating wastewater, they provide an economic return from the reuse of water and nutrients to produce marketable crops or other agriculture products and/or water and fodder for livestock. The water may also be used to support reforestation projects in water-poor areas. In slow-rate systems, either primary or secondary wastewater is applied at a controlled rate, either by sprinklers or by flooding of furrows, to a vegetated land surface of moderate to low permeability. The wastewater is treated as it passes through the soil by filtration, adsorption, ion exchange, precipitation, microbial action, and plant uptake. Vegetation is a critical component of the process and serves to extract nutrients, reduce erosion, and maintain soil permeability.

Overland flow systems are a land application treatment method in which treated effluents are eventually discharged to surface water. The main benefits of these systems are their low maintenance and low technical manpower requirements. Wastewater is applied intermittently across the tops of terraces constructed on soils of very low permeability and allowed to sheet-flow across the vegetated surface to the run-off collection channel. Treatment, including nitrogen removal, is achieved primarily through sedimentation, filtration, and biochemical activity as the wastewater flows across the vegetated surface of the terraced slope. Loading rates and application cycles are designed to maintain active microorganism growth in the soil. The rate and length of application are controlled to minimize the occurrence of severe anaerobic conditions, and a rest period between applications is needed. The rest period should be long enough to prevent surface ponding, yet short enough to keep the microorganisms active. Site constraints relating to land application technologies are shown in Table.

Table. Site Constraints for Land Application Technologies

Feature	Slow Rate	Rapid Infiltration	Subsurface Infiltration	Overland Flow
Soil texture	Sandy loam to clay loam	Sand and sandy loam	Sand to clayey loam	Silty loam and clayey loam
Depth to groundwater	3 ft	3 ft	3 ft	Not critical
Vegetation	Required	Optional	Not applicable	Required
Climatic restrictions	Growing season	None	None	Growing season
Slope	<20 per cent,	Not critical	Not applicable	2%-8%

	cultivated land < 40%, uncultivated land		finished slopes

In rapid infiltration systems, most of the applied wastewater percolates through the soil, and the treated effluent drains naturally to surface waters or recharges the groundwater. Their cost and manpower requirements are low. Wastewater is applied to soils that are moderately or highly permeable by spreading in basins or by sprinkling. Vegetation is not necessary, but it does not cause a problem if present. The major treatment goal is to convert ammonia nitrogen in the water to nitrate nitrogen before discharging to the receiving water.

Subsurface infiltration systems are designed for municipalities of less than 2,500 people. They are usually designed for individual homes (septic tanks), but they can be designed for clusters of homes. Although they do require specific site conditions, they can be low-cost methods of wastewater disposal.

Extent of Use

These treatment technologies are widely used in Latin America and the Caribbean. Combinations of some of them with wastewater reuse technologies have been tested in several countries. Colombia has extensively tested aerobic and anaerobic mechanical treatment systems. Chile, Colombia, and Barbados have used activated sludge plants, while Brazil has utilised vertical reactor plants. Argentina, Bolivia, Colombia, Guatemala, Brazil, Chile, Curaçao, Mexico, Jamaica, and Saint Lucia have successfully experimented with different kinds of terrestrial and aquatic treatment systems for the treatment of wastewaters.

Curaçao, Mexico, and Jamaica have used stabilisation or facultative lagoons and oxidation ponds; their experience has been that aquatic treatment technologies require extensive land areas and relatively long retention times, on the order of 7 to 10 days, to adequately treat wastewater. An emerging technology, being tested in a number of different countries, is a hybrid aquatic-terrestrial treatment system that uses wastewaters for hydroponic cultivation. However, most of the applications of this hybrid technology to date have been limited to the experimental treatment of small volumes of wastewater.

Operation and Maintenance

Operation and maintenance requirements vary depending on the particular technology used.

In mechanical activated-sludge plants, maintenance requirements consist of periodically activating the sludge pumps, inspecting the system to ensure that are no blockages or leakages in the system, and checking BOD and suspended solids concentrations in the plant effluent to ensure efficient operation.In the case of aquatic treatment systems using anaerobic reactors and facultative lagoons for primary wastewater treatment, the following operational guidelines should be followed:

- Periodically clean the sand removal system (usually every 5 days in dry weather, and every 2 to 3 days in wet weather).
- Daily remove any oily material that accumulates in the anaerobic reactor.
- Daily remove accumulated algae in the facultative lagoons.
- Open the sludge valves to send the sludge to the drying beds.
- Establish an exotic aquatic plant removal programme (aquatic plant growth can hamper the treatment capacity of the lagoons).
- Properly dispose of the materials removed, including dried sludge.

A preventive maintenance programme should also be established to increase the efficiency of the treatment systems and prolong their lifespan.

When using terrestrial treatment systems or hybrid hydroponic cultivation systems for wastewater treatment, it is advisable to have two parallel systems, and to alternate applications of wastewater to these systems every 12 hours in order to facilitate aeration and to avoid damage to the system.Care is required to avoid hydraulic overload in these systems, as the irrigated plant communities could be damaged and the degree of treatment provided negated. Periodic removal of sediments accumulated in the soil is also required to improve the soil-plant interaction and to avoid soil compaction/subsidence.

LEVEL OF INVOLVEMENT

Government involvement is essential in the implementation of most of the wastewater treatment technologies. The private sector, particularly the tourism industry, has successfully installed "packaged" or small-scale, self-contained sewage treatment plants at individual sites. In some cases, the installation of these plants has been combined with the reuse of the effluent for watering golf courses, lawns, and similar areas.

The selection and construction of the appropriate wastewater treatment technology is generally initiated and financed, at least partially, by the government, with the subsequent operation and maintenance of the facility being a responsibility of the local community. Nevertheless, despite the large number of well-known and well-tested methods for wastewater treatment, there still exist a significant number of local communities in Latin America which discharge wastewater directly into lakes, rivers, estuaries, and oceans without treatment. As a result, surface water degradation, which also affects the availability of freshwater resources, is more widespread than is desirable within this region.

Costs

Construction costs and operation and maintenance costs for wastewater treatment systems with a capacity of 0.1 to 1 million gallons per day are summarised in Figures 29 and 30. Most of the cost data come from systems implemented in the United States. Similar systems in Latin America might be less expensive, in some cases, owing to lower labour costs and price differentials in construction materials. Nevertheless, the relative cost comparison among technologies is likely to be applicable to all countries.

Figure compares the operating and maintenance costs (labour, energy, chemicals, and materials such as replacement equipment and parts) of the various systems of 0.1 to 1 mgd treatment capacity. All costs were obtained from the USEPA*Innovative and Alternative Technology Assessment Manual.* They have been indexed to the USEPA Operation, Maintenance, and Repair Index of Direct Costs for the first quarter of 1993 (4.3). All costs are presented in dollars per million gallons of wastewater treated. The cost for mechanical systems is significantly larger than for any of the other systems, particularly at smaller flows. The cost of harvesting plants from aquaculture systems is not included; this could be a significant amount for some systems.

Figure compares of the capital cost of the wastewater treatment processes. The cost data are also from the *Innovative and Alternative Technology Assessment Manual,* with the exception of wetland and aquaculture data, which were obtained from more recent sources. All natural systems are assumed to have a facultative lagoon as the primary treatment unit. The cost of chlorination/ disinfection is included for all systems except the slow rate and rapid infiltration systems. The cost of land is excluded in all cases, as is the cost of liners for the aquatic treatment systems. The mechanical treatment plant cost was derived as the cost of an oxidation ditch treatment system, and includes the cost of a clarifier, oxidation ditch, pumps, building, laboratory, and sludge drying beds. These costs also include the cost of engineering and construction management, in addition to the costs for piping, electrical systems, instrumentation, and site preparation. All costs are in March 1993 dollars.

WASTEWATER TREATMENT AND REUSE IN INDIA

In Calcutta, India, systematic reuse of wastewater for aquaculture started in the early 1940s. The sewage-fed fish ponds were initially created on about 4 628 ha in an 8 000 ha wetland area. As of 1987, about 3 000 ha of ponds remained active. The treatment process involves screening the raw sewage prior to it entering the ponds. After twelve days, the ponds are repeatedly netted and manually agitated using split bamboo rods. The agitation enhances the oxidation and mixing of the effluent, and promotes improved water quality. The pond is stocked with fish after 25 days, and additional sewage effluent is applied to the ponds during a 3 hour period in the morning, 7 days/month at an estimated rate of 130 m3/day/ha. Even though the total fish pond area has been reduced over the period of operation, total production and yield of fish has gone up from 0.6 tones/ha in 1948 to between 4 and 9 tones/ha in 1984.

Advanced Wastewater Treatment Systems

Some contaminants, such as inorganic substances and a sizeable portion of microbiological populations, present in the waste stream remain in the effluent after the preliminary and secondary treatment processes. Amongst others, nitrates, phosphates and ammonia radicals may still be found in high concentrations. These pollutants can be removed through advanced treatment processes using autotrophic plants to take up nutrients and selected heavy metals and organic substances, flocculation of colloidal particulate matter with chemical flocculants, and removal of synthetic organic contaminants using absorbents and the oxidising agents. Technologies used in advanced wastewater treatment systems include filtration, carbon adsorption, microstraining, chemical phosphorus removal, and biological nitrogen removal.

Filtration can remove most of the residual suspended solids, BOD5, and bacteria from the secondary effluent using multimedia or microstrainer filters. Multimedia filters contain low density charcoal for removal of particles with large grain sizes, medium density sand for intermediate sizes, and a high density medium for the smallest grain sizes. These filters can decrease the concentration of suspended solids in activate sludge-treated effluent from 25 mg/l to approximately 10 mg/l. The carbon adsorption technique is used also to adsorb persistent organic substances onto activated carbon, the organic removal capacity of which depends on the surface area of the carbon particles within the cartridge. Microstrainers can also be used to remove residual suspended solids. These filters consist of woven steel wire or a special cloth fabric mounted on revolving drums which capture the solids still remaining in the wastewater.

Only about 20 per cent of the phosphorus in domestic wastewater is removed during secondary treatment. With phosphorus removal techniques, phosphorus is removed through chemical precipitation of the phosphorus with aluminum sulphate (alum), ferric chloride, or calcium carbonate (lime). This process requires a reaction basin and settling tank to remove the precipitate. Likewise, nitrogen is removed either chemically or biologically.

The chemical process is called ammonia stripping and the biological process is called nitrification/denitrification. In the ammonia stripping process, nitrogen is removed in two stages by first raising the pH to convert ammonium into ammonia, and then stripping the ammonia by passing large volume of air through the effluent. In the biological process, secondary effluent is further aerated to convert ammonia nitrogen to nitrate nitrogen. During the denitrification processes, nitrate nitrogen is converted to gaseous nitrogen by bacteria under anaerobic conditions. Using these techniques, Chen found that the addition of polyaluminium-chloride (PAC) resulted in a 70 per cent removal of suspended solids at a PAC dosage rate of 30 mg/l. If polyelectrolytes are added (at a rate of about 1 mg/l), the dosage of PAC could be reduced to around 10 mg/l with a similar result. Air flocculation, or dissolved air flotation filtration (DAFF), followed by sedimentation, resulted in the removal of more than 80 per cent of the suspended solids at an aeration rate of 0.5-1.0 Nl air/l. This technology is more effective for smaller solids than for larger solids in wastewater. Organic removal, with either sedimentation or combined air flocculation and sedimentation processes, removed about 15 per cent to 40 per cent of the COD or BOD5. The efficiency of organic removal from wastewater was increased to about 60 per cent by utilising chemical coagulation and sedimentation treatment.

The use of advanced treatment is only recommended where major pollutants are not removed to a sufficient extent by the secondary treatment. Usually, advanced treatment are very complicated and expensive, and its use in developing countries to produce suitable effluent for aquaculture or farm purposes is not recommended.

Reuse of Wastewater in Irrigation

In the face of growing water scarcity, reuse of marginal quality water is the best alternative available for agriculture. Marginal quality water, as defined by FAO refers to water that possesses certain characteristics (such as agricultural drainage water, municipal wastewater, and brackish water) which have the potential to cause problems when the water is used for purposes other than the intended use.

Converting marginal quality water to freshwater that can be used for agricultural purposes requires less complex treatment technologies than those required to produce a multi-purpose quality water. Further, use of wastewater for agriculture mimics the traditional use of night soils for agricultural purposes that has been practised in different parts of Asia from ancient times. Sewage farming was initiated in Bombay, India, as early as 1877, and, in Delhi, from 1913. In modern times, the most intensive use of wastewater for irrigation has been made in Israel. In India, modern use of sewage effluents for irrigation is reported to be about six decades old. China's sewage irrigation systems have developed rapidly since 1958. In Laos, effluent of sewage is used directly for the irrigation of 400 to 500 ha fields.

Reuse of irrigation drainage water provides another important source of water for agricultural purposes. Conventional irrigation methods, such as flood or spray irrigation, result in excess water being applied to agricultural fields. The run-off that results, referred to as drainage water or return flows, may be collected and reused for irrigation purposes downstream. This practice, which is widespread though not well documented, can be found in many farmer-managed irrigation systems of Nepal, India and Thailand. In western hills of Argakhachi, Nepal, five parallel canals run across the base of the hills and successively collect drainage water from the farming areas upslope for reuse downstream. In the Kailai Terhi-Gurgi irrigation system, farmers have constructed parallel drainage networks to collect drainage water in the upper portions of the system for reuse in the lower portions of the area. The exact quantities of water reused through this process are not known. In some cases, rules have been formulated for the allocation of rights to reuse drainage water. In arid zones, such as in Egypt, drainage water is collected by an extensive network of covered and open drains and reused. The quantity of drainage water collected and reused during the 1988/89 hydrological year was estimated to be 2 634 million m^3. The drainage water available for reuse had a salinity content within the limit of 1.5 mS/cm, but the quantity of water available was reported to be decreasing and the salinity increasing.

PRE-TREATMENT AND PRIMARY WASTEWATER TREATMENT PROCESSES

Pretreatment of surface water includes processes such as bankside filtration, presedimentation, off-river storage, roughing filters, and microstrainers. Many pretreatment processes are natural processes, enhanced by design to improve water quality. Pretreatment options may be compatible with a variety of water treatment processes ranging in complexity from simple disinfection to membrane processes. Pretreatment is used to reduce, and/or to stabilise variations in the microbial, natural organic matter and particulate load.

GROUNDWATER

For bacterial and protozoan compliance purposes, the DWSNZ distinguish between secure and non-secure groundwaters, with non-secure groundwaters (which includes springs) being considered equivalent to surface waters. Except where discussing compliance issues, the Guidelines consider groundwater to include all water extracted from under the ground.

Springs flow out of the ground at the surface but may contain water that has been underground for a very short time or distance. However, the types of pretreatment commonly applied to springs means that in this section of the Guidelines, springs are considered equivalent to groundwater.

When surface water enters the ground, changes in its quality occur relatively slowly. For this reason, groundwater sources have a more consistent quality than surface waters. Most well pumps have a fixed output so even the flow rate does not change.

Refer to Chapter 3: Source Waters, Section 3.2 for a detailed discussion on groundwater. This includes the development of the well, screens, corrosion, deposition of iron and manganese.

AERATION

Aeration is a physical process aimed at:

- Increasing the dissolved oxygen of the water; and/or
- Decreasing the dissolved carbon dioxide or other gases.

The first objective is more common in wastewater treatment, where oxygen is required for bacterial respiration. The second is more common in drinking water. The aeration process removes the gas by jostling it out of solution and sending it to the surface.

Surface water such as stream or river water already has a high dissolved oxygen and low dissolved carbon dioxide content. However, this is not usually the case with groundwater.

When surface water goes underground, it often carries organic material with it. This material decays over time, adding to the carbon dioxide content. This is not toxic or even distasteful, lemonade contains very high levels! The problem with carbon dioxide is that it reacts with the water to form carbonic acid, lowering the pH of the water.

If this falls to below 7 (as a guide), the following problems may occur:

- The water will dissolve iron and manganese and, potentially, other metals from the ground itself. These metals stay in solution as long as the pH is low; higher pH levels will normally see them precipitate out as unsightly red, brown or black slimes or encrustations. This pH lift occurs at a tap when the pressure is released and the carbon dioxide comes out of solution
- Metallic fittings, particularly copper, zinc (in brass), and iron will be corroded. This may affect people's health, especially in the case of copper, as well as causing bitter tastes and staining of basins, baths and pans. For discussion on corrosion of plumbing materials, refer to Chapter 10: Chemical Compliance, Section 10.3.4.
- Concrete and other lime-based materials such as plaster pipe linings and asbestos cement pipes will dissolve. This can result in detritus (loose sand etc) and the loss of corrosion-prevention linings.

The simplest test for whether there is a high carbon dioxide content is to measure the pH, then aerate the water (by, for example, shaking a half-full sample bottle) and re-measure the pH. If it increases by one pH unit or so, you can be confident the water has enough carbon dioxide in it to merit aeration.

The laboratory method for analysing carbon dioxide is described in *Standard Methods*. Care is needed when collecting a sample for carbon dioxide analysis; the procedure is described in Chapter 4: Selection of Water Source and Treatment.Some underground waters contain other dissolved gases such as ammonia and/or hydrogen sulphide, or even methane. These will have marked effects on the aesthetics of the water, in both taste and odour. They also can be reduced by aeration, but may be more difficult to treat than carbon dioxide. Laboratory testing is needed to verify their presence, though you may be able to detect it by the smell.

To aerate water, it needs to be split into a thin film or droplets to maximize its exposure to the air. This can be done a number of ways:

Tray aerators consist of a series of, usually, four or five horizontal trays that have small holes at regular intervals. The trays are mounted one above the other, about 150 - 200 mm apart. The water is dropped on to the top tray, splashes over it, and goes down through the holes on to the next tray, where the same thing happens.

The tray area needed is calculated by dividing the flow by the loading rate. The loading rate is between 30 and 70 m^3/h per m^2 of tray area. For example:

For a flow of 150 m^3/h and trays with a loading rate of 50 m^3/h per m^2:

- You would need 150 divided by 50 = 3 m^2 of tray area
- 5 trays works best, so they need to be 3 m^2 divided by 5 = 0.6 m^2 each.

So you would have 5 trays, one above the other, with each tray, say 1m by 0.6m. When in doubt, use extra area, it will do no harm if they are bigger than necessary.

The trays can be made of plastic (uPVC, ABS or polypropylene), stainless steel or hardwood timber. Tanalised timber must not be used because it will leach arsenic and other treatment chemicals. The holes are typically 10 – 12 mm in diameter, about 25 – 40 mm apart. A good working tray aerator can be made from the plastic crates used to carry bread. If they have too many holes (so the water does not spread out), a sprinkling of coarse gravel will help disperse the flow over the whole area.

Fig. Tray Aerator Made from Bread Crates at Pleasant Point

Tray aerators need a good air flow to provide oxygen and remove carbon dioxide. They should have screens to keep sandflies or other insects away, and be shaded to limit algal growth. Shade cloth is simple and cheap.

Some tray aerators will precipitate very fine iron or manganese on to the trays. This will need to be hosed off at regular intervals, usually, at least weekly.

Examples of tray aerators can be found at Hannah's Clearing (Westland) and Waitane Meats (Gore).

Cascade aerators are similar to tray aerators, but the trays are displaced to form steps. This horizontal offset may appeal architecturally and does enable easier cleaning, but it requires a collection system (along a line or at a point) for the water to be redistributed on to the next tray. This is not needed in tray aerators.

The design loading and performance of cascade aerators are very similar to those of tray aerators. Like tray aerators, they need good ventilation and shade, and regular removal of precipitates if iron and/or manganese are present.

No cascade aerators are in use in New Zealand. Spray aerators use jets to spray the water up into the air. The finer the droplets formed, the more aeration is achieved and, consequently, the greater the carbon dioxide reduction. Spray aerators use more electric power (to produce the hydraulic head) than tray aerators and are less common, partly for this reason. Like tray aerators, the jets need to be ventilated, shaded and screened.

Spray aerators need about 15 – 20 metres pressure head to produce velocities in the order of 8 – 10 m/s. With typical loading rates in the order of 10 m/h for a single layer, spray aerators occupy about 25 times as much area as tray aerators.

A design constraint is the need to have multiple jets that do not interfere with each other. There are special nozzles that reduce the jet angle, but they are expensive.

A New Zealand example of a spray aerator was Bulls Water Treatment Plant (Rangitikei District Council) but these have been replaced with tray aerators.Entrained air aerators disperse air through the water to allow the transfer of gas from the water into the air, or vice versa. However, they are rarely used, because the energy cost of running a compressor is usually much higher than that of pumping the water.

Packed tower aerators are towers through which the water flows down against a current of air blown from a compressor. The towers are filled with surface contactors, a little like the plastic media for wastewater treatment, to increase the contact area between the water and the air. These contactors are not there for microbiological reasons; they are there solely to increase the contact area. They may be made of plastic, wood or loose media.

Packed tower aerators are often called aerators for removing gas and air stripping towers for removing volatile carbon compounds. They can achieve good results in removing or oxidising gases such as methane, ammonia and hydrogen sulphide but are not common in New Zealand.

Removal Rates

All aerators typically remove only about 50 per cent of the dissolved gases. Some may reach nearly 75 per cent, but none will remove all of it. The remaining carbon dioxide is usually neutralised with an alkali, such as lime or sodium hydroxide (caustic soda), forming bicarbonate and lifting the pH back to non-aggressive levels.

Cleaning

If not much iron or manganese is being precipitated, aerators may need to be hosed down only every few months or so. However, high precipitation rates may necessitate cleaning several times a week. Insect screens and air inlets also need to be cleaned from time to time.

Filtration

After aeration, filtration may be needed to remove further iron and manganese. If so, the carbon dioxide should be neutralised before the water is filtered. The filters should be fine sand (say, 18/36 grade) and be of the normal rapid sand filter design.

OXIDATION PROCESSES

Aeration is the most common process for removing gases such as carbon dioxide and occasionally methane. Aeration is also used to help in precipitating dissolved iron and manganese. Sometimes the direct addition of an oxidising chemical may be needed to help precipitate the metals and maybe oxidise gases such as ammonia.

The following oxidising chemicals are in common use:

- Chlorine, in any form. For example, it will oxidise ferrous iron to ferric iron, making it insoluble so that it precipitates out
- Potassium permanganate, also known as Condy's crystals. This is very effective at destroying some organic substances and oxidising any manganese bound on to them. Again, the manganese is rendered insoluble and precipitates out.
- Ozone, usually used to oxidise taste and odour compounds because many of these compounds are very resistant to oxidation, but also used to oxidise iron and manganese, and ammonia, which breaks down to nitrogen and water. It is used on one source in Wanganui to reduce the ammonia content.

Solutions and liquid chemicals are usually added by dose pump into carry water. In smaller supplies, the carry water may be the main pipe. If the chemical is a gas (gas chlorine and ozone), it is always added to carry water, separate from the main supply.There is often a feedback control loop that adjusts the dose rate to respond to variations in the raw water. The effect is measured in one of three ways: chlorine level, pH or ORP (oxidation/reduction potential).

SURFACE WATER

NON-SECURE GROUNDWATER AND SPRINGS ARE COVERED

The pretreatment processes for surface water described below aim to improve the quality of the raw water prior to the main treatment process. Many of them seek to reduce the natural variations or extremes in water quality. A typical case is turbidity (measured as NTU), used to indicate how much mud (or similar) is suspended in the water during heavy rainfall and its run-off. Rivers may reach levels of several hundred NTU during such events, tending to overwhelm the filtration processes by the sheer volume of the solids to be removed.Pretreatment processes may reduce the level of these solids by factors of ten or more, allowing a treatment plant to continue functioning rather than shutting down or damaging any treatment components.

BANK FILTRATION AND INFILTRATION GALLERIES

This section covers issues related to the design, operation and maintenance of bank filtration. For discussion on protozoal compliance issues, refer to Chapter 8: Protozoa Compliance.

Experience has shown that both bank filtration and full infiltration galleries can be effective at reducing the concentration of suspended material from surface water that percolates down to the collection area. In some cases the turbidity can be quite low, and consistently so, despite peaks of turbidity in the surface water. Protozoan (oo)cysts are removed too, which is why the bank filtration process can qualify for protozoan log credits.

The consistently lower turbidity can mean that less expensive treatment such as cartridge or diatomaceous earth filtration can be used instead of chemical coagulation in order to achieve protozoan compliance.Bank filtration is the process of using vertical or horizontal wells beside rivers and/or lakes. The idea is to use naturally occurring gravels and sands to filter the water as it flows through to the gathering well or gallery. To gain protozoa removal credits for this process, the amount of gravel and/or sand present needs to be significant.

All sites have their own peculiar restrictions, and advantages. Whilst some simplistic diagrammatic examples are shown below, the well or gallery needs to be designed for each location. Few streams and rivers are sited in alluvial gravels and sands alone, there will be flow-blocking material such as silts, clays and bedrock ledges. These will control where the water may be taken from.

The PHRMP should allow for the risks at each particular site.

Risks may include:

- Flood damage to the structures
- Floods flowing over the site and inundating the (originally clean) water
- Flow levels dropping to such an extent that there is inadequate water available to take.

There will be other risks peculiar to each site.

Some examples of bank filtration and infiltration galleries are:

Vertical wells

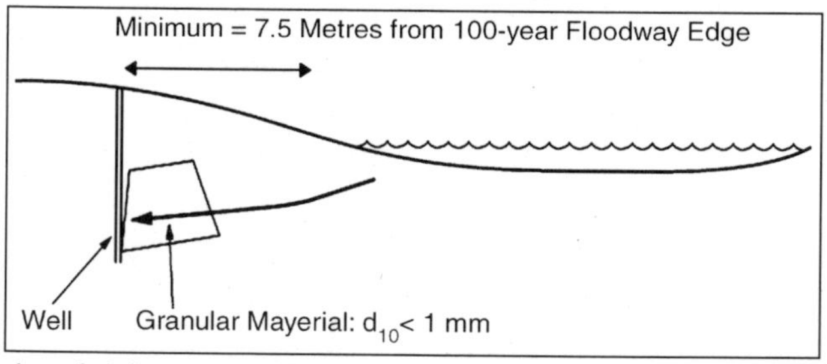

An example of this type of intake is the Taieri bores in Outram, Dunedin.

In some cases, after wet weather, groundwater may flow towards the river, so the water collected may be from adjacent surface land, not the river.

The granular material must have a d_{10} of 1 mm or less. The d_{10} is the size that 10 per cent of the material is smaller than; likewise, d_{60} is the size that 60 per cent is smaller than. A sand ranging between 0.6 mm and 1.2 mm would probably have a d_{10} of about 0.66 mm and a d_{60} of about 0.96. Within a given range, the relative amounts of different-sized grains can vary and the effectiveness of the filter is determined by the fineness of the material overall. This is expressed as the ratio of the d_{60} to the d_{10}, or uniformity coefficient (UC). The UC of the sand described above would be 0.96 divided by 0.66 = 1.45. Good filters usually have a UC of 1.5 or lower.

Vertical wells may use a central screen surrounded by a filter pack. The shaft is dug to a much greater diameter than that of the screen, say 1.1 m for a 200 mm screen, and a temporary liner such as a large steel pipe (for this example, about 1 m in diameter) inserted to stop the country collapsing. The screen is put in place, the filter material inserted in the gap and the liner removed. This technique enables a good thickness of filter material but is more suited to shallower wells of up to 15 m deep; any deeper and it becomes very difficult to remove the liner. Normally, in deep wells, the screen is placed directly against the country of the aquifer. Following this, the well is usually surged to pull small sand through the screen, this process is called well development. It is important that the design prevents dirty water flowing down the sides of the shaft, polluting the cleaner water below. This would be one of the risks addressed by the PHRMP.

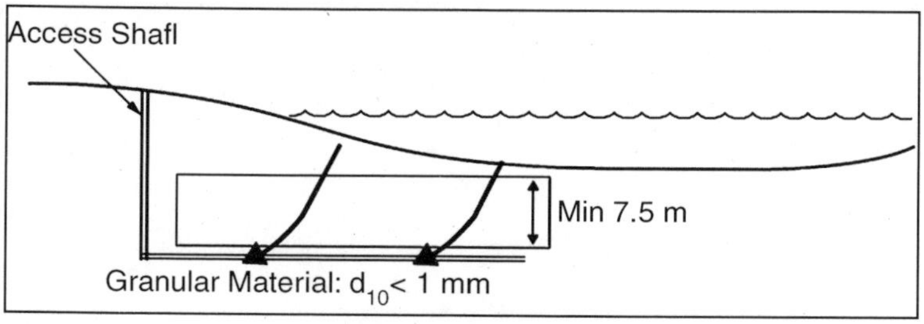

Fig. Horizontal Wells

The protozoa log removal credits allowed for bank filtration systems in the DWSNZ are based on data analysis carried out by the USEPA. For such a system to qualify for 0.5 log removal of *Cryptosporidium* oocysts or *Giardia* cysts, it is required that:

- A horizontal or vertical well is used.
- The ground is unconsolidated gravels or sands that either have a d_{10} of <1 mm or can be demonstrated to be an effective filter.
- The groundwater flow path is at least 7.5 metres long. For vertical wells, this distance is measured from the 100-year flood edge to the well; for horizontal wells, it is measured from the river edge and bed under normal flows to the nearest point of the gallery.
- Turbidity measurements are taken at least 4-hourly. These must show an average daily maximum of less than 1 NTU, unless a satisfactory explanation is given for non-compliance.

To qualify for 1 log removal credit, the groundwater flow path must be at least 15 m.

In New Zealand some infiltration galleries differ from these bank filtration arrangements.

The usual method of constructing these galleries is:

- A horizontal trench is excavated, parallel to the river
- A slotted pipe is laid, usually bedded in a filter pack of carefully graded gravel
- The trench is backfilled, usually with the excavated gravels
- The supply is withdrawn from all of the slotted pipe length
- Many of these galleries do not comply with the criteria for bankside filtration.

Non-Complying Infiltration Gallery

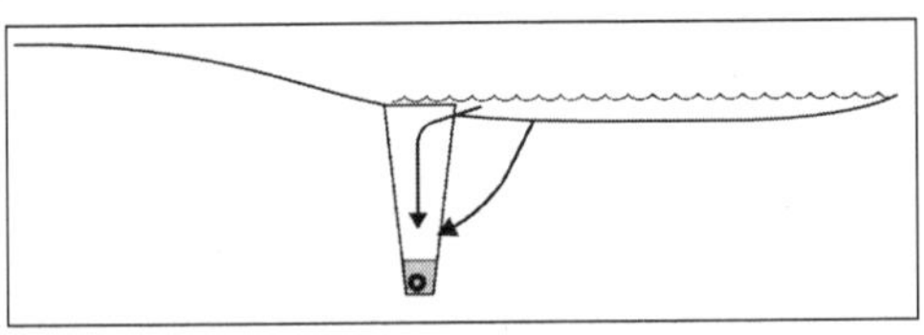

The features that make this non-compliant are:

- The length of the filter path is too short
- The backfill/country material may be too coarse.

Compliant Infiltration Gallery (a)

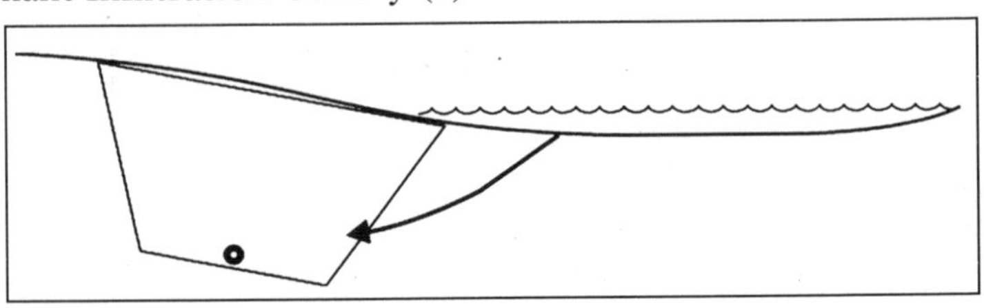

This arrangement complies with the requirements of the DWSNZ, provided the distance is greater than 7.5 m (for 1 log credit) and the country has a d_{10} of <1 mm.An example of this type of gallery can be found on the Opihi River, Timaru City. Compliant Infiltration Gallery (b)

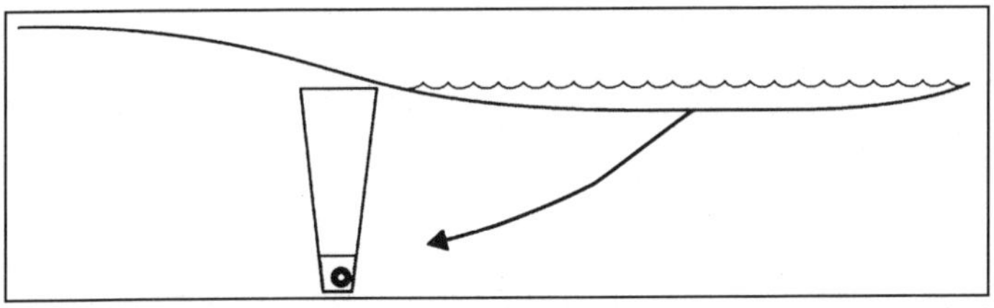

This arrangement also complies with the requirements of the DWSNZ, provided the distance is greater than 7.5 m (for 1 log credit) and the country has a d_{10} of <1 mm.

An example of this type of gallery can be found at Coal Creek, Greymouth.

The bank filtration mechanism operates by the normal methods of filtration:

- Direct mechanical screening, *i.e.* stopping the transport of large particles. The material will stop particles that are 30 per cent of its d_{10} or bigger. So, if the d_{10} is 1 mm, it will stop particles of 300 microns and more
- Settling of particles on to horizontal layers in the sand
- Spinning out particles at tight corners, sending them into quiet areas where they settle out
- Adsorption of particles on to the surfaces of the material.

Bank filtration also significantly reduces the organic material in surface water. This is probably because the organic material sticks to the surface of the filtration material and because tiny aquatic animals graze on the micro-organisms.

As noted above, organic material in surface water decays over time. Thus the ideal time to abstract the raw water is after most of the particles have been removed but before any dissolved organics have produced carbon dioxide, dropping the pH and dissolving iron or manganese from the soil.

Design loading rates for infiltration galleries are usually around 0.2 – 0.5 litres per second per metre of length, with the pipes sized to keep velocities under 0.7 m/s. The general rule is the lower the loading rate, the better the result.

The pipe should also be large enough to allow for maintenance; an internal diameter of at least 150 mm will allow access for cleaning. The size of the slots should allow a maximum hole velocity of 0.3 m/s, as for well screens.

It is normal to pack the gallery pipe into a filter pack with material graded to limit soil migration. This filter pack is usually five times the d_{30} (*i.e.* the size that 30 per cent of the material is finer than) of the parent country. Typical groundwater flow rates into the galleries/wells are around 1.5 – 15 m/d (USEPA

2003a). For the design loading rates of 0.2 – 0.5 L/s/m, this requires a flow path between 3 m and 10 m wide.Maintenance requirements are usually very low. The very high surface area gives low surface loading rates, with little build-up of material on the soil-water interface. Normally, stream erosion is adequate to keep this surface sufficiently clear.

Some galleries are fitted with air scour header pipes to blow air through the filter material. Care should be taken with this, as air bubbles that get trapped act as aquicludes, thus reducing the flow (or increasing the headloss) through the aquifer. Bank filtration can protect a river intake to a very high degree. There are several examples in New Zealand that achieve extremely good turbidity reduction, including the Opihi gallery in Timaru, which reduces about 150 NTU to about 2 NTU under flood flows.

For further information about the advantages and disadvantages of this technique, site selection and aquifer requirements, design, construction and operational consideration.

OFF-RIVER STORAGE

Off-river raw water storage reservoirs are basins located between a water source (typically a river or stream) and the water treatment plant.

It may be possible to avoid taking water from rivers and streams when water quality is poor (*e.g.* following heavy rainfall) in order to reduce risk and to prevent potential problems in subsequent treatment processes. Rather than operate a stop/start treatment plant with its associated uncertainty of continuous supply, it may be more convenient to use off-river storage.

Off-river storage can also be helpful to ensure an adequate supply of raw water in situations where the run of stream flow is very low during dry periods.

By decreasing the contamination of the raw water, the amount of treatment required can be reduced. For example, a stream or river typically may have a turbidity of 1 - 4 NTU, and if the raw water colour is low, chemical coagulation may not be needed. This means diatomaceous earth, slow sand, membrane, cartridge or bag filtration could provide suitable treatment. However, in New Zealand conditions, it is not unusual for the turbidity in streams or rivers to exceed 50 NTU due to heavy rain in the catchment. If the off-river storage system is only filled while the run of stream water is below (say) 5 NTU, these filtration systems should not be compromised.

It is quite common for other determinands to exhibit elevated levels when the turbidity is high, *e.g.* colour, micro-organisms and nutrients. If the nutrient input to the off-river storage is low, the risk of nuisance algal growths should be minimised; in most cases this can be achieved by turbidity control.

Similarly, the off-river storage could be filled selectively to overcome other problems. For example, if the natural organic matter in the run of stream water increases during certain conditions, potentially causing undesirable levels of disinfection by-products (DBPs) in the drinking-water, on-line UV absorbance could be used to determine when to stop filling the off-river storage. Retention of water in reservoirs can reduce the number of micro-organisms through settling and inactivation, including solar (ultraviolet) disinfection. Most pathogenic micro-organisms of faecal origin (enteric pathogens) do not survive indefinitely in the environment. Substantial die-off of enteric bacteria will occur over a period of weeks of storage.

Enteric viruses and protozoa can survive for longer periods (weeks to months) but are often removed by settling and antagonism from indigenous microbes.

Impoundment also allows suspended material to settle, which makes subsequent disinfection more effective and reduces the formation of DBPs.

The USEPA evaluated a number of studies that investigated the removal of *Cryptosporidium* and other micro-organisms and particles in raw water storage lakes. They summarised these studies in a table that has been reproduced here as Table.

Table. Studies of Protozoa Removal from off-river Raw Water Storage

Researcher	Reservoir	Residence Time	Log Reductions
Ketelaars *et al.* 1995	Biesbosch reservoir system: man-made pumped storage (Netherlands)	24 weeks (avg)	*Cryptosporidium* 1.4*Giardia* 2.3
Van Breeman 1998	As above	24 weeks (avg)	*Cryptosporidium et al.* 2.0*Giardia* 2.6
Van Breeman 1998	PWN reservoir	10 weeks (avg)	*Cryptosporidium et al.* 1.3*Giardia* 0.8
Bertolucci *et al.* 1998	Abandoned gravel quarry used for storage (Italy)	10 weeks (avg)	*Cryptosporidium* 1.3*Giardia* 0.8
Ongerth, 1989	Three impoundments on rivers with limited public access	40, 100 and 200 days respectively	*Giardia*: no removal observed

USEPA was unclear why no decrease in cyst levels was observed in the Ongerth study. They felt it was possible that contamination of the water in the impoundments by *Giardia* from animal sources, either directly or through localised run-off, may have occurred.

Protozoa Compliance for a discussion why the USEPA does not award log credits for off-river storage. However, with respect to determining the source water category in the DWSNZ, water suppliers in New Zealand are able to take advantage of any benefits from off-river raw water storage by sampling after the storage.

Off-river storage also provides opportunities for contamination to be introduced, depending on the surrounding topography and land-use. Constructing cut-off drains around the storage lake or reservoir can reduce the effects of contamination from the run-off due to grasing animals. Even contaminated tributaries can be diverted. Ducks, seagulls and shags can find the storage attractive too, so may need to be discouraged.Many of the benefits of off-river storage are related to the hydraulic retention time.

This can vary day-to-day or seasonally, depending on:

- The filling rate allowed by the regional council
- Whether there is a requirement to release water
- The size of the off-river system catchment
- Whether direct hit rain water is significant to the water balance
- The abstraction rate to the water treatment plant
- How much short-circuiting occurs (under different inflow and water level situations)
- Whether the lake stratifies (if it does, its effective retention time is greatly reduced)
- Whether stratification is broken down by artificial aeration.

PRESEDIMENTATION

Presedimentation is used to remove gravel, sand, and other large or gritty or floating material from the raw water, and to dampen particle loading to the rest of the treatment plant. Presedimentation is similar to conventional sedimentation, except that presedimentation may be operated at higher loading rates and may not involve use of chemical coagulants.

Also, some water supplies operate the presedimentation process periodically and only in response to periods of high particle loading.Sometimes a presedimentation basin is no more than a pond that has been dug out between the intake and the water treatment plant. To enhance their performance, a coagulant is sometimes added, often crudely.

Presedimentation without Coagulant

Presedimentation without coagulant is simply settling the raw water before treatment to allow solids to drop out. It is more commonly used for river sources that are subject to very high solids loadings. It is particularly effective in removing or reducing coarse solids such as gravels and sands. Without chemical aids such as coagulation, it is not very effective at reducing silts, clays or organic materials.Presedimentation is usually carried out in a tank or small natural basin. The use of larger basins is called off-river storage.

On its own, presedimentation involves installing a settlement device such as:

- A clarifier tank with a sloping bottom, or
- A clarifier with a scraped bottom, or
- A plate or tube settler.

For any of these, the key loading feature is the overflow rate. This is the flow into the device divided by the plan area of the settlement surfaces. For a tank, this is simply the plan area of the tank; for tube and plate settlers, it is the total plan area of horizontal surfaces. To settle a particular kind of particle, the overflow rate must be no faster than the speed at which that particle falls.

Because small particles tend to settle more slowly than large ones, the slower the overflow rate, the smaller the particles removed. Clays and organic materials are very slow to settle and difficult to remove without chemical additives such as coagulants, or by using very fine filtration such as membranes. Silts settle a little faster, but still very slowly. For example, 2 micron silt particles will settle at about 10 mm/h. If the flow into the tank is 100 m^3/h, the area of the tank will have to be 100 divided by 0.01; *i.e.* 10,000 m^2, or three football fields!

This can be speeded up significantly by coagulation. Alum floc particles typically settle at around 2 m/h. So, at the same flow rate, the area needed would be only 100 divided by 2; *i.e.* only 50 m^2.

Sand settles faster than silt. For example, at 600 microns, it will fall at around 900 m/h. Using the same flow rate, the area required is only 100 divided by 900; *i.e.* 0.11 m^2.

In practical terms, it is feasible to remove sands, difficult to reduce silts, and virtually impossible to remove clays and organics unless very large areas are used.

Presedimentation without storage as well is not common in New Zealand. The few examples include Mt Grand water treatment plant in Dunedin, which has a parallel plate separator with a loading rate of around 10 m/h. This is being replaced with raw water off-river storage, primarily for an emergency supply, not as a settlement step.

Reservoirs commonly cover very large areas, because it is usually uneconomical to make them deep. This means they are very effective as settlement devices. The settlement function of a raw water reservoir combines very well with its other purposes, such as mixing and emergency reserves. Again, the settlement achieved can be estimated from the plan area of the reservoir; the depth (provided it is enough to limit currents) is not important.

To see why the depth does not matter, consider what happens if the depth is doubled. The particle now has to fall twice as far to fall to the floor, but it has twice as long to do it in. Why? Because the horizontal flow through is now half of the original speed. Therefore, depth does not matter.

Presedimentation with Coagulant

The USEPA proposed in its LT2ESWTR to award a presumptive 0.5 log *Cryptosporidium* treatment credit for presedimentation that meets three criteria.Calling this process presedimentation is a misnomer because the USEPA really considers this to be the sedimentation step in the coagulation, sedimentation, rapid gravity sand filtration process, *i.e.* what the USEPA calls conventional full treatment, hardly a pretreatment process. Therefore presedimentation with coagulant is discussed in Chapter 13: Coagulation Processes.

PREFILTRATION

Prefiltration is used to remove large solids from the raw water before it is treated. The process usually involves screens and is purely mechanical; it is not chemically assisted. Prefiltration is often used at river and reservoir intakes, where a coarse screen prevents large solids that can block valves, such as fish, parts of trees and the like, from being taken in. The holes in the screen are usually 1 – 2 mm, although some are significantly coarser than this.

Types of screens include:

- Fixed bar screen, usually with the bars vertical. These require raking from time to time, often manually, sometimes automatically
- Fixed bar screen mounted on to the downstream face of the weir. These allow water flowing over them to fall into the intake chamber inside the weir structure. If debris builds up against the screen and blocks it, the flow of the water tends to push it off again and back into the river. These screens are suitable only for small streams, because the weir has to carry the entire flow. They require very little maintenance
- Fixed mesh screen. These are not very common, as the mesh is below the surface of the water and is therefore very difficult to get at and clean
- Liftable mesh screen. These are often mounted on a steel frame and are quite heavy. Some are lifted by electric winch, others by a towrope off a vehicle. They are relatively common.

Most intake screens are sized based on a surface velocity of less than 0.15 m/s.

Following this initial screening, prefiltration can be by:

- Rotating drum screen, similar to those used in wastewater. These are not commonly used in water supplies. Their cost (both initial and operating) is high and they are not so much more effective as to warrant the expense.
- *Stepping screen*: These resemble, and operate like, escalators. The solids are lifted out of the water by the steps and are washed off before the steps come around again to re-enter the water. They are more common than rotating drum screens and are often used to remove leaves and coarse algal material. An example can be found at Oamaru.
- *Washable disc filters*: These comprise a stack of discs mounted one on top of the other, with the gaps between them set by bumps on

the discs. The water passes through a hole in the middle of the rings. The gaps between the discs range from 100 microns down to about 5 microns, which is still too big to capture micro-organisms, silts, clays or natural organic matter so they do not qualify for protozoa log credits in DWSNZ. For backwashing, they are unclamped to loosen them before the cleaning flow starts. They also need to be stripped down occasionally (six-monthly or yearly) and, for many applications, soaked in acid or caustic soda. They are very useful filters in many applications. The two most common brands available in New Zealand are Amiad and Arkal, both from Israel.

- Washable mobile media such as sand filters. These present a good barrier against a lot of particles, including quite small, turbidity-causing silts and clays. Typically, they act as the equivalent of about 10 micron filters. They are washed by air/water scour in the same way as conventional gravity rapid sand filters. They differ from normal rapid sand filters by having very high loading rates, too high to prevent floc from being sheared and passing through the sand. They do not qualify for protozoa log credits in DWSNZ.

MICROSTRAINERS

Several microstrainers were installed in New Zealand about 40 - 50 years ago. Some may still be operating. They may still be practical for some source waters.

Microstrainers are revolving drum screens made of the finest stainless steel mesh available, with gaps as small as 23 microns. This size is regarded as adequate to remove lumps too big for chlorine to penetrate. Microstraining followed by chlorination therefore can ensure inactivation of bacteria and viruses.

A microstrainer is typically a cylinder with mesh walls, revolving on its side about three-quarters immersed in water. The raw water enters through one end (the other is closed) and flows out through the mesh into the tank that contains it. The solids are caught on the inside surface of the mesh as the cylinder rotates. At the top of the rotation, they are washed off by high-pressure sprays from outside and collected in a trough mounted just above the inside water level. The difference between the inside and outside water levels, the so-called straining head, is around 40-60 mm.

Washwater use is around 1.5 - 2.5 per cent of flow through the plant. No chemicals are used and the washwater can usually be returned directly to a stream. The loading rate varies depending on mesh size and raw water quality, but is about 25 – 40 m/h through the total screen area.

Microstrainers were installed at many sites in New Zealand in the 1960s and virtually all have since been replaced because they do not present an adequate barrier to protozoa such as *Giardia* and *Cryptosporidium* and do not remove colloidal matter. One plant operating in Dunedin is being replaced with a membrane plant, and there are several in private plants such as meat works.

Algae Removal

High populations of phytoplankton can cause problems with different types of water treatment. The numbers of cells per litre in eutrophic lakes can be so high that processes such as cartridge, membrane, diatomaceous earth or slow sand filters can block quickly and become uneconomic.

Chemical coagulation plants can also be affected. During daylight hours, algae produce oxygen, creating bubbles that can cause the sludge blanket to rise and perhaps break up. The resultant floc carryover can put an excessive load on the filters. The algae can also grow on the sand beds, shortening filter runs. Some algae can be difficult to remove during backwash.

Killing the algae with chlorine or ozone could result in taste and odour problems as the organisms decay. Microstrainers can remove a large percentage of the cells without introducing excessive headloss. The drum speed can be increased if the algae begin to block the microfabric. The microfabric is available in a range of mesh sizes, from 23 microns to 65 microns. Laboratory trials can help select the most appropriate mesh. Between a third and a half of reservoir algae (about 700 cells per mL in a New Zealand raw water) have been removed by a 65 micron microstrainer. During trials for a new Auckland water source, about 90 per cent of Waikato River phytoplankton were removed by a 23 micron microstrainer. On average, these organisms were a little larger than the reservoir species. Despite the high removal of phytoplankton, the turbidity did not decrease much, indicating that those phytoplankton did not contribute greatly to the turbidity measurement.

In both cases, colonies and larger cells were removed more effectively than small cells.

ROUGHING FILTERS

According to the USEPA, roughing filtration is a technique used primarily in developing countries to remove solids from high turbidity source waters prior to treatment with slow sand filters. Typically, roughing filters consist of a series of sedimentation tanks filled with progressively smaller diameter media in the direction of the flow. The media can be gravel, plastic, crushed coconut, rice husks, or a similar locally available material. The flow direction in roughing filters can be either horizontal or vertical, and vertical roughing filters can be either upflow or downflow. The media in the tanks effectively reduce the vertical settling distance of particles to a distance of a few millimetres. As sediment builds on the media, it eventually sloughs off and begins to accumulate in the lower section, while simultaneously regenerating the upper portions of the filter. The filters require periodic cleaning to remove the collected silt.Originally it was thought that the USEPA would recommend a 0.5 protozoa log credit towards additional *Cryptosporidium* treatment requirements for roughing filters. However, after review of available literature, the USEPA was unable to identify design and implementation conditions for roughing filters that would provide reasonable assurance of achieving a 0.5 log removal of oocysts. Hence roughing filters do not appear in the DWSNZ or in Chapter 5: Protozoa Compliance of the Guidelines.

OVERLAND TREATMENT OF WASTEWATER

Apart from the use of effluent for irrigation of crops, termed 'slow rate' land treatment in the US Environmental Protection Agency's Process Design Manual for Land Treatment of Municipal Wastewaters, and 'rapid infiltration' or 'infiltration percolation' of effluent discussed as soil-aquifer treatment in a later section of this document, the EPA manual deals with 'overland flow' as a wastewater treatment method. In overland flow treatment, effluent is distributed over gently sloping grassland on fairly impermeable soils. Ideally, the wastewater moves evenly down the slope to collecting ditches at the bottom edge of the area and water-tolerant grasses are an essential component of the system.

This form of land treatment requires alternating applications of effluent (usually treated) and resting of the land, to allow soil reaction and grass cutting. The total area utilised is normally broken up into small plots to allow this form of intermittent operation and yet achieve continuous treatment of the flow of wastewater. Although this type of land treatment has been widely adopted in Australia, New Zealand and the UK for tertiary upgrading of secondary effluents, it has been used for the treatment of primary effluent in Werribee, Australia and is being considered for the treatment of raw sewage in Karachi, Pakistan.

Table. Site Characteristics And Design Features For Overland Flow Treatment Of Wastewater

Grade	Finished slope 2-8%
Field area required (ha)	6.55-44
Soil permeability	Slow (clays, silts and soils with
impermeable barriers)	
Annual application rate (m)	20-Mar
Typical weekly application rate (cm)	Jun-40

Basic site characteristics and design features for overland flow treatment have been suggested by EPA as shown in Table. It was pointed out that steeper land slopes might be feasible at reduced hydraulic loadings. The ranges given for field area required and application rates cover the wastewater quality from raw sewage to secondary effluent, with higher application rates and lower land area requirements being associated with higher levels of preapplication treatment. Although soil permeability is not critical with this form of land treatment, the impact on groundwater should not be overlooked in the case of highly permeable soils.

The application rate for wastewaters will depend principally on the type of soil, the quality of wastewater effluent and the physical and biochemical activity in the near-surface environment. Rational design procedures, based on the kinetics of BOD removal, have been developed for overland flow systems by Middlebrooks *et al.* Slope lengths from 30–60 m are common in the US for overland flow systems.

The cover crop is an important component of the overland flow system since it prevents soil erosion, provides nutrient uptake and serves as a fixed-film medium for biological treatment. Crops best suited to overland flow treatment are grasses with a long growing season, high moisture tolerance and extensive root formation. Reed canary grass has a very high nutrient uptake capacity and yields a good quality hay; other suitable grasses include rye grass and tall fescue.

Suspended and colloidal organic materials in the wastewater are removed by sedimentation and filtration through surface grass and organic layers. Removal of total nitrogen and ammonia is inversely related to application rate, slope length and soil temperature.

Phosphorus and trace elements removal is by sorption on soil clay colloids and precipitation as insoluble complexes of calcium, iron and aluminium. Overland flow systems also remove pathogens from sewage effluent at levels comparable with conventional secondary treatment systems, without chlorination. A monitoring programme should always be incorporated into the design of overland flow projects both for wastewater and effluent quality and for application rates.

MACROPHYTE TREATMENT

Maturation ponds which incorporate floating, submerged or emergent aquatic plant species are termed *macrophyte ponds* and these have been used in recent years for upgrading effluents from stabilisation ponds.

Macrophytes take up large amounts of inorganic nutrients (especially N and P) and heavy metals (such as Cd, Cu, Hg and Zn) as a consequnce of the growth requirements and decrease the concentration of algal cells through light shading by the leaf canopy and, possibly, adherence to gelatinous biomass which grows on the roots.

Floating macrophyte systems utilising water hyacinth and receiving primary sewage effluent in Florida have achieved secondary treatment effluent quality with a 6 day hydraulic retention time, water depth of 60 cm and hydraulic loading 1860 m^3/ha d. The same authors suggested that similar results had also been observed for artificial wetlands using emergent macrophytes. In

Europe, the land area considered to be necessary for treatment of preliminary-treated sewage is estimated at 2–5 m^2 per population equivalent to achieve a secondary effluent quality.

FLOATING AQUATIC MACROPHYTE SYSTEMS

Floating macrophyte species, with their large root systems, are very efficient at nutrient stripping. Although several genera have been used in pilot schemes, including *Salvinia, Spirodella, Lemna* and *Eichornia*, *Eichornia crassipes* (water hyacinth) has been studied in much greater detail. In tropical regions, water hyacinth doubles in mass about every 6 days and a macrophyte pond can produce more than 250 kg/ha d (dry weight). Nitrogen and phosphorus reductions up to 80 per cent and 50 per cent have been achieved.

In Tamil Nadu, India, studies have indicated that the coontail, *Ceratophyllum demersum,* a submerged macrophyte, is very efficient at removing ammonia (97 per cent) and phosphorus (96 per cent) from raw sewage and also removes 95 per cent of the BOD_5. It has a lower growth rate than *Eichornia crassipes,* which allows less frequent harvesting. In such macrophyte pond systems, apart from any physical removal processes which might occur (especially sedimentation) the aquatic vascular plants serve as living substrates for microbial activity, which removes BOD and nitrogen, and achieves reductions in phosphorus, heavy metals and some organics through plant uptake. The basic function of the macrophytes in the latter mechanism is to assimilate, concentrate and store contaminants on a short-term basis. Subsequent harvest of the plant biomass results in permanent removal of stored contaminents from the pond treatment system.

The nutrient assimilation capacity of aquatic macrophytes is directly related to growth rate, standing crop and tissue composition. The potential rate of pollutant storage by an aquatic plant is limited by the growth rate and standing crop of biomass per unit area.

Water hyacinth, for example, was found to reach a standing crop level of 30 tonnes (dry weight)/ha in Florida, resulting in a maximum storage of 900 kg N/ha and 180 kg P/ha. Fly and mosquito breeding is a problem in floating macrophyte ponds but this can be partially alleviated by introducing larvae-eating fish species such as *Gambusia* and *Peocelia* into the ponds. It should be recognised that pathogen die-off is poor in macrophyte ponds as a result of light shading and the lower dissolved oxygen and pH compared with algal maturation ponds.

In their favour, macrophyte ponds can serve a useful purpose in stripping pond effluents of nutrients and algae and at the same time produce a harvestable biomass. Floating macrophytes are fairly easily collected by floating harvesters. The harvested plants might be fed to cattle, used as a green manure in agriculture, composted aerobically to produce a fertilizer and soil conditioner, or can be converted into biogas in an anaerobic digester, in which case the residual sludge can then be applied as a fertilizer and soil conditioner. Maximum removal by water hyacinth was 5850 kg N/ha year, compared with 1200 kg N/ha year by duckweed.

Emergent Macrophyte Treatment Systems

In recent years, natural and artificial wetlands and marshes have been used to treat raw sewage and partially-treated effluents. Natural wetlands are usually unmanaged, whereas artificial systems are specially designed to maximize performance by providing the optimum conditions for emergent macrophyte growth. The key features of such reed bed treatment systems are:

- Rhizomes of the reeds grow vertically and horizontally in the soil or gravel bed, opening up 'hydraulic pathways'.
- Wastewater BOD and nitrogen are removed by bacterial activity; aerobic treatment takes place in the rhizosphere, with anoxic and anaerobic treatment taking place in the surrounding soil.
- Oxygen passes from the atmosphere to the rhizosphere via the leaves and stems of the reeds through the hollow rhizomes and out through the roots.
- Suspended solids in the sewage are aerobically composted in the above-ground layer of vegetation formed from dead leaves and stems.
- Nutrients and heavy metals are removed by plant uptake.

The growth rate and pollutant assimilative capacity of emergent macrophytes such as *Phragmites communis and Scirpus lacstris* are limited by the culture system, wastewater loading rate, plant density, climate and management factors.

NATURAL BIOLOGICAL TREATMENT SYSTEMS

Natural low-rate biological treatment systems are available for the treatment of organic wastewaters such as municipal sewage and tend to be lower in cost and less sophisticated in operation and maintenance. Although such processes tend to be land intensive by comparison with the conventional high-rate biological processes already described, they are often more effective in removing pathogens and do so reliably and continuously if properly designed and not overloaded. Among the natural biological treatment systems available, stabilisation ponds and land treatment have been used widely around the world and a considerable record of experience and design practice has been documented. The nutrient film technique is a fairly recent development of the hydroponic plant growth system with application in the treatment and use of wastewater.

WASTEWATER STABILISATION PONDS

A recent World Bank Report came out strongly in favour of stabilisation ponds as the most suitable wastewater treatment system for effluent use in agriculture. Table provides a comparison of the advantages and disadvantages of ponds with those of high-rate biological wastewater treatment processes. Stabilisation ponds are the preferred wastewater treatment process in developing countries, where land is often available at reasonable opportunity cost and skilled labour is in short supply.

Key:

FC	= Faecal coliforms;
SS	= Suspended slids;
G	= Good;
F	= Fair;
P	= Poor

Wastewater stabilisation pond systems are designed to achieve different forms of treatment in up to three stages in series, depending on the organic strength of the input waste and the effluent quality objectives. For ease of maintenance and flexibility of operation, at least two trains of ponds in parallel are incorporated in any design.

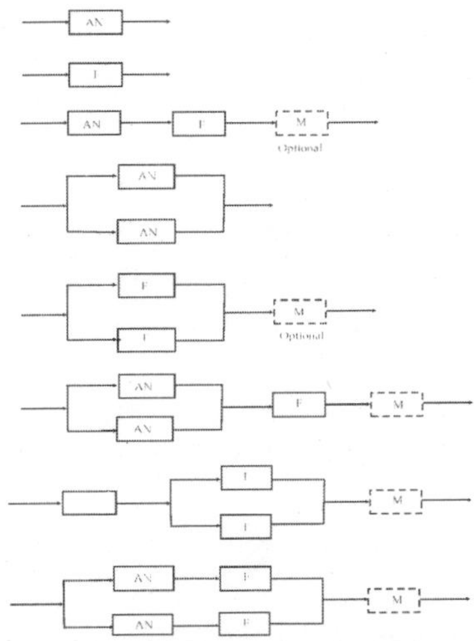

Fig. Stabilisation Pond Configurations AN = anaerobic pond; F = facultative pond; M = maturation pond

Strong wastewaters, with BOD_5 concentration in excess of about 300 mg/l, will frequently be introduced into first-stage anaerobic ponds, which achieve a high volumetric rate of removal.

Weaker wastes or, where anaerobic ponds are environmentally unacceptable, even stronger wastes (say up to 1000 mg/l BOD_5) may be discharged directly into primary facultative ponds. Effluent from first-stage anaerobic ponds will overflow into secondary facultative ponds which comprise the second-stage of biological treatment. Following primary or secondary facultative ponds, if further pathogen reduction is necessary maturation ponds will be introduced to provide tertiary treatment.

Anaerobic Ponds

Anaerobic ponds are very cost effective for the removal of BOD, when it is present in high concentration. Normally, a single, anaerobic pond in each treatment train is sufficient if the strength of the influent wastewater, L_i is less than 1000 mg/l BOD_5. For high strength industrial wastes, up to three anaerobic ponds in series might be justifiable but the retention time t_{an}, in any of these ponds should not be less than 1 day.Anaerobic conditions in first-stage stabilisation ponds are created by maintaining a high volumetric organic loading, certainly greater than 100g BOD_5/m^3 d. Volumetric loading, l_v, is given by:

$$\lambda - = \frac{L_i Q}{V}$$

where:

L_i = Influent BOD_5, mg/l,
Q = Influent flow rate, m^3/d, and
V = Pond volume, m^3

or, since V/Q = t_{an}, the retention time:

$$\lambda_v = \frac{L_i}{t_{an}}$$

Very high loadings, up to 1000g BOD_5/m^{3}d, achieve efficient utilisation of anaerobic pond volume but, with wastewater containing sulphate concentrations in excess of 100 mg/l, the production of H_2S is likely to cause odour problems. In the case of typical municipal sewage, it is generally accepted that a maximum anaerobic pond loading of 400g BOD_5/m^{3}d will prevent odour nuisance.

Table. Bod Removals In Anaerobic Ponds Loaded At 250 G Bod_5/M^{3}d

Retention t_{an} days	*BOD_5 removal per cent3*
1	50
2.5	60
5	70

Anaerobic ponds normally have a depth between 2m and 5m and function as open septic tanks with gas release to the atmosphere. The biochemical reactions which take place in anaerobic ponds are the same as those occurring in anaerobic digesters, with a first phase of acidogenesis and a second slower-rate of methanogenesis. Ambient temperatures in hot-climate countries are conducive to these anaerobic reactions and expected BOD_5 removals for different retention times in treating sewage have been given by Mara as shown in Table. More recently, Gambrill *et al.* have suggested conservative removals of BOD_5 in anaerobic ponds as 40 per cent below 10°C, at a design loading, l_v, of 100 g/m^{3}d, and 60 per cent above 20°C, at a design loading of 300 g/m^{3}d, with linear interpolation for operating temperature between 10 and 20°C. Higher removal rates are possible with industrial wastes, particularly those containing significant quantities of organic settleable solids. Of course, other environmental conditions in the ponds, particularly pH, must be suitable for the anaerobic microorganisms bringing about the breakdown of BOD.In certain instances, anaerobic ponds become covered with a thick scum layer, which is thought to be beneficial but not essential, and may give rise to increased fly breeding. Solids in the raw wastewater, as well as biomass produced, will settle out in first-stage anaerobic ponds and it is common to remove sludge when it has reached half depth in the pond. This usually occurs after two years of operation at design flow in the case of municipal sewage treatment.

Facultative Ponds

and facultative ponds will often be more appropriate than conventional forms of secondary biological treatment for application in developing countries. Primary facultative ponds will be designed for the treatment of weaker wastes and in sensitive locations where anaerobic pond odours would be unacceptable. Solids in the influent to a facultative pond and excess biomass produced in the pond will settle out forming a sludge layer at the bottom. The benthic layer will be anaerobic and, as a result of anaerobic breakdown of organics, will release soluble organic products to the water column above.

Organic matter dissolved or suspended in the water column will be metabolised by heterotrophic bacteria, with the uptake of oxygen, as in convential aerobic biological wastewater treatment processes. However, unlike in convential processes, the dissolved oxygen utilised by the bacteria in facultative ponds is replaced through photosynthetic oxygen production by microalgae, rather than by aeration equipment. Especially intreating municipal sewage in hot climates, the environment in facultative ponds is ideal for the proliferation of microalgae.

High temperature and ample sunlight create conditions which encourage algae to utilise the carbon dioxide (CO_2) released by bacteria in breaking down the organic components of the wastewater and take up nutrients (mainly nitrogen and phosphorus) contained in the wastewater. This symbiotic relationship contributes to the overall removal of BOD in facultative ponds, described diagrammatically by Marais as in Figure.

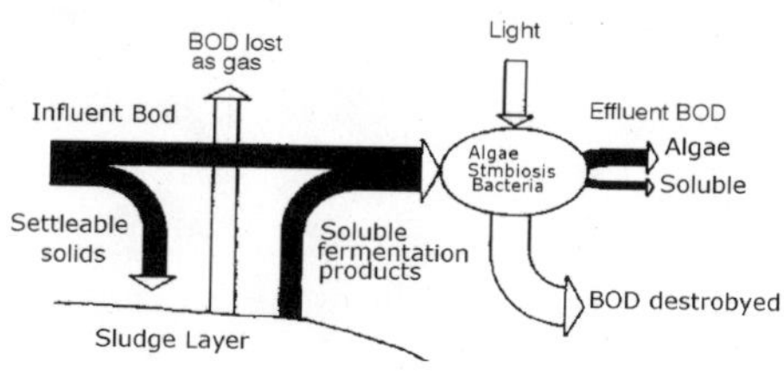

Fig. Energy flows in facultative stabilisation ponds

To maintain the balance necessary to allow this symbiosis to persist, the organic loading on a facultative pond must be strictly limited. Even under satisfactory operating conditions, the dissolved oxygen concentration (DO) in a facultative pond will vary diurnally as well as over the depth. Maximum DO will occur at the surface of the pond and will usually reach supersaturation in tropical regions at the time of maximum radiation intensity.

From that time until sunrise, DO will decline and may well disappear completely for a short period. For a typical facultative pond depth, D_f, of 1.5m the water column will be predominantly aerobic at the time of peak radiation and predominantly anaerobic at sunrise. The pH of the pond contents will also vary diurnally as algae utilise CO_2 throughout daylight hours and respire, along with bacteria and other organisms, releasing CO_2 during the night.

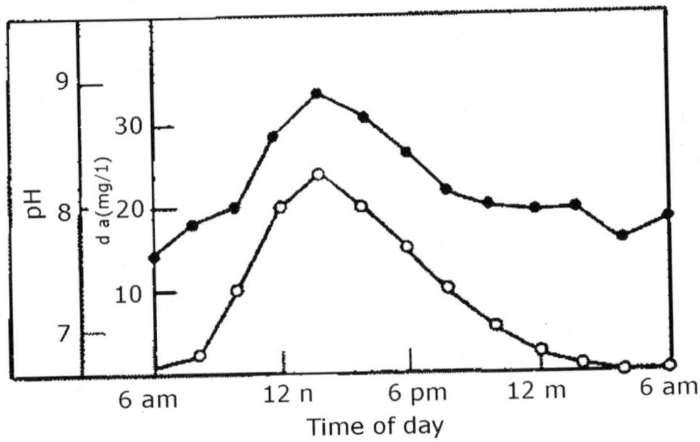

Fig. Diurnal variation of dissolved oxygen and pH in facultative pond, pH:, dissolved oxygen:

Wind is considered important to the satisfactory operation of facultative ponds by mixing the contents and helping to prevent short-circuiting. Intimate mixing of organic substrate and the degrading organisms is important in any biological reactor but in facultative ponds wind mixing is considered essential to prevent thermal stratification causing anaerobiosis and failure. Facultative ponds should be orientated with the longest dimension in the direction of the prevailing wind.

Although completely-mixed reactor theory with the assumption of first-order kinetics for BOD removal can be adopted for facultative pond design, such a fundamental approach is rarely adopted in practice. Instead, an empirical procedure based on operational experience is more common. The most widely adopted design method currently being applied wherever local experience is limited is that introduced by McGarry and Pescod. A regression analysis of operating data on ponds around the world relating maximum surface organic loading, in lb/acre d, to the mean ambient air temperature, in °F, of the coldest month resulted in the following equation (now converted to metric units):

$$\lambda_{s(max)} = 60.3\,(1.099)^T$$

where:

l_s = surface or arael organic loading, kg BOD_5/ha.d

T = mean ambient air temperature of coldest month, °C

Subsequently, Arthur modified this formula and suggested that best agreement with available operating data, including a factor of safety of about 1.5, is represented by the relationship:

$$\lambda_s = 20T - 60$$

This surface (or areal) BOD_5 loading can be translated into a mid-depth facultative pond area requirement (A_f in m^2) using the formula:

$$A_f = \frac{10 L_i Q}{\lambda_s}$$

Thus:

$$A_f = \frac{L_i Q}{2T - 6}$$

and the mean hydraulic retention time in the facultative pond (t_f in days) is given by:

$$t_f = \frac{A_f D_f}{Q}.$$

The removal of BOD_5 in facultative ponds (l_r in kg/ha d) is related to BOD_5 loading and usually averages 70-80 per cent of l_s. Retention time in a properly designed facultative pond will normally be 20-40 days and, with a depth of about 1.5m, the area required will be signficantly greater than for an anaerobic pond. The effluent from a facultative pond treating municipal sewage in the tropics will normally have a BOD_5 between 50 and 70 mg/l as a result of the suspended algae. On discharge to a surface water, this effluent will not cause problems downstream if the dilution is of the order of 8:1 and any live algae in the effluent might well be beneficial as a result of photo synthetic oxygen production during daylight hours. -

Efficiently operating facultative ponds treating wastewater will contain a mixed population of flora but flagellate algal genera such as Chlamydomonas, Euglena, Phacus and Pyrobotrys will predominate. Non-motile forms such as Chlorella, Scenedesmus and various diatom species will be present in low concentrations unless the pond is underloaded. Algal stratification often occurs in facultative ponds, particularly in the absence of wind-induced mixing, as motile forms respond to changes in light intensity and move in a band up and down the water column.

The relative numbers of different genera and their dominance in a facultative pond vary from season to season throughout the year but species diversity generally decreases with increase in loading. Sometimes, mobile purple sulphur bacteria appear when facultative ponds are overloaded and sulphide concentration increases, with the danger of odour production. High ammonia concentrations also bring on the same problem and are toxic to algae, especially above pH 8.0. Maintenance of properly designed facultative ponds will be limited to the removal of scum mats, which tend to accumulate in downwind corners, and the cutting of grass on embankments. To ensure efficient operation, facultative ponds should be regularly monitored but, even where this is not possible, they have the reputation of being relatively trouble-free.

Maturation Ponds

The effluent from facultative ponds treating municipal sewage or equivalent input wastewater will normally contain at least 50 mg/l BOD_5 and if an effluent with lower BOD_5 concentration is required it will be necessary to use maturation ponds. For sewage treatment, two maturation ponds in series, each with a retention time of 7 days, have been found necessary to produce a final effluent with BOD_5 < 25 mg/l when the facultative pond effluent had a BOD_5 < 75 mg/l.

A more important function of maturation ponds, however, is the removal of excreted pathogens to achieve an effluent quality which is suitable for its downstream reuse. Although the longer retention in anaerobic and facultative pond systems will make them more efficient than conventional wastewater treatment processes in removing pathogens, the effluent from a facultative pond treating municipal sewage will generally require further treatment in maturation ponds to reach effluent standards imposed for reuse in unrestricted irrigation.

Faecal coliform bacteria are commonly used as indicators of excreted pathogens and maturation ponds can be designed to achieve a given reduction of faecal coliforms (FC). Protozoan cysts and helminth ova are removed by sedimentation in stabilisation ponds and a series of ponds with overall retention of 20 days or more will produce an effluent totally free of cysts and ova.

Reduction of faecal coliform bacteria in any stabilisation pond (anaerobic, facultative and maturation) is generally taken as following first-order kinetics:

$$N_e = \frac{N_i}{1 + K_b t}$$

where:

N_e = Number of faecal coliforms/100 ml of effluent
N_i = Number of faecal coliforms/100 ml of influent
K_b = First-order rate constant for FC removal, d^{-1}
t = Retention time in any pond, d

For n ponds in series, Eq 8 becomes:

$$N_e = \frac{N_i}{(1+K_b t_{an})(1+K_b t_f)...(1+k_b t_{m0})}$$

where:

t_{mn} = Retention time in the nth maturation pond.

The value of KB is extremely sensitive to temperature and was shown by Marais to be given by:

$$K_{b(T)} = 26(1.19)T^{-2C}$$

where:

$$K_{b(T)} = \text{value of KB at T°C}$$

A suitable design value of N_i in the case of municipal sewage treatment is 1 x 10^8 faecal coliforms/100ml, which is slightly higher than average practical levels.

The value of N_e should be obtained by substituting the appropriate levels of variables in Eq 10 assuming a retention time of 7 days in each of two maturation lagoons (for sewage). If the calculated value of N_e does not meet the reuse effluent standard, the number of maturation ponds should be increased, say to three or more each with retention time 5 days, and N_e recalculated. A more systematic approach is now available whereby the optimum design for maturation ponds can be obtained using a simple computer programme.

Polprasert *et al.* have published an approach to the assessment of bacterial die-off which attempts to take into account the complex physical characteristics of ponds and biochemical reactions taking place in them. A multiple-regression equation involving parameters such as retention time, organic loading, algal concentration and ultra-violet light exposure has been suggested. The Wehner and Wilhelm non-ideal flow equation, including the pond dispersion number, was adopted to predict bacterial survival, in preference to the first order rate equation.

Maturation ponds will be aerobic throughout the water column during daylight hours and the pH will rise above 9.0. The algal population of many species of non-flagellate unicellular and colonial forms will be distributed over the full depth of a maturation pond. Large numbers of filamentous algae, particularly blue-greens, will emerge under very low BOD loading conditions. Very low concentrations of algae in a maturation pond will indicate excessive algal predation by zooplankton, such as Daphnia sp, and this will have a deleterious effect on pathogen die-off, which is linked to algal activity.

Saqqar, in his analysis of the performance of the Al Samra stabilisation ponds in Amman, Jordan, has shown that the coliform and faecal coliform die-off coefficients varied with retention time, water temperature, organic loading, total BOD_5 concentration, pH and pond depth. Total coliform die-off was less than the rate of faecal coliform die-off, except during the cold season. For the series of ten ponds, including at least the first five totally anaerobic, the faecal coliform die-off coefficient, k, for the temperature range 12 - 15°C increased through the pond sequence from 0.11 per day in the first anaerobic pond to 0.68 per day in the final two ponds, which operated as facultative ponds.

BIOLOGICAL OXIDATION IN WASTE WATER TREATMENT

Biological oxidation plants are best suited to treat organic wastes in solution or in non-settleable form. Inorganic compounds such as cyanides, sulphides, thiocyanates and thiosulphates, capable of being oxidised by autotrophic bacteria, may also be treated. The use of the adsorptive properties of the film or flocs to remove settleable matter, or matter not biologically oxidisable, should be strictly limited. Particulate matter is better removed by settlement, possibly preceded by a process of flocculation. Matter in solution which is incapable of biological oxidation, such as metallic salts, may prove injurious to the micro-organisms in the plant. Furthermore, although metals may initially be removed from the liquor by adsorption or even synthesised in the microbial growth, they cannot be broken down like organic matter. They may leave the plant as humus or as excess sludge, in which case they may cause difficulty in the treatment and subsequent disposal of the sludge. Another fraction, however, may return into solution as when the film in a bacteria bed is removed by grasing fauna or when film and activated sludge undergo auto-oxidation under conditions of starvation.

Stone found that with different metals different percentages (60-100) of the applied metal concentration appeared in the effluent. Of this amount different proportions were present in the humus and in solution; with copper, for example, approximately one third was in solution, whilst with nickel the amount was three quarters. Thus a considerable proportion of the applied concentration may appear in the final effluent after the settlement of the humus and result in a toxic effluent. The lethal concentration of the substances to stream life is probably much lower than that for microorganisms and grazers in the biological oxidation plants. Even though the concentration of metals in the waste does not seriously interfere with the biological oxidation, considerable reduction in toxicity may be needed to ensure a non-toxic effluent. It is possible that in treatment plants the metals become complexed with protein molecules, thereby reducing the toxicity. Metals may be present in wastes other than those from metal processing plants. Malt-distillery waste waters, for example, have been reported as containing 66.0 ppm Zn, 6.0 ppm Cu, 4.5 ppm Fe and 2.0 ppm Pb. The metals were probably derived from the vessels and pipelines used in the process.

Radioactive wastes, like metals, cannot be destroyed in biological oxidation plants. Although radioactivity decays at a rate characteristic of the material, this rate, measured in terms of its half-life, cannot be altered by treatment methods. Belcher reported that certain radioactive substances were removed in biological oxidation plants; the radioactivity is transferred to the sludge, however, and the problem of disposal remains. The most widely used method for treating waste waters of low to medium levels of activity is by chemical precipitation, thus transferring the radioactivity to a solid phase, which of course still has to be disposed off. In the latter half of the last century the association of such diseases as cholera and typhoid fevers with the polluted conditions of the rivers was probably a major factor in drawing attention to the need for adequate sewage treatment, even though the causative organisms were then unknown. The removal of pathogens is not now, however, generally considered a primary function of sewage treatment in this country. In America and in Russia much more attention is given to the bacterial standards of effluents. Although some plants effect a high degree of *Bact. coli* removal, some newer processes which are more efficient in the oxidation of organic matter are less efficient in reducing the number of *Bact. coli*. The B.O.D. of an effluent is no true indication of its bacterial content.

THE ASSESSMENT OF PLANT EFFICIENCY

The purpose of waste water treatment is to prevent the pollution of the receiving water. Considering pollution as "an act which changes the natural qualities of the water in a river or stream" the ultimate assessment of the effectiveness of a plant is the extent to which the effluent affects the nature of the receiving water. What constitutes the "natural qualities" of water is the subject of some controversy because of the different interests involved. In this country, fishery interests in preserving the conditions of our streams suitable for fish life, to some extent safeguard the interests of others whose specific requirements may, however, differ. Aesthetic considerations, for example, should largely be satisfied.

Public health requirements, however, are concerned both with the chemical content of the water and with the pathogenic organisms present. Although by satisfying the fishery requirements the chemical aspects are to a large degree safeguarded, the pathogenic content of the water is not usually considered important. Conversely a chlorinated water, satisfactory from a public health aspect, may be quite unsuitable for fisheries. In this country these two needs are satisfied by preserving the streams in as near natural condition as possible and chlorination is carried out after extraction for drinking purposes. The discharge of an effluent may affect the natural quality of a water in several ways. These have been discussed at length elsewhere, and will only be briefly summarised here in discussing methods used to assess them.

Organic Pollution

Biologically oxidisable organic matter discharged to a stream is oxidised by micro-organisms as in the treatment plant. In the process the oxygen dissolved in the stream water is utilised and when the rate of removal is greater than the rate of re-aeration, resulting from surface aeration and the photosynthetic activity of plant life, the oxygen is depleted. Because of the time lag and flow of the stream, this deoxygenation. is greatest some distance below the effluent and is followed by a recovery when the respiratory activity is successively reduced because of the progressive reduction in organic matter present. This phenomenon is known as the "oxygen sag" and the process as "selfpurification". Synthesis, associated with the biological oxidation of the introduced organic matter, results in an increase in saprobic organisms—bacteria, fungi and protozoa—some of which form macroscopic growths attached to the stream bed; these growths are commonly known as "sewage fungus". Both the depletion of oxygen and the development of sewage fungus adversely affect the quality of the stream.

The potential oxygen demand of an effluent can be assessed directly by measuring the rate of oxygen uptake of a given volume of the effluent contained in a sealed vessel under carefully controlled conditions. Such manometric tests can be carried out over long periods and give not only the ultimate oxygen demand of an effluent but also a curve showing the course of oxidation. Techniques involving this principle, reviewed by Jenkins, are unfortunately not practicable as routine tests for large numbers of samples. Manometric methods have, however, been suggested for B.O.D. determinations to eliminate errors due to dilution.

The biochemical oxygen demand (B. O. D.) has been generally accepted as a simple measure of the potential deoxygenating effect of the biologically oxidisable matter present in an effluent. It represents the oxygen uptake over the initial five-day period of the total respiratory curve. As a comparative test of the oxygen demand of effluents it is only strictly valid if the ratio of the B.O.D. to the ultimate oxygen demand is constant, *i.e.* if the respiratory curves are truly exponential. With sewage effluents it has been shown that their oxidation takes place in at least two definite stages representing the oxidation of carbonaceous and nitrogenous matter; sometimes two stages of nitrogenous oxidation are distinguishable—the first representing the oxidation of ammonia to nitrite and the second of nitrite to nitrate. These different stages result in definite steps in the oxidation curve and thus the ratio of B.O.D. to ultimate oxygen demand remaining will vary at different degrees of oxidation.

It can be seen that it is possible for the B.O.D. over a five-day period to be greater when the oxidation is more advanced; compare, for example, the uptake between 5 and 10 days with that between 15 and 20 days when active nitrification was occurring. The implicit faith by some in the B.O.D. test is exemplified by the fallacious argument that nitrification is undesirable in sewage treatment plants because it tends to increase the B.O.D. value! The same effect probably occurs with some effluents from plants treating mixed industrial organic wastes. Thus, although the B.O.D. is a simple and convenient test, care is needed in interpreting the results. Similarly, as pointed out by Wilson, the efficiencies of treatment plants cannot be measured in terms of percentage B.O.D. removal unless the ratio of B.O.D. to the ultimate oxygen demand was the same for influent and effluent.

In contrast to these direct biological tests for assessing the oxygen demand, chemical tests are also used. By determining the oxidisable substances present in the effluent and knowing the course of oxidation, the theoretical ultimate oxygen demand can be calculated. Gameson and Wheatland calculated the ultimate oxygen demand for sewage effluents on this basis, accounting for the complete oxidation of the organic carbon, organic nitrogen, ammonia and nitrite by the equation

U. O. D. = 2.67 (organic carbon) + 4.57 (ammoniacal N + organic N) + 1.14 (nitrous N)

The results they obtained were in good agreement with those obtained by direct manometric methods. The above equation, however, assumes that the carbonaceous matter concerned has a respiratory quotient of unity, *i.e.* one volume of oxygen is utilised to produce one volume of carbon dioxide, one molecule of oxygen being required for each molecule of carbon. With sewage and sewage effluents this was apparently valid but other carbonaceous matter present in industrial wastes may have a different respiratory quotient and the figure of 2.67 C is not, therefore, necessarily generally applicable. Furthermore, although with the sewage effluents used, nearly all the organic matter was biologically oxidisable, with industrial wastes such as gas liquor, some may resist biological oxidation and therefore will not contribute to the oxygen demand, although it may represent a toxic fraction and therefore be of importance in assessing the overall polluting effect.

The determination of organic carbon is a lengthy process and for routine work the amount of oxygen absorbed from acid potassium permanganate or from boiling acid potassium dichromate is used as a measure of the oxygen demand. These two tests, the "Permanganate Value" (4 hr O.A.) and the "Dichromate Value" (Chemical Oxygen Demand) measure the chemically oxidisable matter present. These values do not represent the full oxygen requirements for the oxidation of all the biologically oxidisable organic matter. On the other hand they may include biologically stable organic matter.

The permanganate test is well established in sewage practice and pollution control in this country, probably because of its simplicity and the speed with which it can be carried out. It is, however, essentially an empirical measure and although of undoubted value in the operational control of individual plants continuously treating the same waste, its value as a comparative test is limited except for plants treating similar wastes such as domestic sewage. The relationship between the permanganate value, the B.O.D. and the ultimate oxygen demand varies considerably with wastes having different organic substances present. The dichromate value has been advocated in the *"Recommended Methods for the Analysis of Trade Effluents"* for industrial wastes because of its reproducibility and applicability to a wide variety of wastes. It has been suggested that it is more truly proportionate to the calculated ultimate oxygen demand than is the permanganate value.

For routine plant control one or more of these oxygen demand tests are necessary and where practicable these should be supplemented from time to time by determining the ultimate oxygen demand either by calculation or by manometric means.

However accurately the theoretical oxygen demand of an effluent can be estimated, the prediction of the actual deoxygenating effect of its discharge into a receiving stream is complicated by several factors. Dilution, rate of re-aeration, rate of oxidation, temperature and the biological condition of the stream are inter-related factors which must be taken into consideration. The extent to which the ultimate oxygen demand of an effluent needs to be satisfied by the stream depends on the time of retention of the oxidisable matter in the stream in relation to its rate of oxidation. The pertinent retention time is not necessarily that of the water, for flocculated solids may settle on the stream bed and soluble organic matter may enter food chains and be synthesised into living organisms, thus further complicating the oxygen balance.

In lakes and estuaries with long retention times the full oxygen demand will probably have to be met. In certain circumstances, however, such as the pollution of a small tributary stream which at a short distance below the discharge, is greatly diluted on entering a major river, the rate of oxidation is probably more important in assessing the potential deoxygenation of the tributary stream itself. In such cases the B.O.D., which by involving a time factor (5 days), is to this extent a rate, is probably a more appropriate measure than the ultimate oxygen demand.

The actual deoxygenation resulting from a discharge can, of course, be measured by a series of dissolved oxygen determinations. Because of diurnal and seasonal fluctuations, large numbers of determinations are required, preferably a continuous record. A recording instrument for such determinations has been developed by the W.P.R. Laboratories. Dissolved oxygen determinations, however, only measure one aspect of organic pollution—deoxygenation; the effects on the stream bed, where most of the stream organisms exist, are more difficult to measure analytically. Animal and plant communities living in the beds of streams exist in a dynamic state of balance which is very sensitive to changes in the environment. Organic pollution, by changing the environment, results in a change in the nature of the stream community to an extent depending upon the degree of pollution. Such effects have been studied by many workers and their results have been reviewed by Klein and Hynes. Based on such findings the degree and extent of organic pollution can readily be assessed by the biological examination of the stream bed.

TOXIC AND PHYSICAL POLLUTION

Some effluents affect the natural qualities of the water by addition of toxic or poisonous substances. Where specific substances are known to be involved these can be assessed analytically. In many cases, however, a whole range of substances may be present in the waste and then routine determinations could prove a formidable task. In such cases the biological assay of the toxicity of the effluent, using fish or invertebrate stream fauna as test animals, is sometimes practised. As with organic pollution, difficulties arise in using the results of such tests to predict the effects of discharges on streams. Even when full analysis is practicable, because of interactions involving synergistic and antagonistic effects and other factors such as pH and dissolved oxygen, it is difficult to predict the overall effect on a stream.

Although bio-assay measures more directly the toxicity of an effluent to the animals used in the test, the effect on the complex stream community and environment in which organisms naturally exist is again difficult to forecast. In contrast to organic pollution, which by encouraging the activity of certain species causes a change in the balance of populations, toxic pollution tends to suppress activity generally. Because different species are tolerant of different concentrations of a given poison, it may result in a change in the relative balance of populations and in some cases the more tolerant species may increase in number due to the reduced interspecific competition. The general effect, however, is to reduce the numbers of individuals and the number of species present.

A discharge, although neither toxic nor deoxygenating, may affect the natural quality of the water by changing its physical properties. Discharges which appreciably raise the temperature, increase the turbidity or impart colouration are examples of physical pollution. Inert matter which settles on the stream bed may change appreciably the environment of the bottom dwelling organisms. Such deposits can seriously affect the spawning grounds of some fish, for example. Physical pollution usually has a marked effect on the stream organisms, especially when it affects the physical nature of the stream bed. In such cases it results in a change in the nature of the community; for example the deposition of solids on a naturally stony bed causes the replacement of a stone-loving fauna by one typical of silted conditions. Thus both toxic and physical pollution can readily be detected by biological examination of the receiving stream. Apart from the above mentioned types of pollution the discharge of substances may alter the chemical nature of the water in other ways. Some readily oxidisable inorganic substances, such as sulphites in paper wastes, bring about simple de-oxygenation. Other wastes may alter the degree of hardness of the water or its salinity and thereby affect its natural qualities. Most effluents affect the receiving stream in more than one of the above mentioned ways. The type of tests chosen for operational control of a plant should be determined by the nature of the possible pollution and the nature of the receiving stream. In most cases such tests can with advantage be supplemented by biological examination of the receiving stream.

THE CHOICE OF PROCESS-ACTIVATED SLUDGE OR BACTERIA BEDS

Many factors (and possibly some prejudices) will determine the choice of plant, but here we shall briefly consider the ecological considerations involved in a choice between the two most common biological oxidation processes used in this country—activated sludge and bacteria beds. The theoretical considerations have already been discussed. The sludges resulting from the two processes represent different energy levels, the activated sludge being of a higher energy order than humus sludge. Thus, from a point of view of energy conservation, activated sludge is superior. The "sludge" developed from certain wastes could possibly be exploited as animal feeding stuff; vitamin B.12 is extracted from activated sludge. In sewage practice activated sludge is, however, more difficult to dispose of than humus sludge.

Activated sludge is generally more sensitive to fluctuations in flow and load and to toxic discharges than are bacteria beds. On the other hand, if a bacteria bed breaks down, due to overloading or toxic discharges, the period needed for the recovery of both micro-organisms and grasing fauna is longer than for the re-establishment of the simpler community of activated sludge. Some wastes, such as milk wastes, are not amenable to treatment by activated sludge because they tend to give rise to filamentous sludges which do not readily settle.

Other wastes which are biologically oxidisable but which prevent the establishment of an efficient grasing fauna, or which cause solids to be deposited in the bed which are not readily removed, are best treated by activated sludge. Because of the longer period of maturation needed for bacteria beds, seasonal wastes and those which fluctuate appreciably in volume seasonally are best treated by activated sludge when possible. Since low temperatures have a less serious effect on activated sludge than on bacteria beds, the former are to be preferred for localities subject to severe winter conditions. The more complex community of a bacteria bed, having organisms on several trophic levels, is more stable than that of activated sludge. Because of this, an activated sludge plant is more sensitive to chance changes in conditions, but for the same reason is more responsive to intentional changes imposed by operational control.Thus, with bacteria beds, less frequent adjustment to operating conditions is required. Activated sludge, however, is more amenable to frequent adjustments. The choice of plant will, therefore, be affected by the availability of experienced plant operators. The possibility of a greater degree of variability in bacteria bed operation, such as variable frequency dosing has, however, reduced this difference in the degree of plant control required. Other recognised comparisons between the two processes, such as operational costs, the inevitable linking of fly nuisance with bacteria beds, and the acceptance of non-nitrified effluents from activated sludge, all need to be re-assessed in light of developments in both processes.

APPLICATION OF AUTECOLOGY

DETERMINING A SUITABLE ENVIRONMENT FOR THE REQUISITE ORGANISMS

Although physical adsorption may account for the initial removal of organic matter from the waste water, the breakdown of the organic matter is the result of biological activity. The rate at which this occurs under optimum conditions depends on the readiness with which the micro-organisms are able to utilise the waste, this in turn is determined by the intrinsic properties of the waste and the respective micro-organisms. The theoretical rate at which a waste is oxidised biologically is known as the reaction constant (K) which is a measure of the oxidizability of the waste. In bacteria beds, according to Velz "the rate of extraction of organic matter per interval depth of a biological bed is proportional to the remaining concentration of organic matter measured in terms of its removability". Expressed mathematically,

$$\frac{L_D}{L} = IO^{-KD}$$

where L = Total removable B.O.D. in feed

LD = Remaining removable B.O.D. at depth D

K = The reaction constant.

In practice, this maximum rate may be further limited by different environmental conditions imposed by the waste or by the plant. Such conditions may be studied in isolation in the laboratory but, as mentioned in our study of autecology, care must be taken in the application of the results. For example, the interaction of factors and their effect on the balance of populations must be considered. In this respect, results are probably more applicable to the simpler community of activated sludge than to bacteria beds. Nevertheless such studies provide results which, if intelligently applied, can give a useful guide as to the range of tolerance, if not indicating the optimum conditions for most efficient purification.

Environmental Conditions Imposed by the Waste

Nutrition—For those concerned with waste water treatment it is an encouraging thought that all naturally occurring organic substances and many synthetic ones are attacked by at least one species of micro-organism and that organisms exist which are potentially capable of utilising organic compounds not yet synthesised. The presence of specific substances in the waste, is therefore a most important factor in the environment. Some organisms are able to utilise a wide range of organic substances as their primary food source but others appear to be specific in their requirements. In the latter case it is necessary to ensure a continuous supply of the specific waste, or if this is not practicable, to retain a stock culture of the organisms.

All micro-organisms require certain basic nutrient elements—C, N, P, and S, together with certain trace elements such as K, Ca, Zn, Mg, Fe, Mn, Cu and Co. Furthermore the C, N, and P are required in roughly balanced amounts. In physiology the proportion of C to N is known as the C/N ratio; in waste water treatment practice it is usual to measure the C as B.O.D. and the proportion is then expressed as B.O.D:N or B.O.D:P. The optimum ratio for different organisms varies somewhat but as a basis for experiment a ratio of B.O.D:N:P of 100:6:1.5 could be used. The nitrogen is considered only fully available when present as NH_3 and the phosphorus as soluble PO_4, although other forms of nitrogen which can be converted into NH_4 may thus become available.

Domestic sewage usually provides a nutritionally balanced food with the necessary trace elements and vitamins for bacterial activity. The presence of a large proportion of trade wastes, however, may upset the balance. Some wastes such as those from milk processing and plastic manufacture may increase the carbon-nitrogen ratio and encourage troublesome filamentous growths. Pre-treatment of sewages containing trade wastes, before treatment in biological oxidation plants, may possibly upset the balance; a sewage with a high iron content, if subjected to a process of flocculation, could appreciably reduce the phosphate content for example.

Industrial wastes may be deficient in nitrogen, *e.g.* cider wastes, cotton-kiering wastes; or nitrogen and phosphorus, *e.g.* citrus waste, brewery waste and paper wastes. Such deficiencies can be made good by the addition of the requisite amount of ammonium salt and phosphate to the influent. When practicable, domestic sewage could be used as a source of nitrogen and phosphorus although Wilson found that by supplying the phosphorus, deficient in a chemical manufacturing waste, by the addition of sewage, troublesome slime growths accumulated on the medium in the bacteria beds. As a result, agricultural grade triple super-phosphate was added to provide a concentration of phosphate in excess of that found necessary by laboratory tests—0.4 ppm P. The actual amount of nitrogen and phosphorus required depends to some extent on the method of plant operation. In a plant where the organisms are subjected to endogenous respiration. the autolysis may release nitrogen and phosphorus for further synthesis—the carbon being lost as carbon dioxide. In treating wastes deficient in nutrients, the recycling of effluent containing a higher proportion of such nutrients may be one advantage of recirculation.

Toxicity—The presence of toxic substances in a waste may impair biological activity for reasons previously discussed. The structure of the bacteria bed film or sludge floc affords the embedded bacteria some protection against toxic flushes, but above certain concentrations they seriously affect the activity of the cells. In bacteria beds, toxic matter may be adsorbed on the film in the upper layers of the bed, thus reducing the concentration to which the films lower in the bed is subjected. No such protection is afforded in the activated sludge system where any part or all of the sludge may be subjected to high concentrations of toxic matter. This may be the reason why bacteria beds are usually considered better able to withstand toxic discharges.

It is not possible to state the concentrations at which different toxic substances seriously affect the efficiency of biological oxidation plants. For one thing, the toxicity depends upon such other factors as pH, dissolved oxygen, temperature and the presence of other toxic products. The toxicity of metallic ions in wastes admixed with sewage may be reduced if they are complexed with the proteins present in sewage. A further difficulty is the variation in sensitivity of the different organisms. Because of these complications, reference to the literatures gives a wide range of tolerable toxic concentrations.

The heavy metals are usually considered toxic at above 1-3 ppm, but it has been reported by Ross and Sheppard that the efficiency of bacteria responsible for the oxidation of phenolic wastes from oil refineries was not affected by concentrations of copper up to 100 ppm. Under steady low concentrations of toxic substances it is possible that a more resistant microflora will become established, although this may not be as efficient as the original one. In this respect occasional strong toxic discharges are more serious than a steady weak discharge. As previously mentioned, however, metals are not amenable to biological oxidation and to ensure a non-toxic effluent their concentration in the feed should be reduced to a minimum.

Toxic non-metallic wastes such as phenols, thiosulphates, thiocyanates, cyanides, sulphides, formaldehydes, etc., although inhibiting the biological oxidation of other substances, are themselves subject to attack by specific organisms. Cyanides, for example, although interfering with biological treatment at concentrations between 1 and 2 ppm HCN, can be oxidised to ammonia and oxidised nitrogen in bacteria beds at concentrations of 60 ppm HCN without the addition of sewage. Ware and Painter isolated an organism, an actinomycete, which proved capable of breaking down cyanides and which readily colonised bacteria beds. Where practicable such wastes are best treated separately. Since they are oxidisable to relatively non-toxic compounds they can be accepted in controlled concentrations for treatment at sewage works although this would necessitate a larger oxidation plant, both because of their slower rate of oxidation and their suppressing effect on the biological oxidation of the more readily oxidisable matter.

The partial oxidation of an industrial waste containing mixed compounds before discharging to the sewage works, although greatly reducing the oxygen demand and thereby reducing the load on the sewage works, may contribute a residual waste difficult to oxidise and possibly inhibitory to the biological oxidation of the sewage. Unless sufficiently diluted with sewage effluent it could also impart toxicity to the effluent.

With wastes containing a mixture of such toxic compounds the oxidation of one may be inhibited by the presence of others. In the treatment of gas liquor, for example, the biological oxidation of the thiocyanate was found to be suppressed by the presence of phenols, and in practice did not occur until the phenol oxidation was almost complete. Apart from the toxic effect of the phenol, interspecific competition between the heterotrophic phenoloxidising bacteria and the autotrophic thiocyanate-oxidising organism may be involved, as is thought possible in the carbonaceous. oxidation and nitrification of sewage. To overcome this inhibitory effect, gas liquor is being treated experimentally in a two stage process —the first bed for the oxidation of phenols and the second for thiocyanate oxidation.

pH—Because enzymes are sensitive to pH changes it is not surprising that biological activity is affected by changes in pH. Different organisms have different pH optima and ranges of tolerance, acid conditions generally favouring fungi. Generally a pH between 6.0 and 8.0 is desirable for the biological oxidation of most wastes. Sewage is usually well buffered and its biological oxidation results in only a slight change in pH. With some industrial wastes, however, the change in pH as the waste is oxidised may be appreciable and create conditions unfavourable for further oxidation. The film or activated sludge floc may themselves act as a buffer against some degree of fluctuation in pH but are not able to withstand large changes. When a waste is oxidisable by different organisms, these may require different pH ranges. The bacterial oxidation of phenol is best effected between pH 6.5and 7.8 but it is also successfully attacked by a fungus between pH 2.0 and 2.5.

Environmental Conditions of the Plant

Biological oxidation plants are required to provide an environment in which the organisms and the waste are maintained in intimate contact in the presence of oxygen. In the activated sludge system this is ceffected by creating turbulent conditions in the aeration tanks by mehaical means, which at the same time introduces oxygen into the water from the atmosphere. In the bacteria bed similar requirements are satisfied by allowing the waste water to flow over a static film of microorganisms, the oxygen being supplied by absorption from the air passing through the bed. Factors influencing the degree to which these conditions are provided will be discussed in the respective sections on activated sludge and bacteria beds.

The waste itself, of course, creates environmental conditions within the plant other than those already mentioned. Very strong organic wastes such as cotton-kiering liquor, food processing waste waters and distillery wastes, having a high oxygen demand, result in conditions of oxygen depletion in the immediate environment of the organisms. No matter how efficient is the aeration within the plant, the rate of oxygen uptake would exceed the rate at which it could be transferred from the air through the liquid to the organism. Such liquors should, therefore, be diluted with returned effluent or by other means to produce a liquor having a B.O.D not exceeding 350 ppm, to ensure the prevention of anaerobic conditions within the plant. Wastes having a very high oxygen demand, a B.O.D exceeding 1,000 ppm, are probably best pre-treated by anaerobic processes followed by the biological oxidation of the resultant weaker liquor. It has been found that to induce satisfactory digestion of soluble organic wastes it is necessary to introduce a solid matrix. Pettet and others reported that this can be satisfactorily achieved by adding a massive inoculum of actively digesting sewage sludge which, by gradually increasing the rate at which the waste is added, can be developed into an active anaerobic sludge.

This is retained in the system by settling the effluent from the digester and returning the settled sludge as in the aerobic activated sludge process. In other countries such anaerobic processes have been successfully applied to a variety of strong wastes including those from yeast production and fermentation processes, meat packing, and chewing-gum manufacture in America, and yeast wastes in Denmark and Sweden. Of the wastes investigated in pilot plants by the W.P.R. Laboratories in England, slaughterhouse wastes would appear to be the most amenable to treatment by anaerobic processes, Meat wastes have also been successfully treated anaerobically in pilot plants in New Zealand by Hicks.

Temperature is an important factor determining the rate of biological activity. Each species has a temperature range over which life is possible. Within this range activity increases with increase in temperature to an optimum above which the thermal denaturing of the enzymes occurs at a rapidly increasing rate, resulting in decreased activity and ultimate death of vegetative cells. Some micro-organisms, however, produce highly resistant spores capable of surviving temperatures outside the range of tolerance for the vegetative cell. Organisms active in aerobic treatment plants are usually mesophilic, having optima between 20°C and 40°C, depending on the species. Within the physiological temperature range, biological activity generally changes by a factor of 2 for each 10°C change in temperature. Manometric studies have shown that this factor, known as the

Q10 value, for activated sludge was between 2.0 and 2.06, between temperatures of 0 and 25°C.In view of this significant effect of temperature on biological activity it may be expected that plant efficiency would be similarly affected. In fact, changes in temperature over a wide range of temperatures have little direct effect on the efficiency of treatment plants. There is, for example, little seasonal fluctuation in efficiency of activated sludge plants with changes in temperature, although increased temperatures could be expected to increase the efficiency both by increasing the oxygen transfer rate as well as directly increasing biological activity. Viehl and Meissner considered that the sludge organisms were able to adapt themselves to counteract the effect of low temperatures. Certainly, active sludges can be developed over a wide range of temperatures. Gehm investigated the treatment of waste waters from the pulp and paper industry which had temperatures of between 50 and 53°C; he found that such wastes could be satisfactorily treated at feed temperatures of 52°C by the activated sludge process, the efficiency being much the same as that between 30 and 40°C.

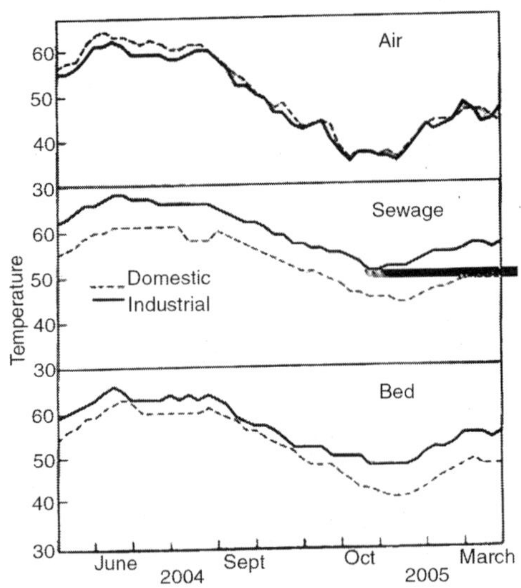

Fig. Comparison of Seasonal Temperature Fluctuations in a Bacteria Bed Treating Domestic Sewage with one Treating Industrial Sewage Together with Corresponding Air and Waste Temperatures.

Although sludges can be developed to operate at different temperatures their efficiency is reduced if they are subjected to violent changes in temperature. In sewage plants such violent changes in temperature do not occur; in plants treating specific industrial wastes, however, they could occur due to sudden discharge of a hot waste. Although the temperature near the surface of a bacteria bed is related to the temperature of the waste and the air temperature, the temperature within the bed is more closely related to that of the waste.

The temperature of the waste is, therefore, an important factor influencing the temperature of bacteria beds. Sewages containing industrial wastes are generally warmer during the winter than purely domestic sewages and thus beds treating sewages containing such wastes are not subjected to as low temperatures as are beds treating domestic sewage. The effect of the waste on the bed temperature depends to some extent on the volumetric loading as well as its temperature. The higher volumetric loadings common with filtration methods used for industrial sewages would, therefore, tend to further enhance this temperature difference.

In bacteria beds marked seasonal fluctuations in efficiency may occur, but these are due to the indirect effect of temperature on the film-fauna balance, as discussed later. In experimental laboratory beds operated at different temperatures over a range 5°C-30°C, although large differences in temperature at different depths within each bed made strict comparisons difficult, it was apparent that there was little difference in the efficiency between 10°C and 30°C; at 7°C, however, there was a marked decrease in efficiency. Grasing fauna were not introduced into the beds and it was found that in the beds of similar efficiency—those operating berween 10°C and 30°C—the amount of accumulated film was the same, whereas in the beds operating at 7°C, less film had been accumulated. Thus this direct effect of temperature was only evident below 10°C.

Ware found that the biological breakdown of cyanide in laboratory experimental beds was little affected by temperature changes in the range 10°-35°C, but that at temperatures outside this range, performance deteriorated rapidly. There was, however, a possibility that an organism other than the actinomyces, with which the bed had been inoculated, was active at the higher temperatures when recognisable actinomycetes were virtually absent. The lack of the expected relationship between temperature and plant efficiency may to some extent be due to adaptation or changes in the microbial flora with changing temperature, but the absence of response to short term temperature changes is more likely to be explained in terms of limiting factors.

The original theory, whereby it was considered that of the many factors influencing a biological process, only one was operative at any time and that this being so, the rate of the process could not be affected by changing the other, non-limiting, factors, is not now fully accepted. Nevertheless, by changing a factor which is not at the time limiting, the resultant effect on the process is only a fraction of that expected theoretically. On the other hand changing a limiting factor causes the expected change in the rate of the process until such times as another factor becomes limiting.

In biological oxidation plants the ultimate factor limiting the rate of oxidation is the intrinsic rate of generation, under optimum conditions, of the micro-organisms concerned. Of the many factors theoretically capable of influencing this rate, the ones most commonly limiting, under normal operating conditions, are probably the rates at which food or oxygen are transferred from the liquid to the cells. With weak wastes such as domestic sewage the transfer rate of the respiratory substrate—the waste—is probably limiting but with strong industrial wastes it is probably the oxygen transfer rate that is limiting, at least in the early stages of oxidation. With the exception of the surface layers of some exposed bacteria beds in winter, temperature is generally not a directly limiting factor and rises in temperature do not, therefore, bring about the increase in rate of oxidation which would be expected from the experimental Q_{10} value.

Temperature, however, affects the transfer rates, but according to Wilson, with oxygen or nutrient availability controlling, a 10°C rise in temperature would increase the transfer rate and, thereby, the rate of biological oxidation, by no more than 20-30 per cent. He suggested that the response of a plant to temperature changes provides a rough indication of the operative limiting factors, the greater the temperature effect the less the control by oxygen and nutrient availability. As an example he quotes data from the Monsanto plant at Ruabon showing that the Q_{10} value for a bacteria bed was 1.4 when the B.O.D. of the feed was 322 but was only 1.08 when the feed B.O.D. was halved.

A knowledge of which factor is limiting in any plant is, of course, important in plant operation. The principle of limiting factors may also explain the discrepancies in operational experiences which result in the advocacy of different and sometimes conflicting practices in treatment methods. The change in a factor, which in one plant produced marked improvements in efficiency, may give disappointing results in an apparently similar plant but where another factor is limiting.

THE APPLICATION OF SYNECOLOGY IN CONTROLLING POPULATIONS

Although autecology enables us to define the conditions of a plant necessary for efficient oxidation, the operation of a plant involves the controlled activity of different populations and the transfer of materials and energy between the waste and the different populations; as such it involves us in synecology. Of the several populations present in plants, the saprobic microorganisms, including the autotrophic bacteria, are the organisms primarily concerned with the breakdown of organic wastes and it is, therefore, the control of their activity that is of primary importance. At first sight it would appear that maximum efficiency would be achieved by maintaining the organisms in a constant state of maximum growth, *i.e.* in the log phase. In the absence of controlling factors, however, this phase cannot be indefinitely maintained. Of the several theoretical factors terminating the log phase, the impedence to the transfer of oxygen and nutrients to the increasing number of cells in the microbial mass—the activated sludge floc or the bacteria bed film—is probably the most important in biological oxidation plants. The accumulation of toxic metabolic by-products may also be a contributory factor.For efficient operation, the organisms must remain physiologically capable of log growth and, therefore, the control of the microbial population is essential. Organisms in this physiological state may, however, be prevented from achieving maximum log growth by such factors as temperature, pH, dissolved oxygen and nutrient concentration. For maximum efficiency, of these factors, nutrient concentration should be the limiting factor, this, of course, being determined by the waste being treated. In practice, providing that the autecological conditions, previously discussed, are satisfied, the synecological considerations involve the establishment and maintainence of a balance between the loading imposed by the waste, the microbial population and the oxygen concentration.

APPLICATION OF BIOLOGICAL PRINCIPLES IN WASTE WATER TREATMENT

Of the three aspects of biology outlined in this introduction, both taxonomy and physiology are necessary for a full appreciation of ecology. For those desirous of further study of any of these aspects of biology a list of appropriate references is given at the end of this chapter. In applying biological principles to waste water treatment, involving the controlled activities of populations of organisms, the ecological approach is probably the most rewarding and will be developed in subsequent sections.

The different aspects of ecology are all reflected in the several studies of the organisms found in sewage-treatment plants. Some resulted in mere systematic lists of organisms associated with different stages of the process and which, although useful as a basis for other workers, are of no practical value to the operator. Other observers have correlated the presence of organisms with different conditions within the plant and such organisms have then been used as "indicator organisms".

For example, early in the history of activated sludge, different associations of protozoa were associated with different levels of sludge efficiency and microscopical examinations of the sludge to determine these associations were found to be of value in plant control. Probably because of the practical nature of the subject, early investigators enquired into the role of the different organisms in the purification process and thus the "synecological" aspect of the subject was introduced at an early stage. An outline of the historical development of our knowledge of the ecology of waste water treatment may be of interest.

HISTORICAL DEVELOPMENTS IN WASTE WATER ECOLOGY

The discovery that micro-organisms were associated with water and putrefying liquids was made possible by the invention in 1675 of the compound microscope by Leuwenhoek. It was not until 1743, however, that Baker indicated the scavenging activities of protozoa and bacteria, and in 1839 Schwann and Schultze demonstrated that micro-organisms were the true agents of decompositio. Following his work, in the middle of the 19th century, on fermentation, Pasteur established the importance of micro-organisms in the process and distinguished between aerobic and anaerobic organisms for the first time.

On the subject of organic wastes he declared "Dead matter which ferments and putrefies is not obedient, at any rate inclusively, to forces of a nature purely physical or chemical. It is life which rules over the work of death and the dissolution of animal and vegetable matter. This constant return to the atmospheric air and to the mineral kingdom of the constituents which vegetables and animals have borrowed from them is an act related to the development and multiplication of organised beings".

The setting up of the River Pollution Commissions of 1865 and 1868, as a result of the foul conditions of the rivers of this country, first initiated a scientific study into the problems of sewage treatment. At that time, the recognised method of disposal was by land treatment, but there appears to have been no appreciation of the role of microorganisms in the process in spite of the findings of earlier workers mentioned above. As a result of his investigations on the purification of sewage, Sir Edward Frankland in the second report of the Commission reported that the process was not merely mechanical filtration but involved chemical oxidation; the role of micro-organisms was apparently not appreciated.

He did appreciate, however, the need for adequate aeration and as a result of his experiments he did devise the "intermittent filtration" method using land. It was truly claimed "that Frankland's conclusions, based as they were on excellent experiments, have formed the foundation for all further progress relating to sewage purification up to the present time". Unfortunately, however, his mechanical-chemical concept of the process has persisted, although in a subconscious form, and still influences the design and operation of biological treatment plants; the term sewage "filter" reflects this attitude to the process!

Later work, however, by Schloesing and Muntz showed that nitrification was brought about by micro-organisms and they concluded that these were essential to the purification process. Warington showed that nitrification proceeded in two stages, each stage being the result of the activity of separate organisms. These were later isolated as *Nitrosomonas* and *Nitrobacter* by Winogradsky. For a more detailed history the reader is referred to Stanbridge.

The role of larger animals in the stabilisation of organic matter was also reported by Dr. Sorby to the Royal Commission on the Metropolitan Sewage

Discharge in 1883. He associated the disappearance of sedimentary faecal matter in the river Thames with the presence of certain crustacea and worms. Thus before the advent of the bacteria bed, which was evolved from soil and sand filters as a result of experiments at the Lawrence Experiment Station, Massachusetts, the role of both the micro-organisms and the macrofauna in sewage purification had been established. The results of the Lawrence experiments stimulated investigations of the principles involved in the bacteria bed method of treatment. In these, although the role of the micro-organisms was appreciated by most, and studies were carried out on the bacterial flora, the activity of the larger organisms, previously associated with purification, was at first overlooked, their presence, when observed, being regarded as incidental, as indeed it still is by some workers.

Johnson, in reviewing the early works, quoted Dunbar as first drawing attention to the function of higher animals and plants and states that later Höfer, in 1907, investigated their activities more closely. Dibdin in 1904 also observed the presence of large numbers of active insects and Annelid worms in his slate beds, and Fowler also reported on the importance of the higher forms of life in sewage purification. About the same time Harrison in his evidence to the *Royal Commission on Sewage Disposa* suggested that the seasonal discharge of solids from beds was probably due to the activity of the macrofauna. Work of a more truly ecological nature was carried out by Johnson when he studied the factors influencing the distribution of different species of macrofauna in bacteria beds at Wakefield. Parkinson and Bell demonstrated the useful activity of the springtail *Achorutes subviaticus* in controlling film accumulation.

Later, the environment of the bacteria bed, supporting a reduced number of species, attracted the attention of more academic workers as a habitat on which to study problems of the dynamics of populations. Notable among such workers were Dr. Lloyd and his succession of students at Leeds University who carried out a series of studies on insect and protozoan populations on sewage works in the West Riding; these are summarised by Lloyd. Nuisance arising from flies leaving the bacteria beds probably attracted attention to the macrofauna before their beneficial activities were fully realised.

Later, this nuisance resulted in many investigations into methods of control using different chemicals. Some such investigations were ecological in that they attempted to relate the populations to environmental conditions but most, although of practical value in assessing the effectiveness of different control measures, did not advance the knowledge of bacteria bed ecology. Another practical problem—the choking of the beds with accumulated film—has led to a series of ecological investigations both in this country. At Birmingham a study on the effect of insecticide treatment on the bacteria bed community and film accumulation has led to the two practical difficulties being studied as one ecological problem. In the ecology of activated sludge both academic interest and practical problems have initiated ecological investigations. Ecological successions of different species of protozoa in the development of a sludge and the relative roles of protozoa and bacteria in the process have been studied by many workers. The development of filamentous growths associated with 'bulking' has also resulted in investigations being carried out.

The outcome of such ecological investigations usually results in the formulation of hypotheses which necessitate laboratory work for their proof or otherwise and much work being carried out at present by the Water Pollution Research Laboratory and others will, it is hoped, assist in the solution of outstanding ecological problems.

ACTIVATED-SLUDGE AND BACTERIA-BED ENVIRONMENTS

The treatment of waste waters by both methods involves similar biochemical processes the two differ in many respects. Firstly, whereas the activated sludge is truly an aquatic environment that of the bacteria bed has been likened to the wrack zone of the sea shore. Also, whereas the fauna of activated sludge is normally restricted to microorganisms, the bacteria bed also supports higher forms of life such as oligochaete worms and insects. Associated with this difference in the fauna is the difference in the degree of control the operator has over the two biological systems. In the activated sludge process the amount of microbial mass is controlled by the withdrawal of excess sludge from the system; in the bacteria bed excess film is removed chiefly by biological agencies.

A fundamental difference in the ecology of the two systems is that in the bacteria bed a succession of communities becomes established at different levels of the bed and associated with the corresponding different degrees of purification; whereas in the activated-sludge process the same community within the floc is at one time associated with the untreated waste and at the other extreme with the purified effluent.

It is of interest to note that many of the advantages and disadvantages of the two systems are attributable to these differences; fly nuisance and the choking of the beds being connected with film control by biological agencies; bulking of the sludge, with resultant settlement difficulties, and the difficulties associated with disposal of the surplus activated sludge are probably connected with its removal from the system as microbial flocs rather than as the faeces of the grasing fauna. Again, the relative sensitivity of activated sludge to toxic discharges may be associated with circulation of the sludge, whereas, in the more robust 5 bacteria beds such discharges are first taken by the surface growths.

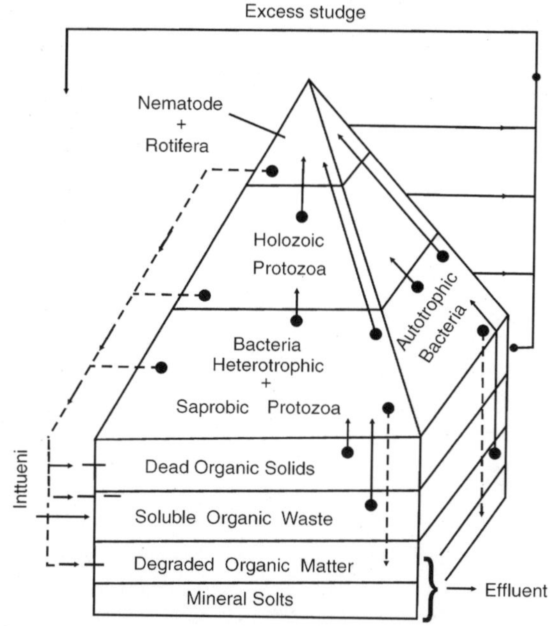

Fig. Diagrammatic Representation of the Main Food-links in the Purification of Organic Wastes by Activated Sludge and by Bacteria Beds.

A highly nitrified effluent is more often associated with an efficient bacteria bed than with activated sludge and this may also be connected with the differences in the relative ecological systems. Nitrifying organisms are more effective in isolation from the other processes of purification, *e.g.* in the cleaner depths of the bacteria bed or in the cleaner secondary bed of the two-stage process. It may be therefore, that in the multipurpose floc of the activated sludge their activity is restricted. Also the length of life of a floc in the system is largely determined by the growth rate of the heterotrophic bacteria and may be such that the floc is discharged before the more slowly growing autotrophic nitrifying organisms become effectively established.

In nature, a succession of different ecological conditions and associated communities concerned with the breakdown of organic matter in waters has been described. A high concentration of complex decomposable organic matter is associated with the absence, or very low concentration, of oxygen—*Polysaprobic* condition. As purification proceeds and the organic matter is broken down first to aminoacids (*á-mesosaprobic*) and then to salts, the oxygen concentration increases until oxidation and mineralisation is completed (*oligosaprobic*). Both the activated-sludge and the bacteria-bed environment differ from these systems in that, because of the aeration, high concentration of complex organic matter may be associated with fairly high oxygen concentrations. Obviously large numbers of organisms gain access to both activated sludge plants and bacteria beds; many find the environment inhospitable and never become established; others may persist but are not very successful, whilst the few species which find the habitat suitable, increase greatly and with reduced competition may become more abundant than in their natural habitat. In the following discussion on the ecologies of the two processes only those organisms considered to have a significant role in the community, *i.e.* those organisms which if removed from the community would appreciably upset its balance of populations.

2

Understanding the Aquaculture

Aquaculture, also known as aquafarming, is the farming of aquatic organisms such as fish, crustaceans, molluscs and aquatic plants. Aquaculture involves cultivating freshwater and saltwater populations under controlled conditions, and can be contrasted with commercial fishing, which is the harvesting of wild fish. Mariculture refers to aquaculture practiced in marine environments and in underwater habitats.

The reported output from global aquaculture operations would supply one half of the fish and shellfish that is directly consumed by humans; however, there are issues about the reliability of the reported figures. Further, in current aquaculture practice, products from several pounds of wild fish are used to produce one pound of a piscivorous fish like salmon.

Particular kinds of aquaculture include fish farming, shrimp farming, oyster farming, algaculture (such as seaweed farming), and the cultivation of ornamental fish. Particular methods include aquaponics and Integrated multi-trophic aquaculture, both of which integrate fish farming and plant farming.

HISTORY

The indigenous Gunditjmara people in Victoria, Australia may have raised eels as early as 6000 BC. There is evidence that they developed about 100 square kilometres (39 sq mi) of volcanic floodplains in the vicinity of Lake Condah into a complex of channels and dams, that they used woven traps to capture eels, and preserve eels to eat all year round.

Aquaculture was operating in China circa 2500 BC. When the waters subsided after river floods, some fishes, mainly carp, were trapped in lakes. Early aquaculturists fed their brood using nymphs and silkworm feces, and ate them. A fortunate genetic mutation of carp led to the emergence of goldfish during the Tang Dynasty. Japanese cultivated seaweed by providing bamboo poles and, later, nets and oyster shells to serve as anchoring surfaces for spores. Romans bred fish in ponds. In central Europe, early Christian monasteries adopted Roman aquacultural practices. Aquaculture spread in Europe during the Middle Ages since away from the seacoasts and the big rivers fish had to be salted in order to not perish. Improvements in transportation during the 19th century made fresh fish easily available and inexpensive, even in inland areas, making aquaculture less popular.Hawaiians constructed oceanic fish ponds. A remarkable example is a fish pond dating from at least 1,000 years ago, at Alekoko. Legend says that it was constructed by the mythical Menehune dwarf people.In 1859 Stephen Ainsworth of West Bloomfield, New York, began experiments with brook trout. By 1864 Seth Green had established a commercial fish hatching operation at Caledonia Springs, near Rochester, New York. By 1866, with the involvement of Dr. W. W. Fletcher of Concord, Massachusetts, artificial fish hatcheries were under way in both Canada and the United States. When the Dildo Island fish hatchery opened in Newfoundland in 1889, it was the largest and most advanced in the world.Californians harvested wild kelp and attempted to manage supply circa 1900, later labeling it a wartime resource.

21ST-CENTURY PRACTICE

About 430 (97%) of the species cultured as of 2007 were domesticated during the 20th century, of which an estimated 106 came in the decade to 2007. Given the long-term importance of agriculture, it is interesting to note that to date only 0.08% of known land plant species and 0.0002% of known land animal species have been domesticated, compared with 0.17% of known marine plant species and 0.13% of known marine animal species. Domestication typically involves about a decade of scientific research. Domesticating aquatic species involves fewer risks to humans than land animals, which took a large toll in human lives. Most major human diseases originated in domesticated animals, through diseases such as smallpox and diphtheria, that like most infectious diseases, move to humans from animals. No human pathogens of comparable virulence have yet emerged from marine species.

Harvest stagnation in wild fisheries and overexploitation of popular marine species, combined with a growing demand for high quality protein, encourage aquaculturists to domesticate other marine species.

DEFINITION

According to the FAO, aquaculture "is understood to mean the farming of aquatic organisms including fish, molluscs, crustaceans and aquatic plants. Farming implies some form of intervention in the rearing process to enhance production, such as regular stocking, feeding, protection from predators, etc. Farming also implies individual or corporate ownership of the stock being cultivated."

SPECIES GROUPS

AQUATIC PLANTS

Microalgae, also referred to as phytoplankton, microphytes, or planktonic algae constitute the majority of cultivated algae. Macroalgae, commonly known as seaweed, also have many commercial and industrial uses, but due to their size and specific requirements, they are not easily cultivated on a large scale and are most often taken in the wild.

FISH

The farming of fish is the most common form of aquaculture. It involves raising fish commercially in tanks, ponds, or ocean enclosures, usually for food. A facility that releases juvenile fish into the wild for recreational fishing or to supplement a species' natural numbers is generally referred to as a fish hatchery. Fish species raised by fish farms include salmon, bigeye tuna, carp, tilapia, catfish and cod.

In the Mediterranean, young bluefin tuna are netted at sea and towed slowly towards the shore. They are then interned in offshore pens where they are further grown for the market. In 2009, researchers in Australia managed for the first time to coax tuna (Southern bluefin) to breed in landlocked tanks.

CRUSTACEANS

Commercial shrimp farming began in the 1970s, and production grew steeply thereafter. Global production reached more than 1.6 million tonnes in 2003, worth about 9 billion U.S. dollars. About 75% of farmed shrimp is produced in Asia, in particular in China and Thailand. The other 25% is produced mainly in Latin America, where Brazil is the largest producer. Thailand is the largest exporter.

Shrimp farming has changed from its traditional, small-scale form in Southeast Asia into a global industry. Technological advances have led to ever higher densities per unit area, and broodstock is shipped worldwide. Virtually all farmed shrimp are penaeids (i.e., shrimp of the family *Penaeidae*), and just two species of shrimp, the Pacific white shrimp and the giant tiger prawn, account for about 80% of all farmed shrimp. These industrial monocultures are very susceptible to disease, which has decimated shrimp populations across entire regions. Increasing ecological problems, repeated disease outbreaks, and pressure and criticism from both NGOs and consumer countries led to changes in the industry in the late 1990s and generally stronger regulations. In 1999, governments, industry representatives, and environmental organizations initiated a Programme aimed at developing and promoting more sustainable farming practices. Freshwater prawn farming shares many characteristics with, including many problems with marine shrimp farming. Unique problems are introduced by the developmental life cycle of the main species, the giant river prawn.

The global annual production of freshwater prawns (excluding crayfish and crabs) in 2003 was about 280,000 tonnes of which China produced 180,000 tonnes followed by India and Thailand with 35,000 tonnes each. Additionally, China produced about 370,000 tonnes of Chinese river crab.

MOLLUSCS

Abalone farming began in the late 1950s and early 1960s in Japan and China. Since the mid-1990s, this industry has become increasingly successful. Over-fishing and poaching have reduced wild populations to the extent that farmed abalone now supplies most abalone meat.

OTHER GROUPS

Other groups include aquatic reptiles, amphibians, and miscellaneous invertebrates, such as echinoderms and jellyfish. They are separately graphed at the top right of this section, since they do not contribute enough volume to show clearly on the main graph. Commercially harvested echinoderms include sea cucumbers and sea urchins. In China, sea cucumbers are farmed in artificial ponds as large as 1,000 acres (400 ha).

AROUND THE WORLD

In 2004, the total world production of fisheries was 140 million tonnes of which aquaculture contributed 45 million tonnes, about one third. The growth rate of worldwide aquaculture has been sustained and rapid, averaging about 8 percent per annum for over thirty years, while the take from wild fisheries has been essentially flat for the last decade. The aquaculture market reached $86 billion in 2009.

Aquaculture is an especially important economic activity in China. Between 1980 and 1997, the Chinese Bureau of Fisheries reports, aquaculture harvests grew at an annual rate of 16.7 percent, jumping from 1.9 million tonnes to nearly 23 million tonnes. In 2005, China accounted for 70% of world production. Aquaculture is also currently one of the fastest growing areas of food production in the U.S.

Approximately 90% of all U.S. shrimp consumption is farmed and imported. In recent years salmon aquaculture has become a major export in southern Chile, especially in Puerto Montt, Chile's fastest-growing city.

OVER REPORTING

China overwhelmingly dominates the world in reported aquaculture output. They report a total output which is double that of the rest of the world put together. However, there are issues with the accuracy of China's returns.

In 2001, the fisheries scientists Reg Watson and Daniel Pauly expressed concerns in a letter to *Nature*, that China was over reporting its catch from wild fisheries in the 1990s. They said that made it appear that the global catch since 1988 was increasing annually by 300,000 tonnes, whereas it was really shrinking annually by 350,000 tonnes. Watson and Pauly suggested this may be related to China policies where state entities that monitor the economy are also tasked with increasing output. Also, until recently, the promotion of Chinese officials was based on production increases from their own areas.

China disputes this claim. The official Xinhua News Agency quoted Yang Jian, director general of the Agriculture Ministry's Bureau of Fisheries, as saying that China's figures were "basically correct". However, the FAO accepts there are issues with the reliability of China's statistical returns, and currently treats data from China, including the aquaculture data, apart from the rest of the world.

METHODS

MARICULTURE

Mariculture is the term used for the cultivation of marine organisms in seawater, usually in sheltered coastal waters. In particular, the farming of marine fish is an example of mariculture, and so also is the farming of marine crustaceans (such as shrimps), molluscs (such as oysters) and seaweed.

INTEGRATED

Integrated Multi-Trophic Aquaculture (IMTA) is a practice in which the by-products (wastes) from one species are recycled to become inputs (fertilizers, food) for another. Fed aquaculture (for example, fish, shrimp) is combined with inorganic extractive (for example, seaweed) and organic extractive (for example, shellfish) aquaculture to create balanced systems for environmental sustainability (biomitigation), economic stability (product diversification and risk reduction) and social acceptability (better management practices).

"Multi-Trophic" refers to the incorporation of species from different trophic or nutritional levels in the same system. This is one potential distinction from the age-old practice of aquatic polyculture, which could simply be the co-culture of different fish species from the same trophic level. In this case, these organisms may all share the same biological and chemical processes, with few synergistic benefits, which could potentially lead to significant shifts in the ecosystem. Some traditional polyculture systems may, in fact, incorporate a greater diversity of species, occupying several niches, as extensive cultures (low intensity, low management) within the same pond. The "Integrated" in IMTA refers to the more intensive cultivation of the different species in proximity of each other, connected by nutrient and energy transfer through water.

Ideally, the biological and chemical processes in an IMTA system should balance. This is achieved through the appropriate selection and proportions of different species providing different ecosystem functions. The co-cultured species are typically more than just biofilters; they are harvestable crops of commercial value. A working IMTA system can result in greater total production based on mutual benefits to the co-cultured species and improved ecosystem health, even if the production of individual species is lower than in a monoculture over a short term period.

Sometimes the term "Integrated Aquaculture" is used to describe the integration of monocultures through water transfer. For all intents and purposes however, the terms "IMTA" and "integrated aquaculture" differ only in their degree of descriptiveness. Aquaponics, fractionated aquaculture, IAAS (integrated agriculture-aquaculture systems), IPUAS (integrated peri-urban-aquaculture systems), and IFAS (integrated fisheries-aquaculture systems) are other variations of the IMTA concept.

NETTING MATERIALS

Various materials, including nylon, polyester, polypropylene, polyethylene, plastic-coated welded wire, rubber, patented rope products (Spectra, Thorn-D, Dyneema), galvanized steel and copper are used for netting in aquaculture fish enclosures around the world. All of these materials are selected for a variety of reasons, including design feasibility, material strength, cost, and corrosion resistance.

Recently, copper alloys have become important netting materials in aquaculture because they are antimicrobial (i.e., they destroy bacteria, viruses, fungi, algae, and other microbes) and they therefore prevent biofouling (i.e., the undesirable accumulation, adhesion, and growth of microorganisms, plants, algae, tubeworms, barnacles, mollusks, and other organisms). By inhibiting microbial growth, copper alloy aquaculture cages avoid costly net changes that are necessary with other materials. The resistance of organism growth on copper alloy nets also provides a cleaner and healthier environment for farmed fish to grow and thrive.

ISSUES

Aquaculture can be more environmentally damaging than exploiting wild fisheries on a local area basis but has considerably less impact on the global environment on a per kg of production basis. Local concerns include waste handling, side-effects of antibiotics, competition between farmed and wild animals, and using other fish to feed more marketable carnivorous fish. However, research and commercial feed improvements during the 1990s and 2000s have lessened many of these concerns.

Fish waste is organic and composed of nutrients necessary in all components of aquatic food webs. In-ocean aquaculture often produces much higher than normal fish waste concentrations. The waste collects on the ocean bottom, damaging or eliminating bottom-dwelling life. Waste can also decrease dissolved oxygen levels in the water column, putting further pressure on wild animals.

IMPACTS ON WILD FISH

Salmon farming currently leads to a high demand for wild forage fish. Fish do not actually produce omega-3 fatty acids, but instead accumulate them from either consuming microalgae that produce these fatty acids, as is the case with forage fish like herring and sardines, or, as is the case with fatty predatory fish, like salmon, by eating prey fish that have accumulated omega-3 fatty acids from microalgae. To satisfy this requirement, more than 50 percent of the world fish oil production is fed to farmed salmon.

In addition, as carnivores, salmon require large nutritional intakes of protein, protein which is often supplied to them in the form of forage fish. Consequently, farmed salmon consume more wild fish than they generate as a final product. To produce one pound of farmed salmon, products from several pounds of wild fish are fed to them. As the salmon farming industry expands, it requires more wild forage fish for feed, at a time when seventy five percent of the worlds monitored fisheries are already near to or have exceeded their maximum sustainable yield. The industrial scale extraction of wild forage fish for salmon farming then impacts the survivability of the wild predator fish who rely on them for food.Fish can escape from coastal pens, where they can interbreed with their wild counterparts, diluting wild genetic stocks. Escaped fish can become invasive, out competing native species.

COASTAL ECOSYSTEMS

Aquaculture is becoming a significant threat to coastal ecosystems. About 20 percent of mangrove forests have been destroyed since 1980, partly due to shrimp farming. An extended cost–benefit analysis of the total economic value of shrimp aquaculture built on mangrove ecosystems found that the external costs were much higher than the external benefits. Over four decades, 269,000 hectares (660,000 acres) of Indonesian mangroves have been converted to shrimp farms. Most of these farms are abandoned within a decade because of the toxin build-up and nutrient loss. Salmon farms are typically sited in pristine coastal ecosystems which they then pollute. A farm with 200,000 salmon discharges more fecal waste than a city of 60,000 people. This waste is discharged directly into the surrounding aquatic environment, untreated, often containing antibiotics and pesticides." There is also an accumulation of heavy metals on the benthos (seafloor) near the salmon farms, particularly copper and zinc.

GENETIC MODIFICATION

Salmon have been genetically modified for faster growth, although they are not approved for commercial use, in the face of opposition. One study, in a laboratory setting, found that modified salmon mixed with their wild relatives were aggressive in competing, but ultimately failed.

ANIMAL WELFARE

As with the farming of terrestrial animals, social attitudes influence the need for humane practices and regulations in farmed marine animals. Under the guidelines advised by the Farm Animal Welfare Council good animal welfare means both fitness and a sense of well being in the animal's physical and mental state. This can be defined by the Five Freedoms:

- Freedom from hunger and thirst
- Freedom from discomfort
- Freedom from pain, disease, or injury
- Freedom to express normal behaviour
- Freedom from fear and distress

However, the controversial issue in aquaculture is whether fish and farmed marine invertebrates are actually sentient, or have the perception and awareness to experience suffering. Although no evidence of this has been found in marine invertebrates, recent studies conclude that fish do have the necessary receptors (nociceptors) to sense noxious stimuli and so are likely to experience states of pain, fear and stress. Consequently, welfare in aquaculture is directed at vertebrates; finfish in particular.

COMMON WELFARE CONCERNS

Welfare in aquaculture can be impacted by a number of issues such as stocking densities, behavioural interactions, disease and parasitism. A major problem in determining the cause of impaired welfare is that these issues are often all interrelated and influence each other at different times.

Optimal stocking density is often defined by the carrying capacity of the stocked environment and the amount of individual space needed by the fish, which is very species specific. Although behavioural interactions such as shoaling may mean that high stocking densities are beneficial to some species, in many cultured species high stocking densities may be of concern. Crowding can constrain normal swimming behaviour, as well as increase aggressive and competitive behaviours such as cannibalism, feed competition, territoriality and dominance/subordination hierarchies. This potentially increases the risk of tissue damage due to abrasion from fish-to-fish contact or fish-to-cage contact. Fish can suffer reductions in food intake and food conversion efficiency. In addition, high stocking densities can result in water flow being insufficient, creating inadequate oxygen supply and waste product removal. Dissolved oxygen is essential for fish respiration and concentrations below critical levels can induce stress and even lead to asphyxiation. Ammonia, a nitrogen excretion product, is highly toxic to fish at accumulated levels, particularly when oxygen concentrations are low.

Many of these interactions and effects cause stress in the fish, which can be a major factor in facilitating fish disease. For many parasites, infestation depends on the host's degree of mobility, the density of the host population and vulnerability of the host's defence system. Sea lice are the primary parasitic problem for finfish in aquaculture, high numbers causing widespread skin erosion and haemorrhaging, gill congestion,and increased mucus production. There are also a number of prominent viral and bacterial pathogens that can have severe effects on internal organs and nervous systems.

IMPROVING WELFARE

The key to improving welfare of marine cultured organisms is to reduce stress to a minimum, as prolonged or repeated stress can cause a range of adverse effects. Attempts to minimise stress can occur throughout the culture process. During grow out it is important to keep stocking densities at appropriate levels specific to each species, as well as separating size classes and grading to reduce aggressive behavioural interactions. Keeping nets and cages clean can assist positive water flow to reduce the risk of water degradation.

Not surprisingly disease and parasitism can have a major effect on fish welfare and it is important for farmers not only to manage infected stock but also to apply disease prevention measures. However, prevention methods, such as vaccination, can also induce stress because of the extra handling and injection. Other methods include adding antibiotics to feed, adding chemicals into water for treatment baths and biological control, such as using cleaner wrasse to remove lice from farmed salmon.

Many steps are involved in transport, including capture, food deprivation to reduce faecal contamination of transport water, transfer to transport vehicle via nets or pumps, plus transport and transfer to the delivery location. During transport water needs to be maintained to a high quality, with regulated temperature, sufficient oxygen and minimal waste products. In some cases anaesthetics may be used in small doses to calm fish before transport. Aquaculture is sometimes part of an environmental rehabilitation Programme or as an aid in conserving endangered species.

PROSPECTS

Global wild fisheries are in decline, with valuable habitat such as estuaries in critical condition. The aquaculture or farming of piscivorous fish, like salmon, does not help the problem because they need to eat products from other fish, such as fish meal and fish oil. Studies have shown that salmon farming has major negative impacts on wild salmon, as well as the forage fish that need to be caught to feed them. Fish that are higher on the food chain are less efficient sources of food energy.

Apart from fish and shrimp, some aquaculture undertakings, such as seaweed and filter-feeding bivalve mollusks like oysters, clams, mussels and scallops, are relatively benign and even environmentally restorative. Filter-feeders filter pollutants as well as nutrients from the water, improving water quality. Seaweeds extract nutrients such as inorganic nitrogen and phosphorus directly from the water, and filter-feeding mollusks can extract nutrients as they feed on particulates, such as phytoplankton and detritus.

Some profitable aquaculture cooperatives promote sustainable practices. New methods lessen the risk of biological and chemical pollution through minimizing fish stress, fallowing netpens, and applying Integrated Pest Management. Vaccines are being used more and more to reduce antibiotic use for disease control.

Onshore recirculating aquaculture systems, facilities using polyculture techniques, and properly sited facilities (for example, offshore areas with strong currents) are examples of ways to manage negative environmental effects.

Recirculating aquaculture systems (RAS) recycle water by circulating it through filters to remove fish waste and food and then recirculating it back into the tanks. This saves water and the waste gathered can be used in compost or, in some cases, could even be treated and used on land. While RAS was developed with freshwater fish in mind, scientist associated with the Agricultural Research Service have found a way to rear saltwater fish using RAS in low-salinity waters. Although saltwater fish are raised in off-shore cages or caught with nets in water that typically has a salinity of 35 parts per thousand (ppt), scientists were able to produce healthy pompano, a saltwater fish, in tanks with a salinity of only 5 ppt. Commercializing low-salinity RAS are predicted to have positive environmental and economical effects. Unwanted nutrients from the fish food would not be added to the ocean and the risk of transmitting diseases between wild and farm-raised fish would greatly be reduced. The price of expensive saltwater fish, such as the pompano and combia used in the experiments, would be reduced. However, before any of this can be done researchers must study every aspect of the fish's lifecycle, including the amount of ammonia and nitrate the fish will tolerate in the water, what to feed the fish during each stage of its lifecycle, the stocking rate that will produce the healthiest fish, etc.

Some 16 countries now use geothermal energy for aquaculture, including China, Israel, and the United States. In California, for example, 15 fish farms produce tilapia, bass, and catfish with warm water from underground. This warmer water enables fish to grow all year round and mature more quickly. Collectively these California farms produce 4.5 million kilograms of fish each year.

AQUAPONICS

Aquaponicsis a sustainable food production system that combines a traditional aquaculture (raising aquatic animals such as snails, fish, crayfish or prawns in tanks) with hydroponics (cultivating plants in water) in a symbiotic environment. In aquaculture, effluents accumulate in the water, increasing toxicity for the fish. This water is led to a hydroponic system where the by-products from the aquaculture are filtered out by the plants as vital nutrients, after which the cleansed water is recirculated back to the animals. The term *aquaponics* is a portmanteau of the terms *aquaculture* and *hydroponic*.Aquaponic systems vary in size from small indoor or outdoor units to large commercial units, using the same technology. The systems usually contain fresh water, but salt water systems are plausible depending on the type of aquatic animal and which plants. Aquaponic science may still be considered to be at an early stage, relative to other sciences.

FUNCTION

Aquaponics consists of two main parts, with the aquaculture part for raising aquatic animals and the hydroponics part for growing plants. Aquatic effluents, resulting from uneaten feed or raising animals like fish, accumulate in water due to the closed-system recirculation of most aquaculture systems. The effluent-rich water becomes toxic to the aquatic animal in high concentrations but these effluents are nutrients essential for plant growth. Although consisting primarily of these two parts, aquaponics systems are usually grouped into several components or subsystems responsible for the effective removal of solid wastes, for adding bases to neutralize acids, or for maintaining water oxygenation. Typical components include:

- *Rearing tank*: the tanks for raising and feeding the fish;
- *Solids removal*: a unit for catching uneaten food and detached biofilms, and for settling out fine particulates;
- *Biofilter*: a place where the nitrification bacteria can grow and convert ammonia into nitrates, which are usable by the plants;
- *Hydroponics subsystem*: the portion of the system where plants are grown by absorbing excess nutrients from the water;
- *Sump*: the lowest point in the system where the water flows to and from which it is pumped back to the rearing tanks.

Depending on the sophistication and cost of the aquaponics system, the units for solids removal, biofiltration, and/or the hydroponics subsystem may be combined into one unit or subsystem, which prevents the water from flowing directly from the aquaculture part of the system to the hydroponics part.

Nitrification

Nitrification, the aerobic conversion of ammonia into nitrates, is one of the most important functions in an aquaponics system as it reduces the toxicity of the water for fish, and allows the resulting nitrate compounds to be removed by the plants for nourishment. Ammonia is steadily released into the water through the excreta and gills of fish as a product of their metabolism, but must be filtered out of the water since higher concentrations of ammonia (commonly between 0.5 and 1 ppm) can kill fish. Although plants can absorb ammonia from the water to some degree, nitrates are assimilated more easily, thereby efficiently reducing the toxicity of the water for fish. Ammonia can be converted into other nitrogenous compounds through healthy populations of:

- *Nitrosomonas*: bacteria that convert ammonia into nitrites, and
- *Nitrobacter*: bacteria that convert nitrites into nitrates.

In an aquaponics system, the bacteria responsible for this process form a biofilm on all solid surfaces throughout the system that are in constant contact with the water. The submerged roots of the vegetables combined have a large surface area, so that many bacteria can accumulate there. Together with the saliency of ammonia and nitrites in the water, the surface area determines the speed with which nitrification takes place. Care for these bacterial colonies is important as to regulate the full assimilation of ammonia and nitrite. This is why most aquaponics systems include a biofiltering unit, which helps facilitate growth of these microorganisms. Typically, after a system has stabilized ammonia levels range from 0.25 to 2.0 ppm; nitrite levels range from 0.25 to 1 ppm, and nitrate levels range from 2 to 150 ppm. During system startup, spikes may occur in the levels of ammonia (up to 6.0 ppm) and nitrite (up to 15 ppm), with nitrate levels peaking later in the startup phase. Since the nitrification process acidifies the water, non-sodium bases such as potassium hydroxide or calcium hydroxide can be added for neutralizing the water's pH if insufficient quantities are naturally present in the water to provide a buffer against acidification. In addition, selected minerals or nutrients such as iron can be added in addition to the fish waste that serves as the main source of nutrients to plants.

A good way to deal with solids buildup in aquaponics is the use of worms, which liquefy the solid organic matter so that it can be utilized by the plants and/or animals.

Hydroponics subsystem

Plants are grown as in hydroponics systems, with their roots immersed in the nutrient-rich effluent water. This enables them to filter out the ammonia that is toxic to the aquatic animals, or its metabolites. After the water has passed through the hydroponic subsystem, it is cleaned and oxygenated, and can return to the aquaculture vessels. This cycle is continuous. Common aquaponic applications of hydroponic systems include:

- *Deep-water raft aquaponics*: styrofoam rafts floating in a relatively deep aquaculture basin in troughs.
- *Recirculating aquaponics*: solid media such as gravel or clay beads, held in a container that is flooded with water from the aquaculture. This type of aquaponics is also known as *closed-loop aquaponics*.

- *Reciprocating aquaponics*: solid media in a container that is alternately flooded and drained utilizing different types of siphon drains. This type of aquaponics is also known as *flood-and-drain aquaponics* or *ebb-and-flow aquaponics*.
- Other systems use towers that are trickle-fed from the top, nutrient film technique channels, horizontal PVC pipes with holes for the pots, plastic barrels cut in half with gravel or rafts in them. Each approach has its own benefits.

Most green leaf vegetables grow well in the hydroponic subsystem, although most profitable are varieties of chinese cabbage, lettuce, basil, roses, tomatoes, okra, cantaloupe and bell peppers. Other species of vegetables that grow well in an aquaponic system include beans, peas, kohlrabi, watercress, taro, radishes, strawberries, melons, onions, turnips, parsnips, sweet potato and herbs. Since plants at different growth stages require different amounts of minerals and nutrients, plant harvesting is staggered with seedings growing at the same time as mature plants. This ensures stable nutrient content in the water because of continuous symbiotic cleansing of toxins from the water.

Aquaculture subsystem

Freshwater fish are the most common aquatic animal raised using aquaponics, although freshwater crayfish and prawns may also be used. In practice, tilapia are the most popular fish for home and commercial projects that are intended to raise edible fish, although barramundi, Silver Perch, Eel-tailed catfish or tandanus catfish, Jade perch and Murray cod are also used. For temperate climates when there isn't ability or desire to maintain water temperature, bluegill and catfish are suitable fish species for home systems. Koi and goldfish may also be used, if the fish in the system need not be edible.

Normal operations

Aquaponic systems do not typically discharge or exchange water under normal operation, but instead recirculate and reuse water very effectively. The system relies on the relationship between the animals and the plants to maintain a stable aquatic environment that experience a minimum of fluctuation in ambient nutrient and oxygen levels. Water is only added to replace water loss from absorption and transpiration by plants, evaporation into the air from surface water, overflow from the system from rainfall, and removal of biomass such as settled solid wastes from the system. As a result, aquaponics uses approximately 2% of the water that a conventionally irrigated farm requires for the same vegetable production. This allows for aquaponic production of both crops and fish in areas where water or fertile land is scarce. Aquaponic systems can also be used to replicate controlled wetland conditions that are useful for water treatment by reclaiming potable water from typical household sewage. The nutrient-filled overflow water can be accumulated in catchment tanks, and reused to accelerate growth of crops planted in soil, or it may be pumped back into the aquaponic system to top up the water level.. The three main inputs to the system are water, feed given to the aquatic animals, and electricity to pump water between the aquaculture subsystem and the hydroponics subsystem. Spawn or fry may be added to replace grown fish that are taken out from the system to retain a stable system. In terms of outputs, an aquaponics system may continually yield plants such as vegetables grown in hydroponics, and edible aquatic species raised in an aquaculture.

HISTORY

Ancient

Aquaponics has ancient roots, although there is some debate on its first occurrence:

- Aztec cultivated agricultural islands known as *chinampas* and are considered by some as the first form of aquaponics for agricultural use where plants were raised on stationary (and sometime movable) islands in lake shallows and waste materials dredged from the Chinampa canals and surrounding cities are used to manually irrigate the plants.
- South China and Thailand who cultivated and farmed rice in paddy fields in combination with fish are cited as examples of early aquaponics. These polycultural farming systems existed in many Far Eastern countries and raised fish such as the oriental loach, swamp eel, Common and crucian carp as well as pond snails in the paddies.

REGIONS

North America

United States

While the development of aquaponics is often attributed to the various works of the New Alchemy Institute and the works of Dr. Mark McMurtry et al. at the North Carolina State University, many papers of initial development of aquaponics concepts pre-date both institutions by nearly a decade.Tom and Paula Speraneo, owners of a small greenhouse operation near West Plains, Missouri, modified the North Carolina State method and raised tilapia in above-ground tanks inside a solar greenhouse. The effluent from the tanks was used to fertigate gravel-cultured vegetables in raised benches.In addition, the Speraneos manipulated the watering cycle, and this methodology of the Speranos forms the basis for the style of "flood and drain" media grow bed aquaponics systems that have been widely adopted in Australia based on models promoted by Joel Malcolm and

Murray Hullam, and are now gaining increasing popularity in the United States.Inspired by the successes of the New Alchemy Institute and the North Carolina State University with aquaponics, other institutes followed suit. Besides the reciprocating aquaponics based on the techniques developed by Dr. Mark McMurtry et al. at the North Carolina State University, Dr. James Rakocy and his colleagues at the University of the Virgin Islands researched and developed the "Deep Water" or "Raft Culture" aquaponics

The system combines tilapia with various vegetables.In 1997 Rebecca L. Nelson and John S. Pade began publishing the Aquaponics Journal, a quarterly scientific journal that brings together research and various applications of aquaponics from around the globe.Recent years have seen a shift towards community integration of aquaponics, such as the Non-profit foundation Growing Power that offers Milwaukee youth job opportunities and training while growing food for their community. The model has spawned several satellite projects in other cities, such as New Orleans where the Vietnamese fisherman community has suffered from the Horizon Disaster and in the South Bronx in New York City. In addition, aquaponic gardeners from all around the world have gathered in online community sites and forums to openly share their experiences and promote the development of this form of gardening.

Canada

The first aquaponics research in Canada was a small system added onto existing aquaculture research at a research station in Lethbridge. Canada saw a rise in aquaponics setups throughout the '90s, predominantly as commercial installations, that for example combine trout with floating lettuce production, or to water fruiting vegetable crops that warm up the water too much to be recirculated back into the fish ponds.

Eels are also known to be raised. A setup based on the deep water system developed at the University of Virgin Islands was built in a greenhouse at Brooks, Alberta where Dr. Nick Savidov and colleagues researched aquaponics from a background of plant science. The team made findings on rapid root growth in aquaponics systems, on closing the solid waste loop, and that because of certain advantages in the system over traditional aquaculture, the system can run well at a low pH level, which is favoured by plants but not fish. The Edmonton Aquaponics Society in Northern Alberta is adapting Dr. Savidov's commercially sized system to a smaller scale prototype that can be operated by families, small groups, or restaurants. They intend to further develop the closed solid waste loop.

South America

Barbados

Barbados is a densely populated island that deals with water scarcity. In

40 years' time, focus has shifted from domestic fruit and vegetable production on small farms to importing 80% of all fruits and vegetables for cost reasons. Aquaponics would make Barbados and other Caribbean islands less dependent on the world food market and reduce stress on the dwindling fish supplies. An inter-organizational project that started in late 2009 sets out to encourage and enable Barbadians to start aquaponics at home, with revenue generated by selling produce to tourists.

Asia

Bangladesh

Bangladesh is the world's most densely populated country. Most of its farmers use/misuse agrochemicals to enhance food production/storage life. However, the country is yet to establish a good monitoring system to ensure chemical-residue free safe food in the market, capable city dwellers are looking for a better way to grow their own safe foods for domestic consumption. To help support mainly this client group and people living in adverse climatic conditions such as salinity-prone southern part and flood-prone haor area in the eastern region, a low-cost model of backyard aqaponics system has been developed by Professor Dr. M.A. Salam at the Department of Aquaculture of Bangladesh Agricultural University, Mymensingh. The initial model has been developed in 2011, and demonstrated and reported to the press in January 2012. Further research may be necessary to develop appropriate models for different situations among client groups.

Taiwan

Taiwan is a densely populated island that is faced with freshwater scarcity. Dispensation of water is government-controlled. The closed-loop system of aquaponics is used by agricultural farmers to save water by also rearing fish, while fish farmers grow plants that filter the water from the fish tanks. Taiwan Aquaponics Association is the not-for profit federation to promote Aquaponics at Taiwan.

Japan

Japan is another highly densely populated country that has a long history of agricultural innovation and technological improvement. Aquaponics however, has only recently been introduced to this country, and there have been some projects to integrate aquaponics into communities in Tohoku, northern Japan, that suffered after the 2011 Tôhoku earthquake and tsunami. Japan Aquaponics, founded by Aragon St-Charles, is a leading social enterprise that is pioneering the use of aquaponics in this country.

Australia

Due to a ban on tilapia in all states, native freshwater fish including Silver

Perch, Jade Perch, Sleepy Cod, Murray Cod and Barramundi are popular in aquaponics and aquaculture systems, along with non-native rainbow trout, brown trout and Crayfish such as common yabby and Redclaw. The law changed in WA June 2012, banning tilapia.

Pros and cons

The unique advantages of aquaponic systems are:

- Conservation through constant water reuse and recycling.
- Organic fertilization of plants with natural fish emulsion.
- The elimination of solid waste disposal from intensive aquaculture.
- The reduction of needed cropland to produce crops.
- The overall reduction of the environmental footprint of crop production.
- Building small efficient commercial installations near markets reduces food miles.
- Reduction of pathogens that often plague aquaculture production systems.
- Reduction of erosion by eliminating the need to plough the soil.

Some conceivable disadvantages with aquaponics are:

- Initial expenses for housing, tank, plumbing, pumps, and grow beds.
- The infinite number of ways in which a system can be configured lends itself to equally varying results, conflicting research, and successes or failures.
- Some aquaponic installations rely heavily on man-made energy, technological solutions, and environmental control to achieve recirculation and water/ambient temperatures. However, if a system is designed with energy conservation in mind, using alternative energy and a reduced number of pumps by letting the water flow downwards as much as possible, it can be highly energy efficient.
- While careful design can minimize the risk, aquaponics systems can have multiple 'single points of failure' where problems such as an electrical failure or a pipe blockage can lead to a complete loss of fish stock.
- Like all aquaculture based systems, stock feed usually consists of fish meal derived from lower value species. Ongoing depletion of wild fish stocks makes this practice unsustainable. Organic fish feeds may prove to be a viable alternative that negates this concern. Other alternatives include growing duckweed with an aquaponics system that feeds the same fish grown on the system, excess worms grown from vermiculture composting, using prepared kitchen scraps, as well as growing black soldier fly larvae to feed to the fish using composting grub growers.

BEST AQUACULTURE PRACTICES

Best Aquaculture Practices (BAP) is a set of farm raised seafood standards developed by Global Aquaculture Alliance (GAA). BAP standards address environmental and social responsibility, animal welfare, food safety and traceability for farms and hatcheries of shrimp, tilapia and catfish as well as seafood processing plants. The Best Aquaculture Practices are organized as a tiered ranking system: certified companies with four-star ratings are considered the most strictly compliant. The BAP are licensed by the Aquaculture Certification Council (ACC) which carries out all the inspections and certification.

COPPER ALLOYS IN AQUACULTURE

Recently, copper alloys have become important netting materials in aquaculture (the farming of aquatic organisms including fish farming). Various other materials including nylon, polyester, polypropylene, polyethylene, plastic-coated welded wire, rubber, patented twine products (Spectra, Dyneema), and galvanized steel are also used for netting in aquaculture fish enclosures around the world. All of these materials are selected for a variety of reasons, including design feasibility, material strength, cost, and corrosion resistance.

What sets copper alloys apart from the other materials used in fish farming is that copper alloys are antimicrobial, that is, they destroy bacteria, viruses, fungi, algae, and other microbes. (For information about the antimicrobial properties of copper and its alloys, see Antimicrobial properties of copper and Antimicrobial copper alloy touch surfaces).

In the marine environment, the antimicrobial/algaecidal properties of copper alloys prevent biofouling, which can briefly be described as the undesirable accumulation, adhesion, and growth of microorganisms, plants, algae, tube worms, barnacles, mollusks, and other organisms on man-made marine structures. By inhibiting microbial growth, copper alloy aquaculture pens avoid the need for costly net changes that are necessary with other materials. The resistance of organism growth on copper alloy nets also provides a cleaner and healthier environment for farmed fish to grow and thrive.

In addition to their antifouling benefits, copper alloys have strong structural and corrosion-resistant properties in marine environments. It is the combination of all of these properties – antifouling, high strength, and corrosion resistance – that has made copper alloys a desirable material for such marine applications as condenser tubing, water intake screens, ship hulls, offshore structure, and sheathing. In the past 25 years or so, the benefits of copper alloys have caught the attention of the marine aquaculture industry. The industry is now actively deploying copper alloy netting and structural materials in commercial large-scale fish farming operations around the world.

IMPORTANCE OF AQUACULTURE

Much has been written about the degradation and depletion of natural fish stocks in rivers, estuaries, and the oceans (see also Overfishing). Because industrial fishing has become extremely efficient, ocean stocks of large fish, such as tuna, cod, and halibut have declined by 90% in the past 50 years.Aquaculture, an industry that has emerged only in recent decades, has become one of the fastest growing sectors of the world food economy. Aquaculture already supplies more than half of the world's demand for fish. This percentage is predicted to increase dramatically over the next few decades.

THE PROBLEM OF BIOFOULING

Biofouling is one of the most important problems in aquaculture. Biofouling occurs on non-copper materials in the marine environment, including fish pen surfaces and nettings. For example, it was noted that the open area of a mesh immersed for only seven days in a Tasmanian aquaculture operation decreased by 37% as a result of biofouling.

The biofouling process begins when algae spores, marine invertebrate larvae, and other organic material adhere to surfaces submerged in marine environments (e.g., fish nets in aquaculture). Bacteria then encourage the attachment of secondary unwanted colonizers.

Biofouling has strong negative impacts on aquaculture operations. Water flow and dissolved oxygen are inhibited due to clogged nets in fish pens. The end result is often diseased fish from infections, such as netpen liver disease, amoebic gill disease, and parasites. Other negative impacts include increased fish mortalities, decreased fish growth rates, premature fish harvesting, reduced fish product values and profitability, and an adversely impacted environment near fish farms.

Biofouling adds enormous weight to submerged fish netting. Two hundredfold increases in weight have been reported. This translates, for example, to two thousand pounds of unwanted organisms adhered to what was once a clean 10-pound fish pen net. In South Australia, biofouling weighing 6.5 tonnes (approximately 13,000 pounds) was observed on a fish pen net. This extra burden often results in net breakage and additional maintenance costs.

To combat parasites from biofouling in finfish aquaculture, treatment protocols such as cypermethrin, azamethiphos, and emamectin benzoate may be administered, but these have been found to have detrimental environmental effects, for example, in lobster operations. To treat diseases in fish raised in biofouled nets, fish stocks are administered antibiotics. The antibiotics can have unwanted long-term health effects on consumers and on coastal environments near aquaculture operations. To combat biofouling, operators often implement costly maintenance measures, such as frequent net changing, cleaning/removal of unwanted organisms from nets, net repairs, and chemical treatment including antimicrobial coatings on nylon nets. The cost of antifouling a single salmon net can be several thousand British pounds. In some sectors of the European aquaculture industry, cleaning biofouled fish and shellfish pens can cost 5–20% of its market value. Heavy fouling can reduce the saleable product in nets by 60–90%.Antifouling coatings are often used on nylon nets because the process is more economical than manual cleaning. When nylon nets are coated with antifouling compounds, the coatings repel biofouling for a period of time, usually between several weeks to several months. However, the nets eventually succumb to biofouling. Antifouling coatings containing cuprous oxide algaecide/ biocide are the coatings technology used almost exclusively in the fish farming industry today. The treatments usually flake off within a few weeks to six to eight months.

Biofouled nets are replaced after several months of service, depending on environmental conditions, in a complicated, costly, and Labour-intensive operation that involves divers and specialized personnel. During this process, live fish in nets must be transferred to clean pens, which causes undue stress and asphyxiation that results in some loss of fish. Biofouled nets that can be reused are washed on land via manual brushing and scrubbing or high-pressure water hosing. They are then dried and re-impregnated with antifouling coatings.

A line of net cleaners is available for in-situ washings where permitted. But, even where not permitted by environmental, fisheries, maritime, and sanitary authorities, should the lack of dissolved oxygen in submerged pens create an emergency condition that endangers the health of fish, divers may be deployed with special in situ cleaning machinery to scrub biofouled nets.The aquaculture industry is addressing the negative environmental impacts from its operations (see aquaculture issues).

As the industry evolves, a cleaner, more sustainable aquaculture industry is expected to emerge, one that may increasingly rely on materials with anti-fouling, anti-corrosive, and strong structural properties, such as copper alloys.

ANTIFOULING PROPERTIES OF COPPER ALLOYS

In the aquaculture industry, sound animal husbandry translates to keeping fish clean, well fed, healthy, and not overcrowded. One solution to keeping farmed fish healthy is to contain them in antifouling copper alloy nets and structures.

Researchers have attributed copper's resistance to biofouling, even in temperate waters, to two possible mechanisms: 1) a retarding sequence of colonization through release of antimicrobial copper ions, thereby preventing the attachment of microbial layers to marine surfaces; and, 2) separating layers that contain corrosive products and the spores of juveniles or macro-encrusting organisms.The most important requirement for optimum biofouling resistance is that the copper alloys should be freely exposed or electrically insulated from less noble alloys and from cathodic protection. Galvanic coupling to less noble alloys and cathodic protection prevent copper ion releases from surface films and therefore reduce biofouling resistance.

As temperatures increase and water velocities decrease in marine waters, biofouling rates dramatically rise. However, copper's resistance to biofouling is observed even in temperate waters. Studies in La Herradura Bay, Coquimbo, Chile, where biofouling conditions are extreme, demonstrated that a copper alloy (90% copper, 10% nickel) avoided macro-encrusting organisms.

CORROSION BEHAVIOUR OF COPPER ALLOYS

Copper alloys used in sea water service have low general corrosion rates but also have a high resistance to many localized forms of corrosion. A technical discussion regarding various types of corrosion, application considerations (*e.g.*, depth of installations, effect of polluted waters, sea conditions), and the corrosion characteristics of several copper alloys used in aquaculture netting is available (*i.e.*, copper-nickel, copper-zinc, and copper-silicon).

EARLY EXAMPLES OF COPPER SHEATHING

Copper sheathing for bio-resistant ship hulls was developed in the late 18th century when the British Navy experimented with the HMS Alarm, which was coppered in 1758. The British navy sheathed all of their ships by 1781, realizing that the high up-front cost of coating the entire hull of a ship with copper sheeting eventually paid for itself in decreased maintenance cost. Around 1800, the technology was stolen and brought to the United States when the government-contracted Paul Revere developed copper sheeting and galvanic-reaction resistant copper fasteners for hull sheathing. These techniques, which the Navy put into ships like the USS New Hampshire and others, greatly increased the speed and durability of the American Navy by deterring barnacles and shipworms.

ENVIRONMENTAL PERFORMANCE OF COPPER ALLOY MESH

Many complicated factors influence the environmental performance of copper alloys in aquaculture operations. A technical description of antibiofouling mechanisms, fish health and welfare, fish losses due to escapes and predator attacks, and reduced life cycle environmental impacts is summarized in this reference.

TYPES OF COPPER ALLOYS

Copper-zinc brass alloys are currently (2011) being deployed in commercial-scale aquaculture operations in Asia, South America and the USA (Hawaii). Extensive research, including demonstrations and trials, are currently being implemented on two other copper alloys: copper-nickel and copper-silicon. Each of these alloy types has an inherent ability to reduce biofouling, pen waste, disease, and the need for antibiotics while simultaneously maintaining water circulation and oxygen requirements. Other types of copper alloys are also being considered for research and development in aquaculture operations.The University of New Hampshire is in the midst of conducting experiments under the auspices of the International Copper Association (ICA) to evaluate the structural, hydrodynamic, and antifouling response of copper alloy nets. Factors to be determined from these experiments, such as drag, pen dynamic loads, material loss, and biological growth – well documented for nylon netting but not fully understood for copper-nickel alloy nets – will help to design fish pen enclosures made from these alloys. The East China Sea Fisheries Research Institute, in Shanghai, China, is also conducting experimental investigations on copper alloys for ICA.

Copper-zinc alloys

The Mitsubishi-Shindoh Co., Ltd., has developed a proprietary copper-zinc brass alloy, called UR30, specifically designed for aquaculture operations. The alloy, which is composed of 64% copper, 35.1% zinc, 0.6% tin, and 0.3% nickel, resists mechanical abrasion when formed into wires and fabricated into chain link, woven, or other types of flexible mesh. Corrosion rates depend on the depth of submersion and seawater conditions. The average reported corrosion rate reported for the alloy is <5 ìm/yr based on 2- and 5-year exposure trials in seawater.

The Ashimori Industry Company, Ltd., has installed approximately 300 flexible pens with woven chain link UR30 meshes in Japan to raise Seriola (*i.e.*, yellowtail, amberjack, kingfish, hamachi). The company has installed another 32 brass pens to raise Atlantic salmon at the Van Diemen Aquaculture operations in Tasmania, Australia. In Chile, EcoSea Farming S.A. has installed a total of 62 woven chain link brass mesh pens to raise trout and Atlantic salmon. In Panama, China, Korea, Turkey, and the USA, demonstrations and trials are underway using flexible pens with woven chain link UR30 and other mesh forms and a range of copper alloys. To date, in over 10 years of aquaculture experience, chain link mesh fabricated by these brass alloys have not suffered from dezincification, stress corrosion cracking, or erosion corrosion.

Copper-nickel alloys

Copper-nickel alloys were developed specifically for seawater applications over five decades ago. Today, these alloys are being investigated for their potential use in aquaculture.Copper-nickel alloys for marine applications are usually 90% copper, 10% nickel, and small amounts of manganese and iron to enhance corrosion resistance. The seawater corrosion resistance of copper-nickel alloys results in a thin, adherent, protective surface film which forms naturally and quickly on the metal upon exposure to clean seawater.The rate of corrosion protective formation is temperature dependent. For example, at

27 °C (*i.e.*, a common inlet temperature in the Middle East), rapid film formation and good corrosion protection can be expected within a few hours. At 16 °C, it could take 2–3 months for the protection to mature. But once a good surface film forms, corrosion rates decrease, normally to 0.02–0.002 mm/yr, as protective layers develop over a period of years. These alloys have good resistance to chloride pitting and crevice corrosion and are not susceptible to chloride stress corrosion.

Copper-silicon alloys

Copper-silicon has a long history of use as screws, nuts, bolts, washers, pins, lag lag bolts, and staples in wooden sailing vessels in marine environments. The alloys are often composed of copper, silicon, and manganese. The inclusion of silicon strengthens the metal.

As with the copper-nickel alloys, corrosion resistance of copper-silicon is due to protective films that form on the surface over a period of time. General corrosion rates of 0.025–0.050mm have been observed in quiet waters. This rate decreases towards the lower end of the range over long-term exposures (*e.g.*, 400–600 days). There is generally no pitting with the silicon-bronzes. Also there is good resistance to erosion corrosion up to moderate flow rates. Because copper-silicon is weldable, rigid pens can be constructed with this material. Also, because welded copper-silicon mesh is lighter than copper-zinc chain link, aquaculture enclosures made with copper-silicon may be lighter in weight and therefore a potentially less expensive alternative.Luvata Appleton, LLC, is researching and developing a line of copper alloy woven and welded meshes, including a patent-pending copper silicon alloy, that are marketed under the trade name Seawire. Copper-silicon alloy meshes have been developed by the firm to raise various marine organisms in test trials that are now in various stages of evaluation. These include raising cobia in Panama, lobsters in Maine, and crabs in the Chesapeake Bay. The company is working with various universities to study its material, including the University of Arizona to study shrimp, the University of New Hampshire to study cod, and Oregon State University to study oysters. Results will be provided when the studies are concluded.

INLAND SALINE AQUACULTURE

Inland saline aquaculture is the farming or culture of aquatic animals and plants using inland (*i.e.* non-coastal) sources of saline groundwater rather than the more common coastal aquaculture methods. As a side benefit, it can be used to reduce the amount of salt in underground water tables, leading to an improvement in the surrounding land usage for agriculture. Due to its nature, it is only commercially possible in areas that have large reserves of saline groundwater, such as Australia.

SYSTEMS

- Extensive culture: Extensive culture aquaculture systems are simple and with low levels of intervention. An example of this would be a salty dam, lake or pond stocked with trout, where no food is needed to be added as the fish can feed off what naturally occus in the water. While they required little capital investment or management time, their productivity is relatively low.
- Intensive culture: Intensive culture requires more capital outlay and greater management time. Often they use purpose-built facilities (*e.g.* tanks), artificial food and aeration and constant monitoring of water quality. It has much higher productivity rates, but associated high levels of feeding, labour, water pumping and capital costs.
- Semi-intensive culture: Semi-intensive culture is in between extensive and intensive culture. It may range from adding some artificial feed to an extensive system or some aeration and waste management. Costs rise as more inputs are added.

SUITABLE SPECIES

Fish:

- Rainbow trout - robust, fast growth, require low water temperatures, may be limited to winter production
- Brown trout - robust, fast growth, require low water temperatures, may be limited to winter production
- Barramundi - needs higher temperatures, tolerant in a large range of salinity levels
- Macquarie perch - wide tolerance over range of salinity and water quality levels, not suitable for commercial quantities
- Silver Perch - suitable for extensive and intensive systems, prefers warmer water
- Snapper

Other species:

- Crustaceans - brine, shrimp, prawns - these can be included as part of a wastewater treatment Programme as some have the capacity to quickly clean water
- Molluscs - mussels
- Algae - both unicellular and "seaweeds" can be used to extract a range of high-value products, including pharmaceutical chemicals.

Mixing species:

- Chain system: Some inland aquaculture systems involve using a range of separated species to increase its productivity. An example of this would be where water is used to culture a fish specifies, which is then diverted to tanks of shellfish which feed on the fine particles

left by the fish, which then is diverted to algae species which remove the dissolved nutrients, and then last of all the water is sent to a horticultural system.
- Poly-culture: Separate from this type of system is poly-culture, where two or more species are cultured in the same water, possibly multiple fish species or a fish and mollusc species.

INTEGRATED MULTI-TROPHIC AQUACULTURE

Integrated multi-trophic aquaculture (IMTA) provides the by-products, including waste, from one aquatic species as inputs (fertilizers, food) for another. Farmers combine fed aquaculture (*e.g.*, fish, shrimp) with inorganic extractive (*e.g.*, seaweed) and organic extractive (*e.g.*, shellfish) aquaculture to create balanced systems for environment remediation (biomitigation), economic stability (improved output, lower cost, product diversification and risk reduction) and social acceptability (better management practices).

Selecting appropriate species and sizing the various populations to provide necessary ecosystem functions allows the biological and chemical processes involved to achieve a stable balance, mutually benefiting the organisms and improving ecosystem health.Ideally, the co-cultured species each yield valuable commercial "crops". IMTA can synergistically increase total output, even if some of the crops yield less than they would, short-term, in a monoculture.

TERMINOLOGY AND RELATED APPROACHES

"Integrated" refers to intensive and synergistic cultivation, using water-born nutrient and energy transfer. "Multi-trophic" means that the various species occupy different trophic levels, *i.e.*, different (but adjacent) links in the food chain.IMTA is a specialized form of the age-old practice of aquatic polyculture, which was the co-culture of various species, often without regard to trophic level. In this broader case, the organisms may share biological and chemical processes that are minimally complementary, potentially leading to significant ecosystem shifts/damage. Some traditional systems did culture species that occupied multiple niches within the same pond, but with limited intensity and management. The more general term "Integrated Aquaculture" is used to describe the integration of monocultures through water transfer. The terms "IMTA" and "integrated aquaculture" differ primarily in their precision and are sometimes interchanged. Aquaponics, fractionated aquaculture, IAAS (integrated agriculture-aquaculture systems), IPUAS (integrated peri-urban-aquaculture systems), and IFAS (integrated fisheries-aquaculture systems) are variations on the IMTA concept.

RANGE OF APPROACHES

Today, low intensity traditional/incidental multi-trophic aquaculture is much more common than modern IMTA. Most are relatively simple, such as fish/seaweed/shellfish.True IMTA can be land-based, using ponds or tanks, or even open-water marine or freshwater systems. Implementations have included species combinations such as shellfish/shrimp, fish/seaweed/shellfish, fish/seaweed, fish/shrimp and seaweed/shrimp.

IMTA in open water (offshore cultivation) can be done by the use of buoys with lines on which the seaweed grows. The buoys/lines are placed next to the fishnets or cages in which the fish grows. This method is already used commercially in Norway, Scotland, and Ireland.

In the future, systems with other components for additional functions, or similar functions but different size brackets of particles, are likely. Multiple regulatory issues remain open.

MODERN HISTORY OF LAND-BASED SYSTEMS

Ryther and co-workers created modern, integrated, intensive, land mariculture. They originated, both theoretically and experimentally, the integrated use of extractive organisms—shellfish, microalgae and seaweeds—in the treatment of household effluents, descriptively and with quantitative results. A domestic wastewater effluent, mixed with seawater, was the nutrient source for phytoplankton, which in turn became food for oysters and clams. They cultivated other organisms in a food chain rooted in the farm's organic sludge. Dissolved nutrients in the final effluent were filtered by seaweed (mainly Gracilaria and Ulva) biofilters. The value of the original organisms grown on human waste effluents was minimal.

In 1976, Huguenin proposed adaptations to the treatment of intensive aquaculture effluents in both inland and coastal areas. Tenore followed by integrating with their system of carnivorous fish and the macroalgivore abalone.

In 1977, Hughes-Games described the first practical marine fish/shellfish/phytoplankton culture, followed by Gordin, et al., in 1981. By 1989, a semi-intensive (1 kg fish/m) seabream and grey mullet pond system by the Gulf of Aqaba (Eilat) on the Red Sea supported dense diatom populations, excellent for feeding oysters. Hundreds of kilos of fish and oysters cultured here were sold. Researchers also quantified the water quality parameters and nutrient budgets in (5 kg fish m) green water seabream ponds. The phytoplankton generally maintained reasonable water quality and converted on average over half the waste nitrogen into algal biomass. Experiments with intensive bivalve cultures yielded high bivalve growth rates. This technology supported a small farm in southern Israel.

SUSTAINABILITY

IMTA promotes economic and environmental sustainability by converting byproducts and uneaten feed from fed organisms into harvestable crops, thereby reducing eutrophication, and increasing economic diversification.Properly managed multi-trophic aquaculture accelerates growth without detrimental side-effects. This increases the site's ability to assimilate the cultivated organisms, thereby reducing negative environmental impacts.

IMTA enables farmers to diversify their output by replacing purchased inputs with byproducts from lower trophic levels, often without new sites. Initial economic research suggests that IMTA can increase profits and can reduce financial risks due to weather, disease and market fluctuations. Over a dozen studies have investigated the economics of IMTA systems since 1985.

NUTRIENT FLOW

Typically, carnivorous fish or shrimp occupy IMTA's higher trophic levels. They excrete soluble ammonia and phosphorus (orthophosphate). Seaweeds and similar species can extract these inorganic nutrients directly from their environment. Fish and shrimp also release organic nutrients which feed shellfish and deposit feeders.Species such as shellfish that occupy intermediate trophic levels often play a dual role, both filtering organic bottom-level organisms from the water and generating some ammonia. Waste feed may also provide additional nutrients; either by direct consumption or via decomposition into individual nutrients. In some projects, the waste nutrients are also gathered and reused in the food given to the fish in cultivation. This can happen by processing the seaweed grown into food.

Recovery efficiency

Nutrient recovery efficiency is a function of technology, harvest schedule, management, spatial configuration, production, species selection, trophic level biomass ratios, natural food availability, particle size, digestibility, season, light, temperature, and water flow. Since these factors significantly vary by site and region, recovery efficiency also varies.In a hypothetical family-scale fish/microalga/bivalve/seaweed farm, based on pilot scale data, at least 60% of nutrient input reached commercial products, nearly three times more than in modern net pen farms. Expected average annual yields of the system for a hypothetical 1 hectare (2.5 acres) were 35 tonnes (34 long tons; 39 short tons) of seabream, 100 tonnes (98 long tons; 110 short tons) of bivalves and 125 tonnes (123 long tons; 138 short tons) of seaweeds. These results required precise water quality control and attention to suitability for bivalve nutrition, due to the difficulty in maintaining consistent phytoplanton populations.Seaweeds' nitrogen uptake efficiency ranges from 2-100% in land-based systems. Uptake efficiency in open-water IMTA is unknown.

FOOD SAFETY AND QUALITY

Feeding the wastes of one species to another has the potential for contamination, although this has yet to be observed in IMTA systems. Mussels and kelp growing adjacent to Atlantic salmon cages in the Bay of Fundy have been monitored since 2001 for contamination by medicines, heavy metals, arsenic, PCBs and pesticides. Concentrations are consistently either non-detectable or well below regulatory limits established by the Canadian Food Inspection Agency, the United States Food and Drug Administration and European Community Directives. Taste testers indicate that these mussels are free of "fishy" taste and aroma and could not distinguish them from "wild" mussels. The mussels' meat yield is significantly higher, reflecting the increase in nutrient availability.

SELECTED PROJECTS

Asia

Japan, China, South Korea, Thailand, Vietnam, Indonesia, Bangladesh, etc. have co-cultured aquatic species for centuries in marine, brackish and fresh water environments. Fish, shellfish and seaweeds have been cultured together in bays, lagoons and ponds. Trial and error has improved integration over time. The proportion of Asian aquaculture production that occurs in IMTA systems is unknown. the 2004 tsunami, many of the shrimp farmers in Aceh Province of Indonesia and Ranong Province of Thailand were trained in IMTA. This has been especially important as the mono-culture of marine shrimp was widely recognized as unsustainable. Production of tilapia, mud crabs, seaweeds, milkfish, and mussels have been incorporated. AquaFish Collaborative Research Support Programme

Canada

Bay of Fundy

Industry, academia and government are collaborating here to expand production to commercial scale. The current system integrates Atlantic salmon, blue mussels and kelp; deposit feeders are under consideration. AquaNet (one of Canada's Networks of Centers of Excellence) funded phase one. The Atlantic Canada Opportunities Agency is presently funding phase two. The project leaders are Thierry Chopin (University of New Brunswick in Saint John) and Shawn Robinson (Department of Fisheries and Oceans, St. Andrews Biological Station).

Pacific SEA-lab

Pacific SEA-lab is researching and is licensed for the co-culture of sablefish, scallops, oysters, blue mussels, urchins and kelp. "SEA" stands for Sustainable Ecological Aquaculture. The project presently aims to balance four species.The project is headed by Stephen Cross under a British Columbia Innovation Award at the University of Victoria Coastal Aquaculture Research and Training (CART) network.

Chile

The i-mar Research Center at the Universidad de Los Lagos, in Puerto Montt is working to reduce the environmental impact of intensive salmon culture. Initial research involved trout, oysters and seaweeds. Present research is focusing on open waters with salmon, seaweeds and abalone. The project leader is Alejandro Buschmann.

Israel

SeaOr Marine Enterprises Ltd.

SeaOr Marine Enterprises Ltd., which operated for several years on the Israeli Mediterranean coast, north of Tel Aviv, cultured marine fish (gilthead seabream), seaweeds (Ulva and Gracilaria) and Japanese abalone. Its approach leveraged local climate, and recycled fish waste products into seaweed biomass, which was fed to the abalone. It also effectively purified the water sufficiently to allow the water to be recycled to the fishponds and to meet point-source effluent environmental regulations.

PGP Ltd.

PGP Ltd. is a small farm in Southern Israel. It cultures marine fish, microalgae, bivalves and Artemia. Effluents from seabream and seabass collect in sedimentation ponds, where dense populations of microalgae—mostly diatoms—develop. Clams, oysters and sometimes Artemia filter the microalgae from the water, producing a clear effluent. The farm sells the fish, bivalves and Artemia.

The Netherlands

In the Netherlands, Willem Brandenburg of UR Wageningen (Plant Sciences Group) has established the first seaweed farm in the Netherlands. The farm is called "De Wierderij" and is used for research.

South Africa

Three farms currently grow seaweeds for feed in abalone effluents in land-based tanks. Up to 50% of re-circulated water passes through the seaweed tanks. Somewhat uniquely, neither fish nor shrimp comprise the upper trophic species. The motivation is to avoid over-harvesting natural seaweed beds and red tides, rather than nutrient abatement. These commercial successes developed from research collaboration between Irvine and Johnson Cape Abalone and scientists from the University of Cape Town and the University of Stockholm.

United Kingdom

The Scottish Association for Marine Science, in Oban is developing co-cultures of salmon, oysters, sea urchins, and brown and red seaweeds via several projects. Research focuses on biological and physical processes, as well as production economics and implications for coastal zone management. Researchers include: M. Kelly, A. Rodger, L. Cook, S. Dworjanyn, and C. Sanderson.

Bangladesh

Indian carps and stinging catfish are cultured in Bangladesh, but the methods could be more productive. The pond and cage cultures used are based only on the fish. They don't take advantage of the productivity increases that could take place if other trophic levels were included. Expensive artificial feeds are used, partly to supply the fish with protein. These costs could be reduced if freshwater snails, such as *Viviparus bengalensis*, were simultaneously cultured, thus increasing the available protein. The organic and inorganic wastes produced as a byproduct of culturing could also be minimized by integrating aquatic plants, such as water spinach.

MARICULTURE

Mariculture is a specialized branch of aquaculture involving the cultivation of marine organisms for food and other products in the open ocean, an enclosed section of the ocean, or in tanks, ponds or raceways which are filled with seawater. An example of the latter is the farming of marine fish, including finfish and shellfish e.g.prawns, or oysters and seaweed in saltwater ponds. Non-food products produced by mariculture include: fish meal, nutrient agar, jewellery (*e.g.* cultured pearls), and cosmetics.

METHODS

Algae

Seawater algae such as kelp can be farmed in at least two ways. It can be grown around a rope that is anchored to the sea floor so that they do not drift away. Off the coast of California, boats with mowers harvest the top few feet of natural kelp beds. Kelp provides alginate, an edible material used in ice cream and cosmetics. The industry also supplies the dietary supplement industry.

Open ocean

Raising marine organisms under controlled conditions in exposed, high-energy ocean environments beyond significant coastal influence, is a relatively new approach to mariculture. Open ocean aquaculture (OOA) uses cages, nets, or long-line arrays that are moored, towed or float freely. Research and commercial open ocean aquaculture facilities are in operation or under development in Australia, Chile, China, France, Ireland, Italy, Japan, Mexico, and Norway. As of 2004, two commercial open ocean facilities were operating in U.S. waters, raising Threadfin near Hawaii and cobia near Puerto Rico. An operation targeting bigeye tuna recently received final approval. All U.S. commercial facilities are currently sited in waters under state or territorial jurisdiction.

Sea ranching

The Japanese apply a principle based on operant conditioning and the migratory nature of certain species. The fishermen raise hatchlings in a closely knitted net in a Harbour, sounding an underwater horn before each feeding. When the fish are old enough they are freed from the net to mature in the open sea. During spawning season, about 80% of these fish return to their birthplace. The fishermen sound the horn and then net those fish that respond.

Seawater ponds

In seawater pond mariculture, fish are raised in ponds which receive water from the sea. This has the benefit that the nutrition (*e.g.* microorganisms) present in the seawater can be used. This is a great advantage over traditional fish farms (*e.g.* sweet water farms) for which the farmers buy feed (which is expensive). Other advantages are that water purification plants may be planted in the ponds to eliminate the buildup of nitrogen, from fecal and other contamination. Also, the ponds can be left unprotected from natural predators, providing another kind of filtering.

ENVIRONMENTAL EFFECTS

Mariculture has rapidly expanded over the last two decades due to new technology, improvements in formulated feeds, greater biological understanding of farmed species, increased water quality within closed farm systems, greater demand for seafood products, site expansion and government interest. As a consequence, mariculture has been subject to some controversy regarding its social and environmental impacts. Commonly identified environmental impacts from marine farms are:

1. Wastes from cage cultures;
2. Farm escapees and invasives;
3. Genetic pollution and disease and parasite transfer;
4. Habitat modification.

As with most farming practices, the degree of environmental impact depends on the size of the farm, the cultured species, stock density, type of feed, hydrography of the site, and husbandry methods. The adjacent diagram connects these causes and effects.

Wastes from cage cultures

Mariculture of finfish can require a significant amount of fishmeal or other high protein food sources. Originally, a lot of fishmeal went to waste due to inefficient feeding regimes and poor digestibility of formulated feeds which resulted in poor feed conversion ratios.

In cage culture, several different methods are used for feeding farmed fish – from simple hand feeding to sophisticated computer-controlled systems with automated food dispensers coupled with *in situ* uptake sensors that detect consumption rates. In coastal fish farms, overfeeding primarily leads to increased disposition of detritus on the seafloor (potentially smothering seafloor dwelling invertebrates and altering the physical environment), while in hatcheries and land-based farms, excess food goes to waste and can potentially impact the surrounding catchment and local coastal environment. This impact is usually highly local, and depends significantly on the settling velocity of waste feed and the current velocity (which varies both spatially and temporally) and depth.

Farm escapees and invasives

The impact of escapees from aquaculture operations depends on whether or not there are wild conspecifics or close relatives in the receiving environment, and whether or not the escapee is reproductively capable. Several different mitigation/prevention strategies are currently employed, from the development of infertile triploids to land-based farms which are completely isolated from any marine environment. Escapees can adversely impact local ecosystems through hybridization and loss of genetic diversity in native stocks, increase negative interactions within an ecosystem (such as predation and competition), disease transmission and habitat changes (from trophic cascades and ecosystem shifts to varying sediment regimes and thus turbidity).

The accidental introduction of invasive species is also of concern. Aquaculture is one of the main vectors for invasives following accidental releases of farmed stocks into the wild. One example is the Siberian sturgeon (*Acipenser baerii*) which accidentally escaped from a fish farm into the Gironde Estuary (Southwest France) following a severe storm in December 1999 (5,000 individual fish escaped into the estuary which had never hosted this species before). Molluscan farming is another example whereby species can be introduced to new environments by 'hitchhiking' on farmed molluscs. Also, farmed molluscs themselves can become dominate predators and/or competitors, as well as potentially spread pathogens and parasites.

Genetic Pollution and Disease and Parasite Transfer

One of the primary concerns with mariculture is the potential for disease and parasite transfer. Farmed stocks are often selectively bred to increase disease and parasite resistance, as well as improving growth rates and quality of products. As a consequence, the genetic diversity within reared stocks decreases with every generation - meaning they can potentially reduce the genetic diversity within wild populations if they escape into those wild populations. Such genetic pollution from escaped aquaculture stock can reduce the wild population's ability to adjust to the changing natural environment. Also, maricultured species can harbour diseases and parasites (*e.g.*, lice) which can be introduced to wild populations upon their escape. An example of this is the parasitic sea lice on wild and farmed Atlantic salmon in Canada. Also, non-indigenous species which are farmed may have resistance to, or carry, particular diseases (which they picked up in their native habitats) which could be spread through wild populations if they escape into those wild populations. Such 'new' diseases would be devastating for those wild populations because they would have no immunity to them.

Habitat modification

With the exception of benthic habitats directly beneath marine farms, most mariculture causes minimal destruction to habitats. However, the destruction of mangrove forests from the farming of shrimps is of concern. Globally, shrimp farming activity is a small contributor to the destruction of mangrove forests; however, locally it can be devastating. Mangrove forests provide rich matrices which support a great deal of biodiversity – predominately juvenile fish and crustaceans. Furthermore, they act as buffering systems whereby they reduce coastal erosion, and improve water quality for in situ animals by processing material and 'filtering' sediments.

Others

In addition, nitrogen and phosphorus compounds from food and waste may lead to blooms of phytoplankton, whose subsequent degradation can drastically reduce oxygen levels. If the algae are toxic, fish are killed and shellfish contaminated.

SUSTAINABILITY

Mariculture development must be sustained by basic and applied research and development in major fields such as nutrition, genetics, system management, product handling, and socioeconomics. One approach is closed systems that have no direct interaction with the local environment. However, investment and operational cost are currently significantly higher than open cages, limiting them to their current role as hatcheries.

POTENTIAL BENEFITS

Sustainable mariculture promises economic and environmental benefits. Economies of scale imply that ranching can produce fish at lower cost than industrial fishing, leading to better human diets and the gradual elimination of unsustainable fisheries. Maricultured fish are also perceived to be of higher quality than fish raised in ponds or tanks, and offer more diverse choice of species. Consistent supply and quality control has enabled integration in food market channels.Offshore aquaculture Offshore aquaculture, also known as open ocean aquaculture, is an emerging approach to mariculture or marine farming where fish farms are moved some distance offshore. The farms are positioned in deeper and less sheltered waters, where ocean currents are stronger than they are inshore.One of the concerns with inshore aquaculture is that discarded nutrients and feces can settle below the farm on the seafloor and damage the benthic ecosystem. According to its proponents, the wastes from aquaculture that has been moved offshore tend to be swept away from the site and diluted. Moving aquaculture offshore also provides more space where aquaculture production can expand to meet the increasing demands for fish. It avoids many of the conflicts that occur with other marine resource users in the more crowded inshore waters, though there can still be user conflicts offshore. Critics are concerned about issues such as the ongoing consequences of using antibiotics and other drugs and the possibilities of cultured fish escaping and spreading disease among wild fish.

BACKGROUND

Aquaculture is the most rapidly expanding food industry in the world as a result of declining wild fisheries stocks and profitable business. In 2008, aquaculture provided 45.7% of the fish produced globally for human consumption; increasing at an mean rate of 6.6% a year since 1970.

In 1970, a National Oceanic and Atmospheric Administration (NOAA) grant brought together a group of oceanographers, engineers and marine biologists to explore whether offshore aquaculture, which was then considered a futuristic activity, was feasible. In the United States, the future of offshore aquaculture technology within federal waters has become much talked-about. As many commercial operations show, it is now technically possible to culture finfish, shellfish, and seaweeds using offshore aquaculture technology.

Major challenges for the offshore aquaculture industry involve designing and deploying cages that can withstand storms, dealing with the logistics of working many kilometers from land, and finding species that are sufficiently profitable to cover the costs of rearing fish in exposed offshore areas.

TECHNOLOGY

To withstand the high energy offshore environment, farms must be built to be more robust than those inshore. However, the design of the offshore technology is developing rapidly, aimed at reducing cost and maintenance.While the ranching systems currently used for tuna use open net cages at the surface of the sea (as is done also in salmon farming), the offshore technology usually uses submersible cages. These large rigid cages – each one able to hold many thousands of fish – are anchored on the sea floor, but can move up and down the water column. They are attached to buoys on the surface which frequently contain a mechanism for feeding and storage for equipment. Similar technology is being used in waters near the Bahamas, China, the Philippines, Portugal, Puerto Rico, and Spain. By submerging cages or shellfish culture systems, wave effects are minimized and interference with boating and shipping is reduced. Offshore farms can be made more efficient and safer if remote control is used, and technologies such as an 18-tonne buoy that feeds and monitors fish automatically over long periods are being developed.

Existing offshore structures

Multi-functional use of offshore waters can lead to more sustainable aquaculture "in areas that can be simultaneously used for other activities such as energy production". Operations for finfish and shellfish are being developed. For example the Hubb-Sea World Research Institutes' project to convert a retired oil platform 10 nm off the southern California coast to an experimental offshore aquaculture facility. The institute plans to grow mussels and red abalone on the actual platform, as well as white seabass, striped bass, bluefin tuna, California halibut and California yellowtail in floating cages.

Integrated multi-trophic aquaculture

Integrated multi-trophic aquaculture (IMTA), or polyculture, occurs when species which must be fed, such as finfish, are cultured alongside species which can feed on dissolved nutrients, such as seaweeds, or organic wastes, such as suspension feeders and deposit feeders. This sustainable method could solve several problems with offshore aquaculture. The method is being pioneered in Spain, Canada, and elsewhere.

Roaming cages

Roaming cages have been envisioned as the "next generation technology" for offshore aquaculture. These are large mobile cages powered by thrusters and able to take advantage of ocean currents. One idea is that juvenile tuna, starting out in mobile cages in Mexico, could reach Japan after a few months, matured and ready for the market. However, implementing such ideas will have regulatory and legal implications.

SPACE CONFLICTS

As oceans industrialise, conflicts are increasing among the users of marine space. This competition for marine space is developing in a context where natural resources can be seen as publicly owned. There can be conflict with the tourism industry, recreational fishers, wild harvest fisheries and the siting of marine renewable energy installations. The problems can be aggravated by the remoteness of many marine areas, and difficulties with monitoring and enforcement. On the other hand, remote sites can be chosen that avoid conflicts with other users, and allow large scale operations with resulting economies of scale. Offshore systems can provide alternatives for countries with few suitable inshore sites, like Spain.

ECOLOGICAL IMPACTS

The ecological impacts of offshore aquaculture are somewhat uncertain because it is still largely in the research stage. Many of the concerns over potential offshore aquaculture impacts are paralleled by similar, well established concerns over inshore aquaculture practices.

Pollution

One of the concerns with inshore farms is that discarded nutrients and feces can settle on the seafloor and disturb the benthos. The "dilution of nutrients" that occurs in deeper water is a strong reason to move coastal aquaculture offshore into the open ocean. How much nutrient pollution and damage to the seafloor occurs depends on the feed conversion efficiency of the species, the flushing rate and the size of the operation. However, dissolved and particulate nutrients are still released to the environment. Future offshore farms will probably be much larger than inshore farms today, and will therefore generate more waste. The point at which the capacity of offshore ecosystems to assimilate waste from offshore aquaculture operations will be exceeded is yet to be defined.

Wild caught feed

As with the inshore aquaculture of carnivorous fish, a large proportion of the feed comes from wild forage fish. Except for a few countries, offshore aquaculture has focused predominantly on high value carnivorous fish. If the industry attempts to expand with this focus then the supply of these wild fish will become ecologically unsustainable.

Fish escapes

The expense of offshore systems means it is important to avoid fish escapes. However, it is likely there will be escapes as the offshore industry expands. This could have significant consequences for native species, even if the farmed fish are inside their native range. Submersible cages are fully closed and therefore escapes can only occur through damage to the structure. Offshore cages must withstand the high energy of the environment and attacks by predators such as sharks. The outer netting is made of Spectra – a super-strong polyethylene fibre – wrapped tightly around the frame, leaving no slack for predators to grip. However, the fertilised eggs of cod are able to pass through the cage mesh in ocean enclosures.

Disease

Compared to inshore aquaculture, disease problems currently appear to be much reduced when farming offshore. For example, parasitic infections that occur in mussels cultured offshore are much smaller than those cultured inshore. However, new species are now being farmed offshore although little is known about their ecology and epidemiology. The implications of transmitting pathogens between such farmed species and wild species "remains a large and unanswered question".

Spreading of pathogens between fish stocks is a major issue in disease control. Static offshore cages may help minimize direct spreading, as there may be greater distances between aquaculture production areas. However, development of roaming cage technology could bring about new issues with disease transfer and spread. The high level of carnivorous aquaculture production results in an increased demand for live aquatic animals for production and breeding purposes such as bait, broodstock and milt. This can result in spread of disease across species barriers.

EMPLOYMENT

Aquaculture is encouraged by many governments as a way to generate jobs and income, particularly when wild fisheries have been run down. However, this may not apply to offshore aquaculture. Offshore aquaculture entails high equipment and supply costs, and therefore will be under severe pressure to lower Labour costs through automated production technologies. Employment is likely to expand more at processing facilities than grow-out industries as offshore aquaculture develops.

PROSPECTS

Norway and the United States are currently (2008) making the main investments in the design of offshore cages.

FAO

In 2010, the Food and Agriculture Organization (FAO) sub-committee on aquaculture made the following assessments:

- "Most Members thought it inevitable that aquaculture will move further offshore if the world is to meet its growing demand for seafood and urged the development of appropriate technologies for its expansion and assistance to developing countries in accessing them [...] Some Members noted that aquaculture may also develop offshore in large inland water bodies and discussion should extend to inland waters as well [...] Some Members suggested caution regarding potential negative impacts when developing offshore aquaculture.caution regarding potential negative impacts when developing offshore aquaculture.

The sub-committee recommended the FAO "should work towards clarifying the technical and legal terminology related to offshore aquaculture in order to avoid confusion."

Europe

In 2002, the European Commission issued the following policy statement on aquaculture:

- "Fish cages should be moved further from the coast, and more research and development of offshore cage technology must be promoted to this end. Experience from outside the aquaculture sector, *e.g.* with oil platforms, may well feed into the aquaculture equipment sector, allowing for savings in the development costs of technologies."

By 2008, European offshore systems were operating in Norway, Ireland, Italy, Spain, Greece, Cyprus, Malta, Croatia, Portugal and Libya.

In Ireland, as part of their National Development Plan, it is envisioned that over the period 2007–2013, technology associated with offshore aquaculture systems will be developed, including: "sensor systems for feeding, biomass and health monitoring, feed control, telemetry and communications [and] cage design, materials, structural testing and modelling."

United States

Moving aquaculture offshore into the exclusive economic zone (EEZ) can cause complications with regulations. In the United States, regulatory control of the coastal states generally extends to 3 nm, while federal waters (or EEZ) extend to 200 nm offshore. Therefore, offshore aquaculture can be sited outside the reach of state law but within federal jurisdiction. As of 2010, "all commercial aquaculture facilities have been sited in nearshore waters under state or territorial jurisdiction." However, "unclear regulatory processes" and "technical uncertainties related to working in offshore areas" have hindered progress. The five offshore research projects and commercial operations in the US – in New Hampshire, Puerto Rico, Hawaii and California – are all in federal waters. In June 2011, the *National Sustainable Offshore Aquaculture Act of 2011* was introduced to the House of Representatives "to establish a regulatory system and research Programme for sustainable offshore aquaculture in the United States exclusive economic zone".

CURRENT SPECIES

By 2005, offshore aquaculture was present in 25 countries, both as experimental and commercial farms. Market demand means that the most offshore farming efforts are directed towards raising finfish. Two commercial operations in the US, and a third in the Bahamas are using submersible cages to raise high-value carnivorous finfish, such as moi, cobia, and mutton snapper.

Submersible cages are also being used in experimental systems for halibut, haddock, cod, and summer flounder in New Hampshire waters, and for amberjack, red drum, snapper, pompano, and cobia in the Gulf of Mexico.The offshore aquaculture of shellfish grown in suspended culture systems, like scallops and mussels, is gaining ground. Suspended culture systems include methods where the shellfish are grown on a tethered rope or suspended from a floating raft in net containers. Mussels in particular can survive the high physical stress levels which occur in the volatile environments that occur in offshore waters. Finfish species must be feed regularly, but shellfish do not, which can reduce costs. The University of New Hampshire in the US has conducted research on the farming of blue mussels submerged in an open ocean environment. They have found that when farmed in less polluted waters offshore, the mussels develop more flesh with lighter shells.

ORGANIC AQUACULTURE

Organic aquaculture has become more important as consumers have become more environmentally aware and concerned about sustainability and harmful impacts of aquaculture. Aquaculture has the highest rate of expansion and Certified organic aquaculture products have increased since the mid 1990s It has become more mainstream, especially in Germany, the United Kingdom and Switzerland, but many consumers can be confused or sceptical about organically labelled product due to conflicting or misleading standards around the world.

The aquaculture industries in general are still figuring out how to be sustainable, what best practices are and what ecological considerations is can or should be implemented. Current standards are often quite strict and some people argue they are unattainable (and therefore should be relaxed). Organic regulations designed around soil-based systems don't transfer well into aquaculture and tend to conflict with large-scale, intensive (economically viable) practices/goals. There is a definite consumer demand for organic seafood, and organic aquaculture may become a significant management option. Currently integrated aquaculture systems may form the base of approved organic farming practice.

PRODUCTION AND MARKETING

Organic aquaculture was responsible for an estimated US$46.1 billion internationally (2007). There were 0.4 million hectares of certified organic aquaculture in 2008 compared to 32.2 million hectares dedicated to Organic farming. The 2007 production was still only 0.1% of total aquaculture production

The market for organic aquaculture shows strong growth in Europe, especially France, Germany and the UK - for example, the market in France grew 220% from 2007 to 2008. There is a preference for organic food, where available. Organic seafood is now sold in discount supermarket chains throughout the EU. The top five producing countries are UK, Ireland, Hungary, Greece and France. 123 of the 225 global certified organic aquaculture farms operate in Europe and were responsible for 50,000 tonnes in 2008 (nearly half global production).Organic seafood products are a niche market and users currently expect to pay premiums of 30-40%. Organic salmon is the top species and retails at 50%. Market demand is drving Danish rainbow trout farmers to switch to organic farming.

Certification

There is some controversy over licensing restrictions, as some seafood companies propose that wild caught fish should be classified as organic. This is a situation where the flexibility and ambiguity of the term 'organic' is clearly demonstrated. While wild fish may be free of pesticides and unsustainable rearing practices, the industry itself is not necessarily environmentally sustainable nor is it actually farming. As a rule, Organic Aquaculture certification follows rather strict requirements and standards. These rules may vary between different countries or certification bodies. This leads to confusion when products are imported from other countries which can result in a backlash from consumers.

Defining accepted practices is also complicated by the variety of species - freshwater, saltwater, shellfish, finfish, mollusks and aquatic plants. The difficulty of screening pollutants out of the aquatic medium, of controlling food supplies and of keeping track of individuals may mean that fish and shellfish stocks should not be classified as 'livestock' at all under regulations.

A number of countries have created national standards and certifying bodies. A number of 'self-labelling' organisations also exist at local scales in many countries. One of the largest certification organisations is IFOAM which supports and connects organic movements internationally. The variation in standards, as well as the unknown level of actual compliance and the closeness of investigations when certifying are all also variable, and are one of the major problems in consistent organic certification. In 2010, new rules were brought in on 30/06/2010 in the European Union to consistently define the organic aquaculture industry. However, aquaculture was only first included in the organic agriculture standards in 2007.

"Organic" has an official definition in the United States, since the 2008 organic aquaculture standards passed by the US National Organic Standards Board. Canada's General Standards Board's (CGSB) proposed standards in 2010 were strongly opposed because they allowed antibiotic and chemical treatments of fish, up to 30 percent non-organic feed, deadly and uncontrolled impacts on wild species and unrestrained disposal of fish faeces into the ocean. These standards would have certified net pen systems as organic, and most of the fish certiified would not have complied with the US organic aquaculture standards. At the other end of the scale, the extremely strict national legislation in Denmark has made it difficult for the existing organic trout industry to develop.

Current and Future Research and Development

There are a number of problems facing organic aquaculture: difficulty of sourcing and certifying organic juveniles (hatchery or sustainable wild stock); 35-40% higher feed cost; more labour intensive; time and cost of the certification process; a higher risk of diseases, and uncertain benefits.

Various methods and complementary processes are being investigated for organic aquaculture, most notably Integrated Multi-Trophic Aquaculture(IMTA) and aquaponics (a land-based outgrowth of aquaculture in many places). Organic methods of farming various species are also topics of interest, particularly shrimps, salmon and Atlantic Cod .

Projects such as the Danish ORAQUA Project are unvestigating areas such as "Availability of relevant organically produced vegetables with a high protein content; The effect of the organic vegetable protein; sources on growth and digestibility in rainbow trout; The influence of organic protein and oil; eating quality; The influence of organic vegetables on the health and welfare of farmed fish; Optimizing the production by case studies of organic production systems"

Potential Alternatives To Non-Organic Treatments and Processes

A great deal of research needs to be done into a number of areas around organic aquaculture. Current requirements usually stipulate a reduction in unsustainable fishmeal, in favour of organic vegetable and fish by-product replacements. A recent study into organic fish feeds for salmon found that while organic feed provide some benefit to the environmental impact of the fishes' life cycles, the loss of fish meals and oils have a significant negative impact. Another study discovered that certain percentages of dietary protein could be safely replaced.

One major issue is finding practical and sustainable alternatives to non-organic veterinary treatments, feeds, spat and waste disposal. Potential veterinary alternatives include homeopathic treatments and production-cycle limited allopathic or chemical treatments

Not only do the fish have to be organically reared, organic fish feeds need to be developed. Research into ways of decreasing the amount on non-sustainable fishmeal in feed is currently focussing on replacement by organic vegetable proteins. Some organic fish feeds becoming available, and/or the option of integrated multi-species systems (*e.g.* growing plants using aquaponics, as well as larvae or other fish). For example, locating a shellfish bed next to a finfish farm to dispose of the waste and provide the shellfish with controlled nutrients.

RACEWAY (AQUACULTURE)

A raceway, also known as a flow-through system, is an artificial channel used in aquaculture to culture aquatic organisms. Raceway systems are among the earliest methods used for inland aquaculture. A raceway usually consists of rectangular basins or canals constructed of concrete and equipped with an inlet and outlet. A continuous water flow-through is maintained to provide the required level of water quality, which allows animals to be cultured at higher densities within the raceway.

Freshwater species such as trout, catfish and tilapia are commonly cultured in raceways. Raceways are also used for some marine species which need a constant water flow, such as juvenile salmon, brackish water sea bass and sea bream and marine invertebrates such as abalone.

SITE SELECTION

The most important factor to consider when selecting a site for a raceway farm is the water supply. Water sources for raceway aquaculture operations are usually streams, springs, reservoirs or deep wells. Trout do best in spring water because it keeps a constant temperature, while catfish need a strong flow, about 80 litres per second for every 0.4 hectares of raceway. A backup water supply should be positioned so, if the water supply or pump fails, it can flow by gravity into the start of the raceway.

CONSTRUCTION

Most raceways are made of reinforced concrete, though some earthen raceways are also built. Earthen raceways with plastic liners cost little and are easy to build, but cleaning and disinfecting them is difficult and plastic linings are fragile. Reinforced concrete is more expensive, but is durable and can be shaped in complex ways. Raceway tanks can also be built from polyester resin. These tanks have smooth walls, and are mobile and easy to service. However, their cost limits them to small sizes, under 5 cubic metres.

SIZE

A raceway is most often a rectangular canal with a water current flowing from a supply end to an exit end. The length to width ratio is important in raceways. To prevent the fish stock from swimming in circular movements, which would cause debris to build up in the centre, a length to width ratio of at least six to one is recommended. If the width is too large this could result in a feeble current speed which is not desirable (see below). The length of a raceway unit is usually constrained by the water quality or by how much stock a unit can hold for ease of management. The average depth of a raceway for fin fish, such as rainbow trout, is about one metre. This means each section in a raceway should be about 30 m long and 2.5–3 m wide. The landscape should sloped to one or two percent, so the flow through the system can be maintained by gravity. The raceway should not be curved, so the flow will be uniform.A raceway farm for freshwater fin fish usually has a dozen or more parallel raceway strips build alongside each other, with each strip consisting of 15 to 20 or more serial sections.

The risk of unhygienic conditions increases towards the lower level sections, and can be kept in check by ensuring there are not too many sections and the water flow is adequate. In order to isolate any diseased section and avoid transmitting the disease back to the upper raceways, each section should have its own drainage channel. Controls, such as weirs, are also needed to ensure individual raceways can't accidentally overflow or empty.

WATER FLOW

The water flow rate in a raceway system needs to be sufficiently high to meet the respiratory (dissolved oxygen) requirements for the species concerned and to flush out metabolic wastes, especially ammonia. In a well designed system, the existing water in the raceway is largely replaced by new water when the same volume of new water enters the raceway. Self-cleaning can sometimes be achieved if the fish stocks density is sufficiently high and the water level is sufficiently low.

For example, if trout are stocked at 20 kilograms per cubic metre, they can keep the raceway unit clean by their swimming movements, preventing waste solids from settling to the raceway floor.However, in most cases it is necessary to frequently clean raceways. The simplest way is to lower the water level in the raceway units, which increases the speed of the water current, and then herd the fish together till they flush the waste from the raceway. Solid wastes which accumulate at the raceway bottom can be removed by pumps. Oxygen levels in the water can be keep high if the raceway units are placed one after the other with intermediate drops over weirs, or by the use of aeration systems such as pumps, blowers and agitators.

Generally the water should be replaced about every hour. This means a typical raceway section requires a flow rate around 30 liters per second. However, the optimum flow through rate depends on the species, because there are differences in the rates at which oxygen is consumed and metabolic wastes are produced. For example, trout and juvenile salmon are less tolerant of degraded water quality and require a more rapid water turnover than catfish or tilapia.

The flow rate necessary to maintain water quality can also change through the year, as the temperature changes and the cultured species grow larger. For reason such as these, continuous monitoring of water quality is important, including measurements of water flow rates, pH levels and temperature, as well as the levels of dissolved oxygen, and suspended and solid waste material.

MAXIMUM LOAD

The maximum load of organisms that can be cultured in a raceway system depends on the species, and particularly on the size of the species. For trout, stocking rates of 30 to 50 kg/m are normal at the end of a rearing cycle, while for marine species, such as sea bass and sea bream, the achievable load is lower, between 15 and 20 kg/m. The total volume required for a raceway is calculated by dividing the total amount of fish in kg by the desired stocking rate in kg per m.

FEED

In most raceway aquaculture food needs to be supplied. The composition of the food, and the amount and time of feeding needs to be adjusted to the specific species. This can be optimised to reduce costs and minimize the amount of waste.

WASTE WATER

The treatment of waste water issuing from raceway farms is a major concern. Fish fecal matter and uneaten feed are typically the major elements of solid waste produced in raceway aquaculture farms. These can adversely impact the environment in the receiving water body. Of particular environmental concern is the waste product phosphorus. Excessive discharge of phosphorus to receiving waters can result in eutrophication. For example, in Korea poor waste treatments in trout farms resulted in reservoirs and rivers developing red tides, which caused wider social problems. Because raceway aquaculture operations discharge large volumes of water, the concentration of discharged solids is low. This means it is not easy to treat and implement practical, cost effective treatments. Technologies for the removal of solids include microscreens, dual-drain tanks, swirl separators, plate separators, baffles, media filters, air flotation, foam fractionation, chemical flocculation, and constructed wetlands. But because of the impracticality and/ or high costs of these methods, most of them are not applicable for commercial aquaculture. As a consequence, sedimentation (settling) is still the most widely applied and cost effective technology. From 1999, regulations in South Korean require that all raceway farms provide waste water treatment facilities covering at least 20% of the farmed area to prevent pollution of the freshwater environment.

AQUACULTURE ENGINEERING

Aquaculture engineering is the branch of agricultural engineering that studies cultured aquatic species and the production systems used in their culture. Study, research and development in this area ranges from low intensity man-made pond systems to highly intensive recirculation systems involving natural lakes. Species cultured in these systems could range from warm to cold water fish and from fresh to sea water. Ponds make up a large portion of aquaculture production systems. The engineering aspect of aquaculture engineering aims to apply mathematical and engineering concepts to the effective development of aquatic production systems with emphasis on the use of computer simulation for water quality and fish production. Environmental conditions and feeding and fertilization factors are important components of these studies and production.

Whether research, teaching, study, or production systems, aquaculture engineering systems generally make use of water treatment operations to ensure the good quality of the environment for the fish life. Water recirculation systems are also an important aspect of this endeavor, with emphasis on water quality, oxygen content, and feed amount.

3

Feed Management

FISH FEED FORMULATION

Feed formulation is essentially applied nutrition. A number of terms and expressions are introduced that will be put to practical use as information is presented on the nature and qualities of various feedstuffs and the information presented on the nutrient requirements of fish.

Precise understanding of these terms is essential to their correct application. One must recognize that some of these terms have a built-in error that cannot be escaped. This does not eliminate their usefulness in feed formulation. However, one must appreciate the fact that some are useful approximations of the values and not true values.

The terms that one needs to understand to formulate practical fish diets are: crude protein level; energy level, either expressed as metabolizable energy (ME) or as digestible energy (DE); specific amino acid levels; crude fibre level; and ash level. Since most complete practical fish diets are supplemented with a vitamin premix at levels in excess of the dietary requirement, this category of nutrients will be ignored temporarily.

The potential problems occur when one fails to recognize that all of the above mentioned terms, except ME and DE, represent the quantity or level of a nutrient in the feed as determined by chemical tests on a specific sample of a feedstuff. These chemical tests generally correlate well enough with biological methods of feed evaluation (growth studies, tissue, levels) to be very useful to feed formulators, but they are still chemical tests that are subject to experimental error during nutrient level determination. For example, the proximate composition of fish meals changes during the spawning season. Generally, the lipid levels increase before spawning and decrease after spawning. This will alter the percent of protein, ash and carbohydrates in fish meal as the seasons change. Similarly, many plant feedstuffs vary in proximate composition with their stage of maturity at harvest, location grown and other environmental conditions, such as the weather.Tabled values represent an average value that is usually close enough to the actual value to allow accurate feed formulation. However, one must be aware that assumptions are being made in order to recognize the potential sources of error that may exist. Metabolizable, energy and digestible energy values are obtained biologically and, thus, should accurately represent the true energy value of feedstuffs to fish.However, ME values may be obtained in different ways (faeces collection methods) and thus may be subject to experimental error.

It has recently been reported that the digestibility of feed by rainbow trout was lower at 7°C than at 11°C or 15°C. At 11°C and 15°C body size (18.6 g, 207.1 g or 585.7 g) did not affect feed digestibility.

The digestibility of carbohydrate and energy was slightly reduced by meal size in rainbow trout fed at 1.6 percent body weight. Protein and lipid digestibility was not reduced by meal size. Obvious differences exist between fish species in nutrient digestibility, especially in the carbohydrate fraction of feed.Herbivorous and, to a lesser extent, omnivorous fish have longer digestive tracts than do carnivorous fish and are able to obtain more digestible energy from carbohydrates. An awareness of these facts will prevent misuse of ME and DE values. Each feedstuff in any diet formulation should be present for a specific reason; *i.e.*, it is a good energy source, it is rich in a limiting amino acid, etc.

In addition, each feedstuff in a particular diet formulation should be the least costly ingredient available for its particular function in the diet. This leads to another assumption in feed formulation; that is, any nutrient in a particular feedstuff, such as an amino acid, is just as valuable as the same nutrient in any other feedstuff. This allows feed formulators to interchange one feedstuff with another as cost and availability change. Thus, it is assumed that there is no "ideal formulation", but rather an almost infinite number of possible feed formulations that met the nutritional needs of the fish equally well. While this assumption may not be entirely valid and some nutritional judgement must be employed in any feed formulation, it does seem to be valid in most cases. As with the previously mentioned assumption, an awareness of the potential pitfalls involved is necessary for the fish feed formulation so that allowances can be made in diet formulation and problems can be anticipated and avoided.

BALANCING CRUDE PROTEIN LEVEL

In most animal diets, protein is the most expensive portion and is usually the first nutrient that is computed in diet formulation. The energy level of the diet is then adjusted to the desired level by addition of high energy supplements) which are less expensive than protein supplements.The square method is an easy way to determine the proper dietary proportions of high and low protein feedstuffs to add to a feed to meet the dietary requirement of the animal to be fed. For example, suppose rice bran and soybean meal were available as feedstuffs to prepare a diet for carp that was 25 percent crude protein.

A square is constructed and the two feedstuffs are put on the two left corners along with the protein content of each. The desired protein level of the feed is placed in the middle of the square. Next, the protein level of the feed is subtracted from that of the feedstuffs, placing the answer in the opposite corner from the feedstuff. Ignore positive or negative signs.

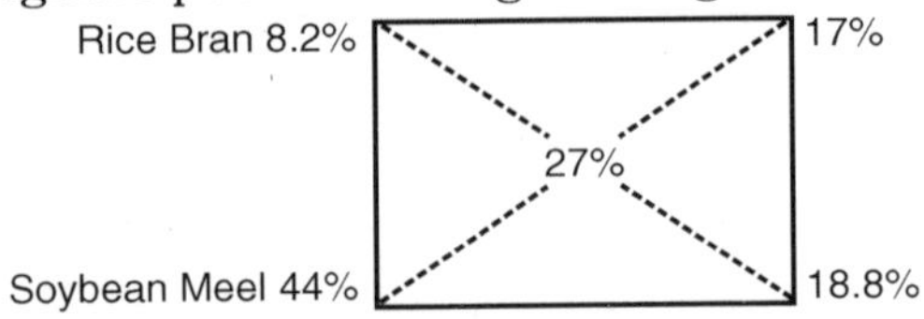

To make the 27 percent crude protein carp feed, we must mix 17/35.8 of rice bran with 18.8/35.8 soybean meal.

Rice Bran 17/35.8 = 47.5%

Soybean meal 18.8/35.8 = 52.5%

So to make 100 kg of this feed we must mix 47.5 kg of rich bran with 52.5 kg of soybean meal.

If more than two feedstuffs are used in a feed, they may be grouped into basal feeds (CP < 20 percent) and protein supplements (CP > 20 percent), averaged within each group and plugged into the square method. For example, suppose shrimp meal and corn were also available for the carp feed. The crude protein levels of the shrimp meal (52.7 percent) and of corn (10.2 percent) are averaged with soybean meal and rice bran, respectively.

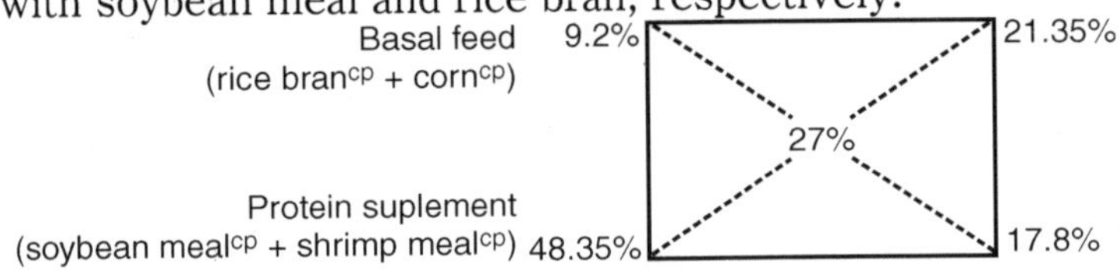

Basal feed = 21.35/39.15 = 54.53%

Protein supplement = 17.8/39.15 = 45.47%

Thus, to make 100 kg of this feed one would mix the following:

Rice bran	27.265 kg
Corn	27.265 kg
Soybean meal	22.735 kg
Shrimp meal	22.735 kg

The square method is helpful to novice feed formulators because it can get them started in diet formulation without the need to resort to trial and error. The square method can also be used to calculate the proportion of feeds tuffs to mix together to achieve a desired dietary energy level as well as a crude protein level. If one wanted to make a feed containing 2 500 kcal ME/kg using wheat middlings (1663 kcal ME/kg) and anchovy fish meal (4 371 kcal, ME/kg) a square could be constructed as follows: The square method cannot be used to simultaneously solve for both crude protein level and ME level.

STEPS IN FEED FORMULATION

The first step in diet formulation is balancing the crude protein and energy levels. This can be accomplished by trial and error, by the square method for either crude protein level or energy level and then adjusting, or by solving simultaneous equations. At first, it is helpful to use at least three feeds tuffs during the initial balancing of protein and energy levels: one high in protein and high in ME, one low or intermediate in protein and high in ME and one low or intermediate in both protein and ME. Once practice makes one more proficient at diet formulation any number of feedstuffs can be used. One must remember to reserve room in the formulation for any feed additive, such as a vitamin or mineral pre-mix.

The second step in diet formulation is to check the levels of indispensable amino acids in the formulation to be sure the dietary levels meet the requirements of the animal to be fed. The requirements of fish for indispensable amino acids is expressed as the dietary level (as a percent of the diet) or as a percent of the dietary protein level. To convert an amino acid level from the percent of diet to percent of protein, divide the dietary level of each amino acid by the dietary protein level. It might be of interest to calculate the dietary levels of all of the indispensable amino acids, but it is not practical to do it all of the time. If the levels of arginine, lysine, methionine and tryptophan meet the dietary requirements of the fish to be fed, the levels of the other six indispensable amino acids will most likely be above required levels. When using unconventional protein supplements, the levels of all ten indispensable amino acids should be checked.

If the diet formulation is low in any amino acid, a feedstuff that contains high levels of that amino acid must be added to the diet at the expense of another ingredient. Once the amino acid requirements are met, the dietary protein and energy levels must be rechecked to, see if any substitution of ingredients has imbalanced the formulation.

A diet mixing sheet should be constructed to standardize diet formulation. The amino acids listed are for illustration purposes only and may be changed to suit different circumstances. In practical feed formulation, pellet quality and acceptability must be considered in addition to nutrient levels and cost. These considerations will vary from species to species and with the type of pellet being made and are dealt with in other sections of this manual.

BEST-BUY TECHNIQUES

The price of the feedstuffs used in diet formulations must be considered to formulate a cost-efficient diet. Feedstuffs can be compared with one another on the basis cost per unit of protein, energy, or amino acid. For example, suppose one has wheat middlings and wheat millrun available for a fish diet, which feedstuff would be the least expensive source of energy? Wheat millrun costs US $ 0.0858/kg and contains about 1200 kcal ME/kg.

$$\text{Cost/kcal} = \frac{.08758}{1200} = \text{US \$ 0.0000/15/kcal.}$$

Wheat middlings cost US $ 01883/kg and contain 1663 kcal ME/kg.

$$\text{Cost/kcal} = \frac{0.1883}{1663} = \text{US \$ 0.0001132/kcal.}$$

Thus, the wheat millrun which has a lower ME value for fish is the better buy because it costs less per kcal.

To compare oat groats and wheat middlings on a cost per unit ME basis one would do the following:

Wheat middlings = US$ 0.0001132/kcal and

Oat groats cost US $ 0.2652/kg and contain about 2450 kcal ME/kg.

$$\text{Cost/kcal} = \frac{0.2652}{2450} = \text{US \$ 0.001082/kcal.}$$

Oats groats, although costing more than wheat middlings, constitute a better buy on an energy basis.

The cost of protein is often the greatest part of the cost of a fish diet. Therefore, substantial savings can be made by using best-buy techniques to determinate least expensive protein supplement. To compare anchovy meal and herring meal, the following calculations are made:

Anchovy meal costs US $ 0.5357/kg and contains 70.9 percent protein.

$$\text{Cost per kg protein} = \frac{0.5357}{0.709} = \text{US \$ 0.7556/kg protein.}$$

Herring meal costs US $ 0.4709/kg and contains 76.7 percent protein.

$$\text{Cost per kg protein} = \frac{0.4709}{0.767} = \text{US \$ 0.61395/kg protein.}$$

On the basis of cost per unit protein, herring meal is less expensive as a dietary ingredient than is anchovy meal.

To compare feedstuffs on the basis of cost per unit of an amino acid, one can calculate the best buy in the same way as before. For example, sesame oil cake which has twice as much methionine content as does groundnut cake on a per unit protein basis would be a more attractive buy at comparable prices. These kinds of comparisons are only valid if the nutrient in one feedstuff is as valuable or available to the animal as the same nutrient in another feed. Such comparisons should be made whenever prices change.

DIET FORMULATION AND GENERAL NOURISH ELEMENTS

The abundance of feed grains in the U.S.A. often obscures the importance of industrial sources of raw materials for the domestic feed industry. Published data from research in poultry and swine nutrition show corn or wheat as the predominant dietary component and soybean meal or fish meal as the only dietary protein source in test diets. Often a test ration would consist of just a cereal grain and soybean meal plus mineral and vitamin supplements.

Increasing competition in the feed industry plus rising cost of traditional feed ingredients in recent years are forcing feed manufacturers to increase the use of less expensive feed ingredients such as by-products from the milling and food processing industries. Feed formulations have been aided by more complete and reliable information of the nutritive value of these products as well as by the development of linear programming techniques for least cost diet formulation based on a large selection of ingredients. Some of the more commonly used industrial by-products currently used for fish diet manufacture in the U.S.A. are listed below.

MILL BY-PRODUCTS

Wheat Germ Meal

Wheat germ meal consists chiefly of the germs of the wheat berry, with some bran and middlings. A rather wide variety of wheat germ grades are produced, depending on the area and regional demand. The quality of wheat germ meal is influenced by the presence of screenings and the level, of other wheat products, which can lower the fat or protein level. Excessive storage time may result in rancidity, due to the oxidation of the rather high level of fat. Generally, I wheat germ meal may have a crude protein level of 25-30 percent; crude fat, 7-12 percent; and crude fibre, 3-6 percent.

Wheat Red Dog

Wheat red dog consists of offal from the tail of the mill, together with some fine particles of wheat bran, wheat germ and wheat flour. Wheat red dog is a fine, floury, light-coloured feed ingredient. The colour may range from creamy white to light brown or light red, depending on the type of wheat being milled. Wheat red dog can act as a pellet binder, as well as contributor of protein, carbohydrates, minerals and vitamins to the feed. Wheat red dog may have a crude protein level of 15.5-17.5 percent; crude fat, 3.5-4.5 percent; and crude fibre, 2.8-4.0 percent.

Wheat Bran

Wheat bran is the coarse outer covering of the wheat berry. Sometimes screenings are ground and added to the bran. Wheat bran contributes bulk to ruminant feeds and is a source of carbohydrates, protein, minerals and vitamins. Generally, wheat bran will have a crude protein level of 14-17 percent; crude fat, 3-4.5 percent; and crude fibre, 10.5-12 percent.

Wheat Middlings

Wheat middlings consist of fine particles of wheat bran, wheat shorts, wheat germ, wheat flour and some of the offal from the tail of the mill. The quality of middlings may be influenced by the amount of screenings added and the fineness of grind. A moisture level over 14 percent may cause problems, especially in hot, humid climates. Wheat middlings have a crude protein, level of 15-17 percent; crude fat, 3-4 percent; and crude fibre, 8.5-9.5 percent.

Wheat Mill Run

Wheat mill run consists of wheat bran, fine particles of wheat bran, wheat shorts, wheat germ, wheat flour and the offal from the tail of the mill. Ground run of the mill screenings are normally added.

Wheat mill run is available in the west, but not normally found in other areas of the country where mill run is separated into bran, middlings and red dog. Wheat mill run may have a crude protein level of 14-17 percent; crude fat, 3-4 percent and crude fibre, 8.5-9.5 percent.

Corn Gluten Feed

Corn gluten feed is that part of commercial shelled corn that remains after the extraction of the larger portion of the starch, gluten and germ by the process employed in the wet milling manufacture of corn starch or syrup. Generally, corn gluten feed will contain a crude protein level of 21-22 percent; crude fat, 1-2.5 percent; and crude fibre, 8-10 percent.

Corn Gluten Meal

Corn gluten meal is the dried residue from corn after the larger part of the starch and germ have been removed and the bran separated by the process employed in the wet milling manufacture of corn starch or syrup, or by enzymatic treatment of the endosperm. Corn gluten meal may contain fermented corn extractives and/or corn germ meal. The crude protein level may range from 41-43 percent; crude fat, 1-3 percent; crude fibre, 4-6 percent; and xanthophyll, 140-220 mg/kg.

Rice Groats

Rice groats are the ground rice kernel, without the bran. Rice groats contain approximately a crude protein level of 8.6 percent; crude fat, 1.2 percent; and crude fibre, 1.0 percent.

Rice Bran with Germ

Rice bran with germ is primarily the pericarp or bran layer and the germ portions of the rice kernel. The crude protein level is approximately 13.3 percent; crude fat, 15.0 percent; and crude fibre, 11.0 percent.

Rice Polishings

Rice polishings are the outer portions of the rice kernel produced by the process used in producing polished white rice. Rice polishings have a crude protein level of 11.8 percent; crude fat, 13.2 percent; and crude fibre, 3.0 percent.

Oat Groats

Oat groats are the cleaned oats with their hulls removed. Generally, oat groats will contain a crude protein level of 16 percent; crude fat, 6 percent' and crude fibre, 2 percent.

OIL EXTRACTIVES

Soybean Meals

Soybean meal is produced in two major protein levels by different processes. Soybean meal must be heated (toasted) sufficiently to destroy, the trypsin inhibitor. Forty-four percent soybean meal is usually mechanically extracted to produce a meal of 44 percent crude protein; crude fat, 4.7 percent; and crude fibre, 6.0 percent. Forty-eight percent soybean meal is dehulled and solvent extracted to yield meal with a crude protein I level of 48 percent; crude fat, 0.9 percent; and crude fibre, 2.8 percent.

Cottonsead Meals

Cottonseed meal is produced in three protein levels. Cottonseed meals are classified as low gossypol if they contain less than 0.04 percent free gossypol. Forty-one percent cottonseed meal has a crude protein. level of 41 percent; crude fat, 2 percent; and crude fibre, 12 percent. Forty-eight percent cottonseed meal has a crude protein level of 48 percent; crude fat, 1 percent; and crude fibre, 8 percent. Although all three cottonseed meals are solvent extracted, the lower protein meals may contain some hulls.

Peanut Meal

Peanut meal is obtained by grinding the cake, chips or flakes obtained by removal of most of the oil from peanut kernels by a mechanical or solvent extraction process. Peanut meal will have a crude protein level of about 46 percent; crude fat, 1-7 percent; and crude fibre, 10-13 percent.

ANIMAL BY-PRODUCTS

Hydrolyzed Poultry Feathers

Feathers from poultry are collected, ground and hydrolyzed under pressure. During processing the cystine linkage is broken, increasing the value of the meal.

Not less than 75 percent of the crude protein must be pepsin digestible. Generally, hydrolyzed poultry feathers will range in crude protein from 80 to 85 percent; crude fat, 2.5 percent; crude fibre, 1.5 percent; phosphorus, 0.75 percent; and ash about 3 percent.

Meat and Bone Meal

Meat and bone meal is the dry rendered product derived from mammalian tissue, exclusive of hair, hoof, horn, manure and paunch contents.

The relatively high fat content of meat and bone meal apparently helps protect the lysine content during rendering. Meat and bone meal generally contain a crude protein of 50 percent; fat, 8-11 percent; crude fibre, 3 percent phosphorus, 4.4-5 percent; calcium, 8.8-10 percent; and ash, about 30 percent.

Fish Meals

Fish meals are manufactured by cooking fish, pressing to remove water and oil and then drying. Fish meals are often made from a single species of fish, *e.g.*, herring meal, but meals containing more than one species, such as bottom fish meal, are also sold. Generally, the single species meals are more uniform in quality.

Some typical analyses follow: Blood meal is produced from clean fresh animal blood, exclusive of all extraneous material such as hair, stomach contents, etc. Blood meal may be dried by several processes, but most often by spray drying.

Spray dried blood meal has approximately a crude protein level of 85 percent; crude fat, 0.5-3 percent; crude fibre, 2.5 percent; ash, 6 percent; and lysine, 9-11 percent, with an availability of 80-90 percent.

	Crude *protein (per cent)*	Crude *fat (per cent)*	Crude *fibre (per cent)*	Sodium *(per cent)*
Herring meal	70	7.5	1	0.5
Menhaden meal	60	7.5	-0.3	
Anchovy meal	64	6	1	0.8
Carp meal, whole	75	9.5	--	
Shrimp meal	36-48	3	11	-
Crab meal	30	1.7	11	-
Bottom fish meal	60	-	--	
Blood Meal				

MISCELLANEOUS

Corn Distillers Dried Grain with Solubles

This product is obtained by removal of ethyl alcohol by distillation from the yeast fermentation of corn or corn mixture by condensing and drying at least 75 percent of the solids of the resultant whole stillage. Corn distillers dried grain with solubles has a crude protein level of 26-27 percent; crude fat, 7-8 percent; and crude fibre, 8.5-9.5 percent.

Brewers Yeast

Brewers dried yeast is the dried, sterilized, unextracted yeast (Saccharomyces) resulting as a by-product from the brewing of beer and ale. Brewers dried yeast has a crude protein level of 45 percent; crude fat, 1 percent; and crude fibre, 2.7 percent.

Brewers Dried Grains

Brewers dried grain is the extracted dried residue of barley malt alone or in mixture with other cereal grain or grain products resulting from the manufacture of work or beer and may contain pulverized spent hops in an amount not to exceed 3 percent. Brewers dried grains may have a crude protein level of 24-30 percent; crude fat, 4.5-8.5 percent; and crude fibre, 10.5-16 percent.

Dried Whey

Dried whey is the residue obtained by drying whey, a by-product of cheese manufacturing. It contains at least 65 percent lactose; crude protein, 13 percent; about 0.8 percent crude fat; and no crude fibre.

ANTIOXIDANTS IN COMPOUNDED FEEDS

Just as oxidation can cause rust and deterioration in metals, a similar type of oxidation occurs in feeds and feedstuffs, resulting in rancidity of fats, destruction of vitamins A, D and E, pigmenters (carotenoids) and amino acids with resultant lowered biological energy values for the diet. If this destruction is allowed to proceed unchecked in a feed, or even in a single ingredient, a lowered feed consumption may result in disastrous nutrient deficiencies. Researchers in a number of fields have been studying the various problems resulting from uncontrolled oxidation and have been developing ways of bringing these oxidation processes under control.

It is noted that oxidative rancidity or lipid peroxidation, as contrasted with hydrolytic rancidity, results in a serious decrease in the energy value of a fat or oil. Undesirable oxidation in feeds may be combatted in several ways. Care should be used to make certain that the ingredients included in the feeds provide adequate margins of safety of vitamins A, E and other natural antioxidants; *e.g.*, lecithin. The use of unstable fats and oils or other pro-oxidants in the feed should be minimized whenever possible.

CRITERIA FOR SELECTING FEED ANTIOXIDANTS

Antioxidants have been used in commercial fish foods in the USA for over 20 years. Although hundreds of chemicals have been tested, only a few have shown the qualifications necessary to make them suitable for use in preventing undesirable oxidations in feedstuffs, in finished feeds and in the guts and carcasses of animals. In order for an antioxidant to be useful in animal feeding, it must have the following qualifications:

- It must be effective in preserving animal and vegetable fats, vitamins and other feed qualities subject to oxidative destruction;
- It must be non-toxic to man and to farm animals (*i.e.*, chickens, swine, fish, etc.);
- It should be effective at very low concentrations; and
- It must be low enough in cost to be economically practical.

COMMONLY USED FEED ANTIOXIDANTS

Of the chemical compounds that have been investigated thus far, three have been found to be outstandingly effective antioxidants for feeds and feed ingredients and can be used both efficiently and economically. They are:

- Ethoxyquin (generic term: 1,2-dihydro-6-ethoxy-2,2,4-trimethylquinoline)
- BHA (butylated hydroxyanisole);
- BHT (butylated hydroxytoluene)

Ethoxyquin, however, has been demonstrated to be the most efficaceous, followed closely by BHT and BHA. With the advent of rations containing a high level of animal and vegetable fats, the requirement for antioxidant protection has become very apparent. The majority of studies over the last few years has focused on ethoxyquin as a preservative or antioxidant. Other chemical preservatives are: ascorbic acid, propionic acid, benzoic acid, citric acid and their various salts. There are technological problems (*i.e.*, moisture level, etc.) associated with the use of these preservatives. Economics, however, remain the most important consideration which limit their use in fish foods.

FUNCTIONAL EFFECTS OF ANTIOXIDANTS

The functional effects of the antioxidants as they affect nutrients and the animals' utilization of them can be summarized as follows:

Nutrient Deficiency Prevention

Dietary deficiencies of vitamins A and E seem to be ameliorated in certain circumstances and ethoxyquin promotes higher levels of vitamin A storage in the liver. Repletion/deletion experiments show that in both monogastric and ruminant animals, a diet containing an anti-oxidant protects fat soluble vitamins throughout ingestion and metabolism. The important benefit of antioxidants most probably lies in their conservation of essential nutrients and their improved utilization by the animal.Altogether too often, it is the practice to use levels of vitamin E far above the animals' nutrient requirement and the result is economically unfavourable. It has been shown in diets designed for chicken and turkey breeders that ethoxyquin has a vitamin E sparing effect.

Prevention of Rancid Oxidation of Fats

In lipid peroxidation, the unsaturated fatty acids undergo a loss of hydrogen, resulting in the formation of a free radical at the site of unsaturation. If the feed material in which this reaction is taking place does not contain vitamin E or some other effective antioxidant, the free radical is quickly converted to a fatty acid peroxide free radical and finally to a fatty acid hydroperoxide. An antioxidant can block this peroxidation by supplying a hydrogen in the first free radical formed, thereby reconverting it to the original fatty acid.

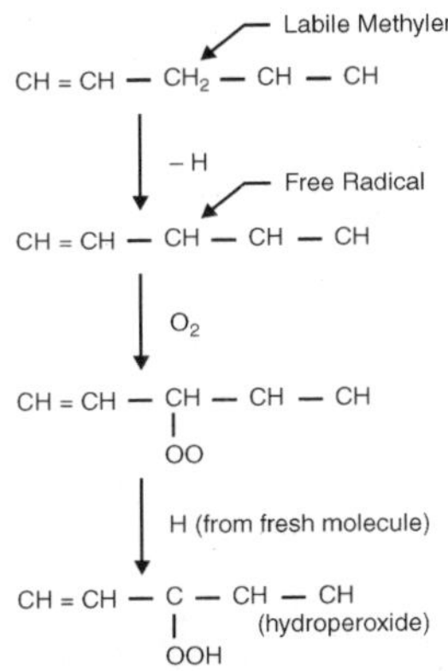

Fig. Peroxidative Rancidity

If the hydroperoxides are allowed to form, they continue to decompose by breaking down into a variety of aldehydes and ketones. Antioxidants prevent oxidative losses of vitamins A and E and pigmenters in stored mixed feeds. Antioxidants stabilize critical oxidation-susceptible nutrients that are naturally present in a fish feed composed of several feedstuffs so that losses are minimal from mixing and storing. If pigmenting substances are used, the anti-oxidants are definitely needed. The benefits of adequate, consistent use span all facets of fish production which include the processing and handling of feedstuffs, formulation and fish cultural practices.

OTHER EFFECTS OF ANTIOXIDANTS IN FEED

It was recently found that BHT, which is widely used as a poultry food additive, prevented mortality in chickens exposed to virulent Newcastle disease virus but prevented an antibody response when chickens were exposed to a virulent Newcastle disease virus.

This in vivo effect of BHT in rations containing 100-150 ppm suggests a possible explanation for vaccination failure sometimes encountered with live vaccines and possibly a new direction in the development of antiviral therapy.Researchers are now trying to define the mechanism of the apparent antiviral action of BHT and extending their attention to other viruses using both BHT and other antioxidants.

ADVENTITIOUS TOXINS IN FEEDS

Toxic materials affecting fish feed ingredients are my cotoxins, toxic products in plant materials, bacteria, residues of pesticides, drugs and agricultural chemicals.

CHEMICAL CONTAMINANTS

Work at the Columbia Pesticide Laboratory of the U.S. Fish and Wildlife Service shows that the most common dietary contaminants have been the organo-chlorine pesticides DDT, DDE, dieldrin, endrin and industrial chemicals such as polychlorinated biphenyls (PCB), phthalate esters and hexachlorobenzene (HCB). Most, but not all, residues are due to the presence of contaminated fish oils and/or meal in the finished diets.

Pesticides and Chlorinated Hydrocarbons

The effects of the chlorinated hydrocarbons have been reasonably well documented in the published literature. Detrimental effects on egg hatchability and survival of young fish are quite striking.

Toxaphene has been shown to interfere with vitamin C metabolism which affects collagen synthesis and ultimately causes skeletal problems.

Dieldrin affects the phenylalanine hydroxylase enzyme which is involved in amino acid metabolism and glutamine synthetase which is involved in protein metabolism.The inhibition of phenylalanine hydroxylase increased the level of blood phenylketo acids, a condition similar to the genetic metabolic disorder phenylketonuria (PKU) that is associated with mental retardation in humans; inhibition of glutamine synthetase might interfere with ammonia detoxification which would be physiologically deleterious. A stimulation of thyroid activity has been associated with DDT and its analogues and endrin has been implicated in the inhibited mobilization of liver glycogen.It would appear that serious attention should be focused on pesticides and other industrial contaminants in dietary feedstuffs and finished fish feeds. The Columbia National Fishery Laboratory has made recommendations that fish oils used at 3-5% of the diet should contain no more than 2 ppm and fish meals no more than 0.1 ppm of organochlorine contaminants, including PCBs.

In addition, the U.S. Food and Drug Administration (FDA) regulations state that the maximum amount of PCBs permitted in animal feed for food producing animals is 0.2 ppm and the amount permitted in animal feed components is 2.0 ppm.

Seed Dressings

Another serious group of contaminants of the ration that may gain entry anywhere from the ingredient supplier to the fish feeder include the newer seed disinfectants and fungicides and certain drugs and chemicals added to feeds. Most seed treatments include organic mercury compounds which are very chemically reactive and readily absorbed. The compounds are highly toxic and the amounts absorbed in light, accidental, contacts accumulate and build up their effects on fish.

Cross Contaminants

Cross contamination occurs when trace amounts of a drug or chemical used as a feed additive are incorporated into a feed which is not supposed to contain the drug. The feed usually affected is the batch immediately following a medicated feed which legally contains the drug in question.

Measurable amounts of the drug-containing feed may remain in the production system and become intermingled with other feed being manufactured. It may occur in one piece of equipment or it may be a combination of residues throughout the entire system. To discover the source, all equipment from the point of addition of the drug to the point of shipping must be checked. Although there are some partial remedies that can be taken in scheduling feed production or with special procedures between feed mixings, experience has shown they are less than adequate answers to the problem. The safest procedures would involve having a completely separate production line or, better yet, a feed manufacturing facility devoted exclusively to fish feed.

OTHER SOURCES OF CHEMICAL CONTAMINANTS

Boiler water additives, hydraulic fluids, lubricants (on pellet mill rolls and other processing equipment), fumigants, defoaming agents and coating materials are examples of materials that could become feed additives during feed manufacture. Misuse of these materials could conceivably lead to human loss as well as fish mortality.

BIOLOGICAL TOXINS

Salmonella

A very serious source of contamination usually overlooked is that of rodent contamination of stored ingredients. There are approximately 1200 serotypes of salmonella organisms that inhabit a great variety of warm and cold blooded animals, including fish. Of these, approximately 80 serotypes have been definitely associated with diseases. Animal by-product derivatives are the worst offenders and the primary source of salmonella contamination in mixed feeds. Fish meals represent the next highest source of salmonella contamination.

Attention has thus far focused on feeds being the means of perpetuating the salmonella cycle (rat to faeces to by-products to feed to fish). Attention also should be given to the role of soil, water, contaminated feeding equipment and wild birds and terrestial animals as contributing to the presence of salmonella organisms in a given environment.

ANTINUTRITIONAL FACTORS

Adventitious toxins has to include the antinutritional factors in plant proteins. Nearly all sources of plant protein possess associated factors which must be eliminated by special processing techniques to make them of maximum nutritional value. For example, for over fifty years, soybeans have been known to be improved by heat treatment.

Soybeans

Factors present in raw soybeans can markedly affect the intestinal tract of animals and influence the digestion and utilization of many nutrients. Inclusion of small amounts of soybean meal will depress growth rate, increase pancreas size, decrease fat absorption and metabolizable energy of the diet of fish. In addition, it has also been found that the feeding of raw soybeans causes the gall-bladder to contract, increases excretion of bile acids, lowers intestinal proteolytic activity and affects methionine metabolism.

Raw soybeans also contain protein which cause agglutination of red blood cells in vitro. It appears that the haemaglutinins (lectins) of raw soybeans are responsible for some of the decrease in efficiency of feed utilization. Apparently these lectins are all capable of causing release of intestinal membrane-bound lipases and amylases, causing these digestive enzymes to be eliminated in the faeces, thereby reducing their digestive capabilities. Fortunately, all of these factors can be destroyed or reduced to minimal levels by proper heat treatment.

Other Vegetable Protein Sources

Groundnuts contain trypsin inhibitor activity which is not destroyed by heat treatment. This is thought to be due to the fact that the inhibitor activity resides in the tannins which are a part of the groundnut skin. cottonseed contains gossypol and fatty acids with a cyclopropene ring which causes nutritional difficulties.

Linseed meal contains an antipyridoxine factor and a cyanogenic glycoside. Rapeseed meal contains glucosides which when hydrolyzed yield a goitrogen. Most all plant protein sources contain phytic acid which can interfere with both mineral and protein availability.

Particular attention must be paid to mineral supplementation when formulating fish diets containing fairly high amounts of plant protein. Even though the temperature which destroys heat labile growth inhibitors is high, minimum protein damage is involved at these high temperatures provided the optimum peak of temperature is reached for a very brief period of time. The brevity of the period at peak temperature is more important than any other factor in obtaining! optimum protein quality.

UNCONVENTIONAL FEED INGREDIENTS FOR FISH FEED

The rapid world-wide expansion of aquaculture and livestock production strongly indicates that a crisis will be precipitated in the livestock and aquaculture feed industries in the near future. Food for humans is not included in this consideration because, generally speaking, livestock, fish and humans can all eat the same basic food commodities and, in emergencies or times of scarcity, feedstuffs are eaten first by humans. The consumption of cultured livestock, fish, etc., is linked to human affluence.

For example, per capita meat consumption in the United States before 1941 was about 50 lb; it is now over 150 lb. Similar percentage increases have occurred in Japan and the other industrialized countries of the world. Fish are probably the most efficient converters of feed to flesh, requiring from 2 to 4 lb of basic feedstuffs to produce 1 lb of fish. In contrast to land animals, however, fish are fastidious eaters in that they require higher levels of dietary protein. In addition, the amino acid requirements to promote rapid growth of fish appear to be more rigid than for land animals.

As an extreme example, the protein in the diets of fish such as eels and yellowtail need to be almost entirely of animal origin, while the protein requirements of ruminants can be partially satisfied from non-protein nitrogen sources such as urea and biuret.

PROBLEMS ASSOCIATED WITH UNCONVENTIONAL FEEDSTUFFS

Faced with the food supply problem for cultured fish, nutritionists have done more work evaluating alternate protein sources in aquaculture diets during the last 7 years than during the previous 50 years. Feedstuffs of vegetable origin are as a whole lower in protein content when compared with those of animal origin. In addition, the presence of high amounts of carbohydrates, fibre and other organic molecules such as glucosides, phytates and cyclopropenes in these sources present the nutritionist with problems that are generally not encountered with sources of animal origin.

Despite these problems, practically all the commercialized feedstuffs are being used to some extent in commercial aquaculture diets. None, however, perform as well as fish meal on a protein equivalent basis. Of the non-commercial items, only single cell protein (SCP), krill meal and possibly leaf protein concentrate have a chance of becoming commercial within three to five years. Because SCP are synthetic products and because the biomass of krill, the volume of either could dwarf everything except soyabean meal.

Formulation Problems

As previously mentioned, the nutritive requirements of fish are such that they probably offer less flexibility in diet formulation than do those of most land animals. First of all, several species of fish selected for culture are almost pure carnivores, requiring a diet of high protein content. These fish have very poor utilization of carbohydrate as an energy source and some evidence is developing that the inclusion of certain types of carbohydrates in the diet is, in fact, detrimental. Hence, it has been recently reported that soyabean meal extracted with alcohol performed better in rations than unextracted meal. This improvement has been attributed to the removal of low molecular weight carbohydrates by the alcohol.

SELECTED UNCONVENTIONAL INGREDIENTS FOR FISH FEED

Soyabean Meal

With regard to its composition, soyabean meal appears to be a reasonably good feed component for aquaculture diets. It contains about 47-50 percent protein, 5-6 percent ash, 1 percent lipid and about 40 percent carbohydrates. It has a lysine content that approaches that of fishmeal, but for most aquaculture diets it is deficient in the sulphur-containing amino acids and in tryptophan. Because of these amino acid deficiencies, soyabean meal cannot be used as the only source of protein and is generally compounded with other feed-stuffs when it is used in aquaculture rations.

Although the inclusion of soyabean meal in fish diets presents no manufacturing problems, problems relating to palatability and availability of nutrients have been encountered. Several investigators have demonstrated that heating soyabean meal rather severely not only increases its acceptability to fish, but also improves the availability of nutrients. This is first accomplished by deactivating trypsin inhibitors and second by denaturing the protein and thus making it more digestible and third by detoxifying natural toxicants. Several studies, however, indicate that when soyabean meal is heated, its performance is still below the expected considering its amino acid composition.

Recent studies with catfish, for example, show that even when soyabean meal was fortified with lysine, cystine and methionine, growth was significantly lower than when menhaden meal was used as the source of protein. It was concluded, among other things, that soyabean meal must contain some anti-growth factors or that catfish do not utilize free amino acids. Other workers, however, have shown that the latter conclusion might not be true, since the carp and salmon can utilize free amino acids. Work done at the Northwest and Alaska Fisheries Center has shown that when soyabean meal substituted for fishmeal in Oregon moist pellet (OMP) type diets, growth is significantly reduced when these diets are fed to rainbow trout. It is believed that part of the problem is due to the phytate content of soyabeans.Chemically, phytic acid is the hexametaphosphate of myoinositol. In plant material it is supposed to occur as the Ca-Mg salt of the hexametaphosphate. The Merck Index states that there are five calcium (Ca) ions and one magnesium (Mg) ion per mole. This may vary with plant species as well as with the maturity of the plant seed. Its chemical structure is still open to question because:

- Theoretically, it can have several isometric forms (*i.e.*, nine stereoisomers and two mesoforms),
- It can form crosslinks between phosphate groups,
- The mechanism for Ca and Mg binding remains uncertain and
- Its ability to bind other ions depends on how much Ca is present.

In plants, phytic acid occurs as the Ca-Mg salt of the acid. Even after all of the phosphate groups have reacted with Ca and Mg, the phytate is still capable of chelating iron, zinc and copper; its ability to chelate is enhanced in the presence of excess calcium. Phytate has been experimentally removed from soyabean meal or other vegetable-based ingredients by reaction with the enzyme phytase Feeding untreated soyabean meal resulted in 25 percent reduction in growth when compared to the OMP control.

Dephytinized soyabean meal gave growth values of about 8-10 percent better than untreated soyabean meal. Total substitution with untreated soyabean meal produced high mortality in the fish after about 90 days. When the zinc, iron and copper contents of the blood of the dead fish were determined, it was found that Zn and Fe levels were significantly lower than normal.

Furthermore, feeding purified (casein-based) diets containing 0.5 percent of either Ca or Na phytate to rainbow trout resulted in a 10 percent reduction in growth. No other obvious physiological defects were noted in the diets containing phytates.

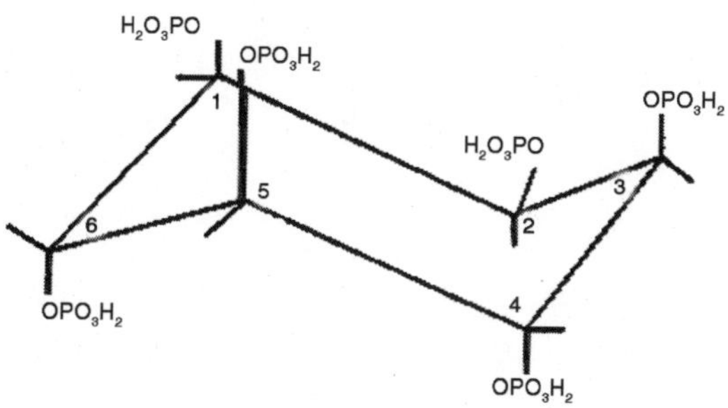

Fig. Structure and Configuration of Phytic Acid (Myoinositol)

Compounds other than phytates present in soyabean meal are also capable of binding trace nutrients. Recent works by other investigators strongly suggest that diets containing proteins of vegetable origin will perform more satisfactorily if they are fortified with Zn and Fe. Recent laboratory data show, for example, that heating soyabean meal also reduces its iron-binding ability so that the improvement in performance of heated soya is not only related to inactivation of trypsin inhibitors and the denaturing of protein but also to the increased bio-availability of micro-nutrients.The problems of soyabean meal, particularly those concerning phytates, are associated more or less with all cereal-based components. The absence of phytase in the soyabean, however, may make the problem with soyabean meal more severe than with other cereal-based feedstuffs such as wheat.

Single Cell Protein

Predictions of future protein shortages have spurred research on non-agricultural methods of protein production. As a result, certain types of high protein products suitable for feeding to livestock and fish are now produced on an industrial scale. Notable among these new protein sources, are single-cell organisms. Man has used single-cell organisms in connexion with his food supply for centuries, primarily in fermentation processes. The direct consumption by man and animals of single-cell organisms, however, is a recent innovation. More commonly referred to as single-cell protein (SCP) because they are high in protein contents this class of feeds tuffs are mainly derived from unicellular organisms such as yeast and bacteria, but can also include fungi and algae. SCP have reasonably well balanced amino acid profiles. When added to the diet of trout, some produced growth comparable to fish meal. SCP are an excellent source of some vitamins and minerals and also possess usable lipids and carbohydrates. Two commercial processes have been developed for large-scale production of SCP. One uses n-paraffin (C-10:C-23 range) as a substrate for growing yeast. The yeast organism most commonly cultured is Candida lipolytica. The other process uses methanol as a substrate and the organism of preference is Methylophilus methylotrophus.

- Bacterial SCP produced from methanol is more easily purified. Yeast plants in-Italy and Spain have been closed down because of substrate contamination of the product.
- The bacteria process operates at higher temperatures, 40-50° vs 37° for the yeast process. Since refrigeration is a major cost in operating the plants, operating temperature becomes a vital economic factor in SCP production. Bacterial SCP also has a higher protein content, about 70 percent as compared to 60 percent for yeast.

SCP has been extensively evaluated with animals such as pigs, poultry and calves; excellent results have been reported. Tests with fish have been conducted primarily with salmonids with variable results.

To date, inconclusive results have been reported with SCP of bacterial origin. Successful substitutions have been made of the entire fish meal component with SCP bacterial in OMP-type diets for rainbow trout in fresh water. When these same diets were fed to coho salmon in salt water, it was possible to substitute only 25 percent of the fishmeal without adverse effects on growth and feed conversion.Fortification of the diets with iron, phenylalanine and/or methionine did not alter growth or feed conversion. At present, tests are in progress to determine whether the dietary requirements for coho are different from rainbow trout or whether the salt water environment is a factor in the tests. One possible factor relating to the poorer performance of the bacterial SCP is the Ca/Mg ratio of the diet. OMP control diet contains a Ca/Mg ratio of about 6:1. Analyses of the bacterial SCP diets (100, percent substitution) show them to contain a Ca/Mg ratio of 16:1.

It is often used in small quantities in preparing trout rations. Brewer's yeast contains about 45 percent protein, 8 percent fat, 13 percent ash, 10 percent water and about 23 percent fibre and carbohydrates. It has an excellent amino acid profile, being deficient only in methionine. Recent work at the University of Colourado, however, has shown that when brewer's yeast contributed over 15 percent of the protein in the diet, growth of fish began to decrease. At the 50 percent level, growth was reduced by 30 percent as compared to fishmeal. These investigators showed that the primary problem was that of digestibility, the BY being only 53 percent digestible, with only 76 percent of the protein being digestible. Here again, the problem of the digestibility of a feed ingredient is an important factor in assessing the value of alternative feed sources.

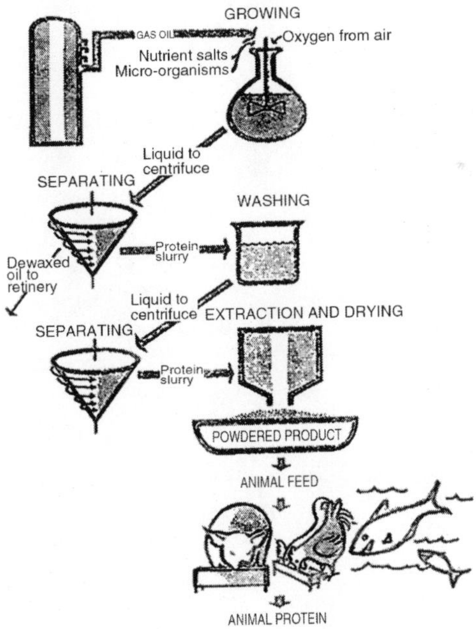

Fig. Process for Production of BP Protein Concentrate

SCP in the form of dried sludge has been evaluated. The sludge consisting of microorganisms grown on paper mill waste had a composition of 43 percent protein, 0.43 percent fat, 3.0 percent ash, 28 percent fibre and 15 percent carbohydrate. When substituted at the 25 percent level in trout diets containing

32 percent herring meal, growth and feed efficiency were significantly lowered; *i.e.*, feed conversion went from 1.45 to 1.58 while percent daily weight gains fell from 18,2 to 13.7 percent, respectively. Amino acid analysis of the SCP used in these tests showed it to be deficient in phenylalanine, methionine and arginine. SCP grown on potato wastes, on the other hand, could be substituted for about 40 percent of the total protein of OMP diets without significant loss of performance. A danger associated with the use of SCP grown on industrial wastes is the possible risk of contamination with mycotoxins-aflatoxin, of course, being the more serious.

Krill

Of all the unconventional feedstuffs of animal origin, krill represents the largest potentially available source. Various estimates made by fishery biologists show that perhaps over 300 million metric tons could be harvested annually. The problem here, of course, is that krill is widely dispersed, primarily in the Arctic and Antarctic regions and would require a tremendous outlay in capital and energy for the harvest. Recently, however, there has been a renewed interest in krill fishery with the principal investigations being carried out by the Federal Republic of Germany, Japan, Poland and the U.S.S.R. Krill meal has been prepared and its reported composition is 55 percent protein, 10 percent moisture, 10-15 percent fat and 15.2 percent ash.

The remaining portion of the composition is probably chitin. Analysis of Russian and Polish krill show krill to contain about 1.5 percent chitin on an as-received basis. Krill has been fed in diets to rainbow trout. A diet containing 35 percent starch and 7.0 percent oil was almost (within 5 percent) equal to a control diet containing 70 percent fishmeal, 24.4 percent starch and 4.1 percent oil. When krill was the sole source of protein, growth of the fish was reported to be lower than the fishmeal-fed control. When isocaloric diets in which the protein contents were solely derived from krill or fishmeal were fed to channel catfish for 19.0 weeks, growth on the krill-fed fish was about 80 percent of those fed the fishmeal diet Analysis of the amino acid contents of krill meal and zooplankton such as red crab show that amino acid profiles of these zooplankton approximate those of herring meal.

A possible explanation for the inferior performance of krill is probably related to its high ash content. A process for reducing ash content of crustacea meals by a milling and screening technique has been reported that not only reduces the ash but also increases the protein level of the final product. High ash content, particularly calcium, has been shown to reduce the bio-availability of manganese and zinc in animal feeding studies.

If krill could be economically harvested and processed at sea,-the Ca content could be reduced by a procedure developed for the processing of red crab In this process, the chitinous fraction is reduced by extracting the krill through a flesh-separating machine, or any other appropriate device such as a Beehive separator. The objective in developing this process which yields several products (*i.e.*, meal, chitosan and protein) was to improve the economic prospects of krill utilization. In feeding salmonids, krill provides an important source of carotenoids. Salmonids are incapable of synthesizing carotenoids and their characteristic red colour can only be derived from ingested carotenoids. In both Europe and the USA, carotenoids are formulated into the diets by the addition of crustacea and/or synthetic carotenoids such as canthaxanthine. Although canthaxanthine offers ease of formulation, about three times more is needed in the diet to impart the same degree of pigmentation as compared to astaxanthine and astaxanthine esters. It also has a stronger tendency to fade during cooking.

Carotenoids are labile compounds and are prone to degradation by heat, acids, alkalis and oxidation. They can be somewhat stabilized by antioxidants such as ethoxyquin. A process for producing a stabilized carotenoid extract was developed.

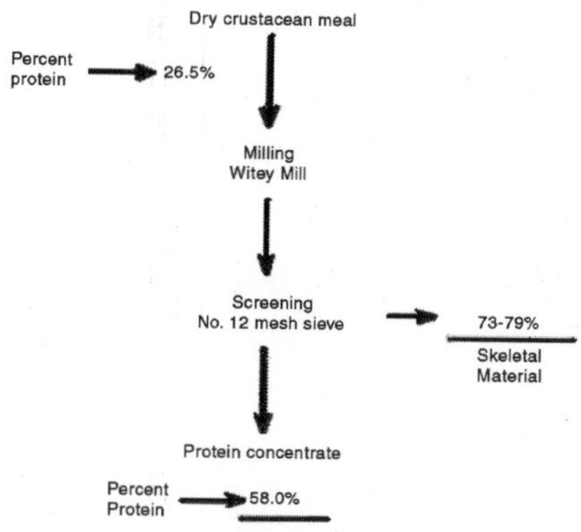

Fig. Decalcification of crustacean meals

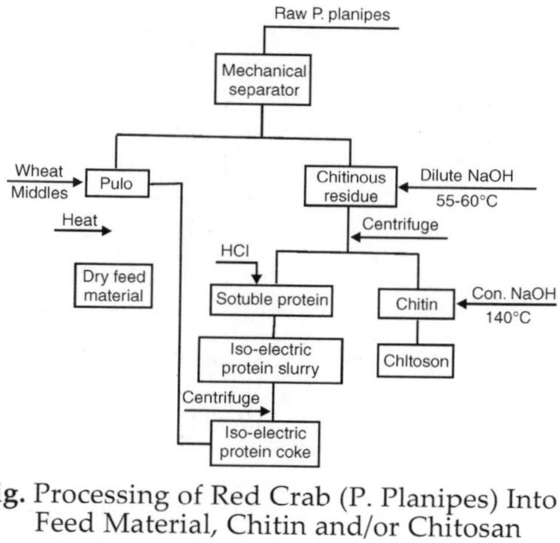

Fig. Processing of Red Crab (P. Planipes) Into a Feed Material, Chitin and/or Chitosan

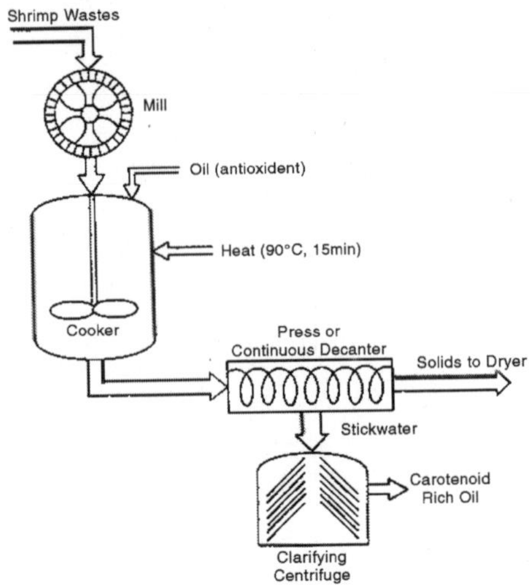

Fig. Preparation of Carotenoid Concentrates by Extraction of Shrimp Waste with Soy Oil

Poultry By-Products and Feather Meal

Poultry by-products and feather meal have long been articles of commerce and their methods of production. Poultry by-products are primarily used by the pet food industry and feather meal is a dietary ingredient, in poultry rations. They appear to be excellent protein and lipid sources containing 69 percent crude protein, 10-21 percent lipid and about 10 percent ash. The so-called hydrolyzed feather meal is really a misnomer. It is at best a slightly hydrolyzed product produced by cooking feathers in the presence of calcium hydroxide to increase its digestibility.

Little, if any, free amino acids are found in this product. Feather meal has been evaluated in both fresh and salt water species. It contains about 80-85 percent protein and is a relatively good source of sulphur-containing amino acids. Whether these amino acids are completely available has not been demonstrated. Good results have been reported when it is used in catfish diets at the 15 percent level. Poultry by-products are lower in lysine than fishmeal and trout diets containing over 75 percent poultry by-products would be deficient in the amino acid. That some of the sulphur-containing amino acids in hydrolyzed feather meal are available to salmonids was recently demonstrated in feeding tests at the University of Washington. In these experiements, trout grew much better on mixes of feather meal and yeast SCP than when either of these commodities was used as the sole source of protein.

Other Potential Feedstuffs

Other feeds tuffs that may be potential dietary ingredients for fish include rapeseed meal and sunflower meal. These commodities contain from 37 to 43 percent protein. Rape-seed meal has been evaluated recently. Inclusion of up to 22 percent in a basic diet containing 40 percent fishmeal produced satisfactory results in the coho salmon. Because rapeseed meal replaced brewer's yeast, cottonseed meal and wheat shorts in the test diet, the results were no indication of its true performance as compared to fishmeal.

The same study also showered that, although the amino acid profile of rapeseed meal was similar to soya-bean meal, its higher fibre, tannin and phytic acid contents could pose a distinct deterrent to its use in aquaculture diets. The high glucoinoleate content in rapeseed could also give rise to isothiocyanates and gastrin formation in the gut thereby interfering with thyroid function. Feeding tests indeed showed that, although the diet containing 22 percent rapeseed meal did not markedly decrease growth and feed efficiency, it did result in a compensatory increase in thyroid activity. A new approach to a rather old process was recently tried; *i.e.*, it was thought that perhaps insect larvae, if properly cultivated and formulated, could serve as an alternate source of nutrients in aquaculture diets. House fly (Musca domestica) larvae were cultivated and tests were made to evaluate their use in OMP-type salmonid and tilapia diets. The larvae were grown by incubating fly eggs on a substrate composed of two parts wheat middlings, one part alfalfa and three parts water The fly egg-substrate mixture was placed in partially covered aluminium containers to a depth of 15 cm and incubated at 30°C for 4-5 days.After incubation, the substrate and larvae were dried in a rotary vacuum dryer (735 mm of vacuum, 85-100°C) to a moisture content of approximately 8 percent. After drying, the larvae were separated from the substrate by screening. The mean proximate composition of the dried fly larvae was:

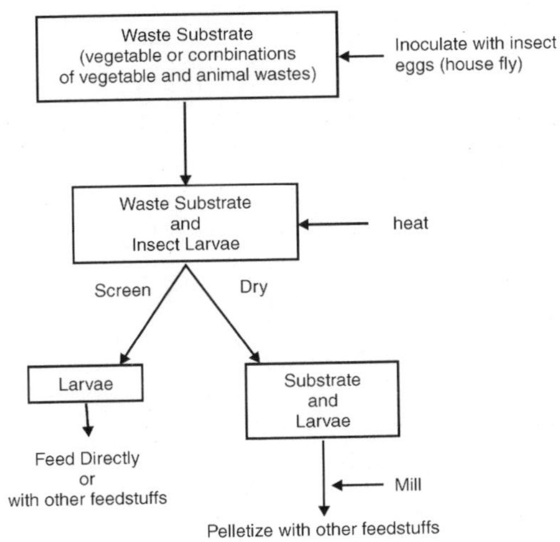

Fig. Use of Insect Larvae in Aquacultural Rations

Amino acid composition was about the same as fishmeal. Prior to formulation into diets, the larvae were milled by passing them through a hammer mill equipped with a 16-mesh (1 mm) screen. Isonitrogenous diets were prepared by replacing appropriate quantities of fishmeal in the control diets with purified chitin. These diets were tested on coho salmon and rainbow trout. Coho salmon were fed a ration ranging from 1 percent to 2 percent of body weight (depending on water temperature) per day. The ration was adjusted every seven days to produce, as closely as possible, the optimum feed conversion. Rainbow trout were fed a ration of 2.5 percent of body weight per day. Ration adjustment was made every seven days. Rainbow trout fed diets containing 25 percent, 50 percent and 100 percent substitution of fly larvae for fishmeal produced the same growth and feed conversion were determined on two sets of diets (*i.e.*, those in which fly larvae was substituted on an equivalent weight basis and another in which the OMP control diet was prepared on an isonitrogenous basis by the substitution of fishmeal with purified chitin. In all experiments, growth and feed conversion (calculated on an isocaloric basis vs direct measurement) were identical.

In another project, diets were prepared by drying the larvae and substrate without separation. After drying, the mixture was milled and directly pelletized without the use of any binders. The pellets produced were very hard and even tilapia did not readily eat the pellets. The project was not pursued further, but it does show promise as a method for upgrading various wastes for use in aquaculture rations. The possibility of using the larvae directly for larval and immature species, such as shellfish species, also presents interesting possibilities. For example Macrobrachia found fly larvae very much to their liking.

WASTE RECYCLING AND DIRECT FERTILIZATION

The potential of producing natural feeds for aquaculture by direct fertilization of the aqueous environment or by recycling wastes in the following publications:

- Problems and Potential of Recycling Wastes for Aquaculture,
- Effects of Feeding, Fertilization and Vegetation in the Production of Red Swamp Crayfish (c) Preliminary Results with a Pilot Plant Waste Recycling Marine Aquaculture Systems.

The above works relate to processes that are designed to supply surplus nutrients through fertilization, either directly or indirectly via waste recycling. Theoretically, these processes are most efficient if they are applied to polyculture systems. In actual commercial practice, however, a combination of prepared foods and naturally grown feed may provide the best promise for this system. Experiments conducted in Louisiana, USA, showed that optimum yield with crayfish was obtained when ponds were fertilized with N-P-K (8:8:8) at 130 kg/ha in combination with pelleted feeds. The fertilizer was applied in November, in March and again in May and June. The vegetation in the pond consisted of phytoplankton, algae, weeds and smart weed. Fertilizing the pond gave a 13-27 percent higher yield than without fertilizer. Whether the crayfish consumed phytoplankton or fixed plant growth did not seem to matter since other zooplankton consumed the plants which, in turn, were eaten by the crayfish. Both Kildow (1974) and Ryther (1975) presented similar systems. Actually, these processes were developed to produce a biological tertiary sewage treatment process capable of removing all inorganic nutrients (primarily N and P) from secondary effluents prior to discharge into the environment (in this instance, the sea). The secondary objective was to develop a marine aquaculture system of a primary crop of shellfish and a secondary crop of other commercially valuable marine organisms such as finfish or shellfish. As for the reality of toxin, metal absorption or accumulation, work done at this laboratory has shown that toxic metals such as mercury are rapidly deposited in the flesh and kidney of salmonids when diets containing the metal are fed. In experiments using dogfish meal as the mercury source, it was shown that this accumulation occurred very rapidly in the fish consuming diets at all replacement levels of fish meal by dogfish meal.

With respect to other toxic compounds, work in the UK has demonstrated that effluents of pea silage could be extremely toxic to fish (carp and goldfish). Several types of toxic compounds have been found in these effluents; *e.g.*, volatile fatty acids, sulphites, phenols and ammonia. Strangely enough, the most toxic factor was the fatty acids which it is assumed were probably found in the greatest amounts.There was also a high percentage of p-cresol which might have been formed by the action of bacteria on tyrosine. Although the purpose of the UK study was to determine the possible toxic effects of pea silage effluents that might find their way into streams, this study would strongly suggest that if direct fertilization of ponds is planned, care should be exercised as the risk of introducing toxic substances via the wastes can be great.

FACTORS AFFECTING THE ORGANOLEPTIC PROPERTIES OF FISH

Diet does influence the organoleptic quality of cultured fish. The factors that contribute to the organoleptic quality of fish are as numerous and complex as those associated with other foods. Unlike other foods, however, little attempt has been made to control these organoleptic factors. With wild fish, flavour and general overall acceptance have been associated with its texture, flesh composition (fat and protein contents) and the characteristic flavour constituents that are associated with the respective species.In marine species, it has long been recognized that their flavour is influenced by seasonal and environmental influences. Environmental factors include not only dissolved substances and plant nutrient but also the available animal food supply (food chain), the abundance of which is influenced by such factors as water temperature, water cycle and photo period. Water temperature influences the fatty acid composition in muscle tissue of teleost fishes. At lower water temperature the fish tend to accumulate more unsaturated fatty acids in their tissues and organs.

The relation of fatty acid composition to flavour and textural alteration during processing and storage, of course, is well recognized. It becomes quickly apparent, therefore, that changes in environmental factors need to be minimized when attempts are made to control the organoleptic qualities of cultured fish via the diet. A broad prospective of the environ-ment as it relates to probable influences on the organoleptic properties of cultured fish. It should be noted that two environments must be considered: the total environment, consisting of oceans, lakes, rivers, ponds, etc. and the local environment (*i.e.*, the location of the fish with respect to the total environment). The type of fish culture, of course, will determine the degree that these environments influence the organoleptic properties of the fish. A review of the literature reveals that most of the research on the flavour attributes of cultured fish is related to environmental influences rather than specific nutrients. This is particularly true of pond-cultured species where, for obvious reasons, fish are reared in near static water conditions. In this type of environment, the growth of flavour-producing organisms, decomposing plant growth and unconsumed feed all contribute to the organoleptic characteristic of the fish.

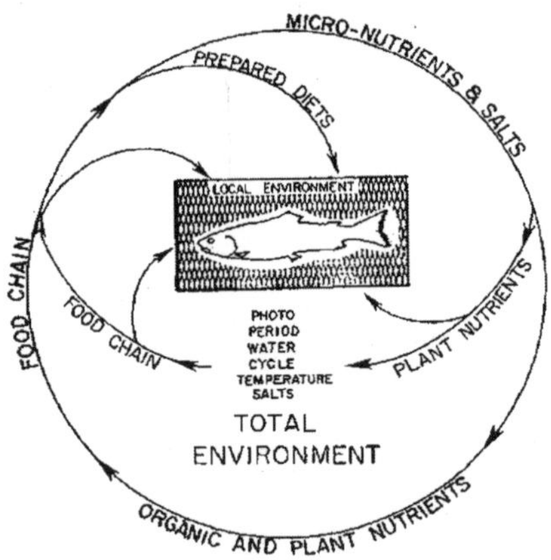

Fig. Environmental and Dietary Influences Affecting the Organoleptic Properties of Cultured Fish

The organoleptic characteristics of European carp grown in ponds fertilized by liquid manure and those grown on diets of grain or high protein pellets have been compared. Carp grown in the fertilized tanks were claimed to be superior in flavour and colour to those fed the prepared diets. However, fish grown in fertilized ponds had a flesh fat content of about 6 percent, while those fed prepared diets ranged from 14 percent to 22 percent fat. It would appear that, all other factors being equal, the compositional constituents in the flesh more strongly influenced the eating quality of the fish than did the flavour components contributed by the diets. In contrast to pond-reared fish, marine species such as salmonids are generally grown in non-static water, thus decreasing the effects from environmental factors. But even in a marine environment, it must be kept in mind that environmental factors can easily override dietary factors when attempts are made to control the organoleptic qualities by dietary means. Fish have one aquaculturist as biological sponges that can absorb many inorganic and organic materials through their gills as well as their intestinal tract. Indeed, catfish have been shown to absorb 2-pentanone and dimethylsulphide within 10 minutes after being exposed to their solutions. Hence, while rigid quality control measures are needed to. prevent off-flavours, desirable flavours could be induced by exposing live fish to potentially beneficial solutions. A recent review of environmental and seasonal factors that influence the flavour attributes of the cod claimed that these qualities are related not only to the amount and type of feed consumed, but also to the composition of the lipid that, in turn, is influenced by feed and sexual maturity of the fish. An interesting point brought out by the author is that the post-mortem pH of the muscle is influenced by food supply, lipid content and swimming activity of the fish.

Work covered in the literature definitely shows that consumers and taste panelists can distinguish between wild and cultured fish. Ignoring "off" or foreign flavours, however, it has not been established whether the consumer prefers wild fish, bearing in mind that this is true only when high quality cultured fish of the same size are compared against the wild fish. In this respect, it had been shown that wild trout possessed distinctly different organoleptic properties from hatchery trout fed a diet consisting of beef, pork spleens and horse meat. While the wild fish were consistently rated superior in flavour and aroma and texture to the hatchery trout, there was also a considerable difference in the organoleptic factors of the hatchery fish as they were taken from different hatcheries. Again, this indicates environmental conditions playing a major role in governing organoleptic qualities of fish.

In conclusion, it is believed that one can safely say that organoleptic changes, whether they be due to diet or environment, are readily detectable, but whether these changes affect acceptability of the fish has yet to be assessed. No doubt efforts need to be made to maintain some of the basic organoleptic characteristics, such as colour and texture traditionally associated with wild species. One for example, would not question that salmon possessing the characteristic red colouration would have a much wider acceptance than those having no colour. Similarly, but not as pronounced, we would expect that a firm texture in finfish such as salmonids or carp would enjoy a wider degree of acceptance than those possessing a mushy texture. The effect of different flavour notes, except when they are distinctly foreign to the fish, however, are as difficult to assess as they are to control. The seemingly simple expedience of controlling the fat content may prove to be difficult. Some work already shows that changing the lipid content in a salmonid diet does not always significantly or uniformly change the lipid content in the salmon. It would appear that water temperature, growth rate, stocking density and sexual maturity are all involved in controlling this variable. Practically all the results in the literature today show that the quality of cultured fish with respect to wild fish should be improved. It is unlikely, however, that this quality improvement will be achieved with the feeds that are in current use. These feeds, after all, were developed to provide optimum growth and feed conversion in the fish. As more is learned about fish nutrition and feeding, it may turn out that supplementary dietary rations will have to be developed that have as their specific function the ability to favourably alter the organoleptic characteristics of the cultured fish. In this area, it is expected that a primary use for alternate or unconventional feedstuffs -for cultured fish will be developed.

4

Marketing Aquaculture Products

MAKING WISE CHOICES WHEN DIRECT MARKETING YOUR AQUACULTURE PRODUCTS

Producers who are ready to harvest their first crop of fish often ask, "Where will I sell them?" It is surprising how many producers ask this question so late in the production cycle. To avoid certain failure of an aquaculture enterprise, producers must determine the marketplace as an initial step. Midwest farmers in traditional farming enterprises, such as corn, soybeans, beef, swine, and poultry use well-established markets that have been highly refined over the last few decades. For example, go to any rural town, regardless of its size, and you'll find a grain elevator or stockyard. Aquaculture, unlike other agriculture enterprises, is in its infancy, especially in the mid-west. As a result large-scale processing plants in the Midwest, solely for farm raised products are nonexistent. Fish and shellfish farmers are forced to use alternative, poorly established, markets for their products. Alternative markets have advantages and disadvantages. The main advantage in alternative markets is that producers become wholesalers, and in some cases, retailers.

Consequently, producers have more control over the prices they set for their products, and they retain some portion of the profit, that would otherwise have gone to the "middleman." The main disadvantage is that considerable time must be spent analyzing and developing their markets, and in some cases, processing their product. Many beginning aquaculturists believe that with the huge markers for fish in Chicago, St.

Louis, Louisville and Indianapolis selling their products will not be difficult, But even though massive quantities of fish are marketed in those areas, such markets generally serve high-volume producers. New farmers producing, relatively small quantities of food species should develop markets within a fifty-mile radius of their farm This bulletin discusses general marketing concepts as applied to food fish and presents ideas for selling aquaculture products locally; Food fish production offers the greatest potential for aquaculture expansion in the Midwest: therefore 'most of this bulletin specifically addresses the process of marketing food fish. Many of the same marketing principles apply to other types of aquaculture products such as sport fish, bait fish, and ornamental fish.

The producer of any aquaculture product should receive from this bulletin assistance in answering the following five questions.

- Who will purchase my product?
- Where do I have to deliver the product?
- When do I have to deliver?
- Upon what criteria will the product be evaluated?
- What price will I pay I be paid and how is it determined

KNOW YOUR COMPETITION

If you are considering aquiculture as a business because you think there is limited competition, then you should reconsider. The seafood industry is well-established and very competitive. It will become more competitive in the near future. Midwestern producers have to compete with wild-caught and farm-raised products of both domestic and foreign origin. By understanding current sources of seafood, the new producer will be more knowledgeable and better able to build a production and marketing Programme responsive to market opportunities. One which is able to endure adverse price fluctuations.

Native Wild

Competition from commercial fisheries for yellow perch, walleye, and crawfish exists in states around the Great Lakes region. Yields from each of these fisheries are being reduced due to overfishing. Overfishing of these species is allowing farm-raised yellow perch to compete with commercially caught yellow perch. Wildcaught minnows from Minnesota, particularly fathead minnows, also compete with farm-raised bait fish. Ocean-caught fish such as cod, pollock, perch, and salmon account for a significant portion of U.S. seafood consumption. They are often cheaper or more reliable than those from lake fisheries or fish farms. Once again, overfishing is resulting in more opportunities for farmraised products.

Foreign Wild

Competing with seafood imports poses the greatest challenge to the aquiculture industry. In 1990, total edible seafood imports (mostly wild-caught) totaled $5.2 billion. Competition between farm-raised species and foreign-wild is greatest for catfish from South America and walleye from Canada. Farmers considering walleye production should also realize that Canada has a well-established commercial walleye fishing industry. They have hundreds of lakes which can be opened to fishing in order to maintain the country's market share by "flooding" the market. Due to this situation, prices could be reduced to the point that marginal farmers who cannot compete with the wild imports would be forced out of business.

Farmed Domestic

Midwest farmers raising food species such as channel cattish and rainbow trout must overcome competition from the large industries in southern and western states which have been producing these species for several decades. Even though pond-side prices of these two species will be lower than for locally grown products, higher shipping costs can make mid-western products more competitive. Shipping costs usually adds 10-20 cents per pound to the wholesale price. Established aquiculture industries will force small scale producers in the Midwest to rely more heavily on direct marketing until significant infrastructure is established to provide Midwest farmers competitive feed and fingerling prices. In the meantime, small-scale producers rely heavily on the premiums that maybe paid for locally grown fish.

Farmed International

The competition from species farmed internationally is greatest for salmon from Europe, and increasingly from Peru and Chile. Tilapia, another product being farmed internationally in large numbers, could be imported cheaply into the United States. Organizations that are currently working to establish markets for tilapia in the U.S. maybe benefiting South American and Asian producers who could take advantage of the growing U.S. market for tilapia.

Selecting a Species

When deciding which species of fish or shellfish to farm, keep three things in mind.

- Choose a marketable species. A good example is common carp. Common carp is easy to farm and is a candidate species if you have a strong ethnic market in your area. Otherwise, it is advisable to consider other more widely accepted species. Of the 60 or so candidate species for food fish aquaculture, only channel catfish and rainbow trout have established industries in the U.S. Other species such as Atlantic salmon, Coho salmon, crawfish, hybrid striped bass, tilapia, yellow perch and various sunfishes offer considerable potential in the Midwest.
- Attempting to farm species with incomplete production information or production peculiarities is a very risky venture. Although species such as walleye, shrimp, and lobster have wide public appeal and are very marketable in the Midwest, each have production problems. For example, walleye producers recognize that fingerling survival is low due to cannibalistic Behaviour. Furthermore, most walleye are produced by state fisheries agencies and released as fingerlings. Few have been grown to marketable food size using commercial diets. This has resulted in little available nutritional information for market-

size walleye. Faced with these production obstacles, a producer could easily fail to reach production goals. In any new business, and especially in aquaculture, cumulative short-term setbacks could lead to loss of farm investment. Tilapia maybe a better choice because they are extremely hardy. They tolerate poor water quality and are easy to spawn. The only production drawback is their intolerance to low temperatures. A second, more serious problem with tilapia is in marketing. Because tilapia grown in the U.S. are non-native, the consumer does not recognize them as being a high-quality food fish. If adequate marketing research and promotion is provided, then tilapia offers great potential.

- Finally, variety is extremely important to small-scale producers. Many marketing outlets prefer buying small quantities of more than one species. Production of more than one species may offer a competitive advantage over single species operations. If production of a variety of species is not feasible, then pooling your resources with other producers will allow you to offer a variety of species.

THE FOUR P'S IN MARKETING

Product price, promotion, and place, classical points in traditional marketing, also apply to marketing aquiculture products. After potential producers determine their competition and choose the "best" species for their area, serious planning and research needs should be conducted on each point.

Product

Market size and product form are two important aspects of an aquaculture product. Different customers often have different ideas on what is the "best" size or form. Market size varies according to species and product form. Channel catfish are usually sold after reaching 1.25 to 1.5 pounds. At this size, one fish will yield two 4.5 oz. fillets.

Yellow perch or bluegill are examples of fish served at a smaller size. Market size for either of these species may range from 0.25 to 1.0 pound. The following is a discussion of definitions and specific information concerning various product forms. Your local and/or state department of health officials should be contacted to determine permit requirements before any type of processing is undertaken.

- Live fish are those sold to live haulers who stock fee-fishing lakes or farm ponds or to consumers who prefer alive product.
- Fish in the round are put on ice and sold just as they come out of the water.
- Drawn defines fish with entrails removed.
- Dressed fish are sold completely cleaned but with the head intact.

- Headed and gutted fish means the head and entrails are removed. Fins and tails may be removed or left intact. Species such as channel catfish have the skin removed. For trout and other salmonids, the skin and scales are left intact, but gills are removed.
- Chunks are cross sections of large dressed fish having backbone as their only bone. They are similar to beef or pork roasts.
- Steaks are cross sections of dressed fish around one inch thick. Larger catfish (2 lbs. or more) are usually sold as steaks.
- Nuggets come from the belly flap after it is cut free from the fillet. Channel catfish nuggets are becoming more common in supermarkets. Part of this popularity may be a result of their lower cost.
- Fillets are boneless pieces of fish obtained either as flank or butterflies.
- Flank fillets are the two sides of the fish cut away from the backbone. Usually, they have no bones or waste.
- Butterfly fillets are the two flank fillets held together by the belly flap. (Trout or yellow perch are examples of fish which are sold as butterfly fillets). Fillet size will depend on a individual preference. Usually, fillets are cut to prescribed proportions which yield a single serving (6-8 oz.) from one or two fillets. As a rule, whole fish need to average around 1.25 to 1.5 pounds depending on the dress-out percentage of each species. For example, 1.5 pound channel catfish, hybrid striped bass, tilapia, or trout will have a fillet dress-out percentage from 35-40 percent.
- Deboned fish means the rib bones and backbones are removed, but the body is left intact. Breaded fish, canned fish, fish sticks, and fish strips are value-added products and were not defined because all involve a high degree of processing.
- Smoked fish is a value-added processing method which is becoming more popular, especially during holidays. Two smoking methods (hot and cold smoking) have been commonly used. Hot smoking is more like Flavourful cooking of brined fish, and never produces enough drying to ensure safekeeping. Hot smoking involves temperatures of 250° to 300°F for a period of 4-5 hours. Cold smoking, on the other hand, preserves fish by drying. Cold smoking times may be as short as 24 hours or as long as three weeks at temperatures never exceeding 80°F.

Regardless of the product form you choose to offer, it is very important to establish and maintain a reputation as a reliable supplier.

Be sure to gain an accurate understanding of each customer's needs regarding both volume and frequency of purchases before you deliver the first fish. Remember, a customer is taking more of a risk when depending on you for fish supplies rather than relying on a local fish wholesaler or other established business. You need to take extra care to be reliable. This can be a major problem area for new or small-scale producers. One approach could be to join with other producers in forming a marketing cooperative.

Price

How much do you charge for your product? Real life implementation of this seemingly simple calculation is not so easy. Pricing a product is an agonizing, lengthy decision, and will likely need priodic adjustments to reflect new market realities. So where do you start in thinking about price? Well, the lowest price you would ever want to charge is the price per pound that just equals your costs per pound, including both fixed and variable costs. The highest price would be what you could talk one or two people at most into paying for your fish. Your base price needs to be somewhere in between. Take into consideration such factors as:

- How are you going to position your product in the food fish market? (Is it a Chevy Chevette or a Cadillac Eldorado?)
- Who are your customers? (Are they individual consumers, upscale restaurants, or food wholesalers?)
- What species and prices are competitors offering?
- What are the quality perceptions, if any, associated with your chosen species? As you may have figured out, pricing is not an independent decision. Many systems for determining the price of your product are used by marketers. From a small scale retail marketing standpoint, only the most relevant systems will be discussed.
- Cost-plus pricing simply adds a constant percentage of profit above the cost of producing a product.
- Return on investment is similar to cost-plus pricing. It begins with an estimate of your annualized investment in your fish enterprise. Add on the percentage return you want to achieve and divide this new total by your annual production or expected sales, or the lower amount. Thus if you sell all you expect at the calculated price you will earn your target profit.
- Competitive pricing is probably the easiest, and in retail marketing, the most common form of pricing. In this system, producers gather market information on the prices and quantities of competing products and then price their products accordingly.
- Penetration pricing is offering a product at a relatively low price in order to gain wide acceptance in the marketplace quickly. This pricing method is often used for unfamiliar species such as tilapia.
- Skimming is the opposite of penetration pricing. It involves the introduction of a product at a relatively higher price for more affluent, quality-conscious customers. Then, as the market becomes saturated, the price is gradually lowered.

- Discount pricing offers customers a reduction from advertised prices for specific reasons. For example, a fish farm advertises in the local newspaper that prices will be 25 percent cheaper if they bring the advertisement from the paper. Another example would be when a producer, who advertises on local radio, offers customers who mention the radio announcement a preset discount.
- Loss-leader pricing is the offering of a portion of the product at a reduced price for a limited time. The goal is to attract more customers to the producer's place of business so that they might also buy non-discounted products as well. This pricing method is seen in farmers markets and supermarkets to introduce a new product or to create consumer excitement.
- Psychological pricing involves establishing prices that look better or convey a certain message to the buyer. For example, instead of charging $3.00 per pound, the producer charges $2.99 per pound. This will make the product appear to be more of a bargain. Or, instead of charging a price close to production costs, the producer charges in a higher price range that buyers associate with higher quality or more desirable fish species.
- Perceived-value pricing is positioning and promoting a product on non-price factors such as quality, healthfulness (clean water, no contaminants), or prestige. Then the producer must decide on a price that reflects this perceived value. This would be a good strategy for farm-raised versus wild-caught fish or any species you could pitch as having non-price attributes.
- Product line pricing, in the case of food fish, means offering a variety of species and/or product forms. You would price the various products based more on targeting certain customers with certain products than on production costs.

Promotion

Producers promote their products to create customers. The best way to attract new customers is to develop excellent promotion strategies. Regardless of the advertising method chosen, producers should expect to spend a portion of time promoting the product(s). There are two general metheds of promoting aquiculture products, generic and. personal promotions. Generic promotion is commonly performed by large marketing groups such the Catfish Institute. Because such groups are involved in generic advertising of specific species and do not focus on products from individual farms, we will discuss personal promotions. What are the methods available to a small-scale retail marketer? One of the most common promotional methods is word of mouth. You convince a friend to buy from you. If satisfied with the quality and or price of your product the buyer will pass along word of your operation to others. The multiplying effect can be tremendous, but often word of mouth is not sufficient. Other common channels for advertising include: radio, newspaper ads, magazines, handbills, flyers, and posters. Promotion is usually aimed at the consumer. The promotional message must be clear, to the point and focused. A small ad under farm produce in the classified section of your hometown newspaper might read, "farmraised striped bass raised indoors free of contaminants; they are highly nutritious and priced to sell State Road 38 and 900 East in Tipp City, Sat. 8-12." The form of promotion you choose as a new producer will depend on the scale of your operation, available resources, availability of your product and geographic location of your farm. In addition to public advertisement, you should consider some on-site product promotion, both visual and verbal. This should accent the non price attributes of your product which will convince the consumer to be a repeat customer.

Place

There are many places and many markets for farmers to sell their food fish products directly, both at the retail and wholesale levels. A number of considerations must be taken into account when deciding what type of market, and specific business or locations, to market your fish (*e.g.*, species, form, location, ability to process. Farmers may sell their product directly to the consumer (retailing) or to other businesses which then sell to the consumer (wholesaling). The promotional principles discussed earlier usually apply to retailing directly to the consumer since most establishments where you may wholesale have their own forms of promotions.

- Direct Retail Sales to consumers area good place to start if supplies are small or availability uncertain.
- Farm side sales are appropriate if the farm is located within a short distance from urban areas. Specific dates can be reserved to make transactions. For example, Saturday morning from 8:00 am. until noon. Equipment such as change, accurate scales, ice, and transport bags, if not essential, will certainly make the purchases more convenient for the consumer.
- Farmers markets are widely utilized in southern states by fish producers. Here in the Midwest the concept of selling fish at existing farmers markets is new. As a result, the producer usually has to convince the manager of the farmers market that the product will be live and safe for human consumption. Hauling tanks are positioned to create a stand or booth similar to other forms of produce. Equipment needs will be similar to farm side sales.
- Roadside stands are often used to capture a percentage of the traffic flow along busy highways. Several states including Illinois and Indiana

offer directories of roadside stands. Obviously location is important in alerting upcoming traffic to the opportunity they are approaching. Offering fresh produce either yourself or in conjunction with another farmer may increase your flow of customers. Equipment needs are similar to those for farmers markets.

- Direct Wholesaling your fish to other wholesalers or retailers may bean excellent route to market your fish depending on the species and your supply situation. Some wholesale customers such as restaurants have a strong preference for steady, year-round supplies, but even this generalization does not hold true in every individual case, It would probably be worth your while to set up an appointment with the manager of every restaurant, grocery store, and food wholesaler within a 50-mile radius of your production site. Find out beforehand, if possible, individual preferences for species, product form, size, volume, availability, and prices. Have a strong sales pitch prepared and a sample of your product. Pricing in the wholesale market will likely be based on individual negotiation so be sure to be armed with a negotiation strategy. Some managers will immediately be interested, others won't. For those who are interested, the accurate, sensitive positioning of your product to fit their individual needs and your ability to convince them of your product's attributes could gain you a customer on a trial basis. Once that customer is gained, you must be extremely diligent to be a reliable supplier of the size, form, quantity, and quality of product that the customer expects. You must work hard to earn a good reputation and to maintain it.
- Live haulers can be used if the producer does not own a transport tank. Live haulers will sell fish to fee fishing operations, farm pond operators, or processing plants. Selling in this manner would not require any investment, on your part, of time or equipment to process, transport, or sell your fish.
- Restaurants can be a good place to start with smaller quantities of fish because they are usually willing to try new species (and suppliers) to fill their "catch of the day" specials or other such short-term menu items. The typical restaurant will take 10 to 80 pounds of a fish species per week depending on product form (another advantage of producing multiple species and/or product forms). For regular menu items, restaurant managers tend to prefer weekly deliveries over monthly. When deciding which businesses to contact, remember that many businesses serving food are not necessarily identified as "restaurants." Don't overlook the country club, VFW, or comer pub. You might be just the right type of supplier or have just the right fish they want. Once you successfully gain a restaurant customer, you should continually educate your

client. How many times have you asked your waiter, when dining out, about the seafood item on special and were given a "song and dance" because he or she knew nothing about it? Educating the head of the serving staff and providing a short brochure or other printed information may be a key to continued success.

- Supermarkets offer one of the best places to sell larger volumes of food fish products. It is quite likely that unless a supermarket is locally owned and operated that you would have to supply all or many of the stores in a chain. This may be a good situation for some producers, but others may have difficulty with the volume or other requirements. But here again, it is probably best to research and interview to determine the compatibility of the producer and the potential customers.

Staff education is likely to also be important in this case also in keeping customers. Staff may need information not only about how the fish are produced and their quality and such, but also about how to prepare them. Since the average consumer simply puts "fish" on the shopping list, a knowledgeable and enthusiastic staff member will be extremely valuable in marketing your product. Specialty stores include ethnic groceries and health food stores.

Fish is a more important part of people's diets in some cultures than in other cultures. Health food stores may be willing to try your product because the perceived quality of farm-raised products is usually higher than wild-caught. Ethnic markets usually are more willing to purchase whole fish.

REGULATIONS

Regulations for marketing aquiculture products locally will focus either on permits allowing retail marketing at the farm, farmers markets, roadside stands, or the setup and use of small-settle processing facilities. In either case you should contact your local health department. The accreditation procedure for small-scale processing plants will involve assurances by the processor that the equipment can be disinfected and that the product can be properly refrigerated.

INSPECTION

Mandatory seafood inspection will soon occur. The United States Department of Agriculture (USDA), National Marine Fisheries Service (NMFS), and the Food and Drug Administration (FDA) have been competing for the right to perform the seafood inspection.

As the inspection procedure is currently proposed. FDA will share the inspection Programme with NMFS and will use the Hazard Analysis Critical Control Point (HACCP) method for inspecting seafood. This method involves random inspections at critical points along the processing line instead of inspecting every fish.

PROCESSING AND MARKETING AQUACULTURE PRODUCTS ON A SMALL SCALE

Aquaculture, or farming of aquatic plants and animals, is a fast growing agricultural industry in the United States. While large-scale production, processing and marketing of species such as catfish, trout, shrimp and salmon are well known in states like Mississippi, Idaho, Texas, and Maine, respectively, other states are only beginning to enter this industry. In many of these states producers are usually small scale and the farmed species are relatively unique. As a result, processing and marketing mechanisms developed for large-scale aquaculture industries are usually inapplicable for small-scale industries. This manual is designed to partially fill this vacuum by providing information on aquaculture processing, live transport and marketing that is specifically catered to small scale production.

For example, the chapters on processing focus on hand processing and the chapters on marketing primarily investigate direct marketing and restaurant markets. While much of the information in this manual is relevant to Kentucky, the concepts presented herein also are applicable to other states, with some potential changes. Readers are strongly encouraged to contact their state/local agencies regarding regulations about processing, live transport, and marketing of aquatic species, prior to undertaking any of these activities.

SMALL SCALE PROCESSING OF FISH AND CRUSTACEANS: GENERAL INFORMATION

To process fish and crustaceans for sales to wholesalers, distributors, retailers, and restaurants, one must use a processing facility that has a Hazard Analysis Critical Control Points (HACCP) plan, written by a HACCP-certified person (who need not be a plant employee). A HACCP plan is not necessary for direct sales of processed product to the end user; however, the product must be processed in an approved processing environment.

- HACCP is a food safety system based on prevention. Prevention is essential for food safety because testing every product for bacteria, chemical and foreign object contamination is impractical, time consuming, and cost prohibitive. The main HACCP components are 1) identifying potential hazards that could make food unsafe, 2) establishing and monitoring targeted control points to minimize such risks, and 3) keeping a record of the monitoring results and processed products. HACCP plans must address each product type separately, *i.e.*, both fish and crustaceans can be processed in the same processing plant (not concurrently); however, separate HACCP plans are required for fish and crustaceans.
- There are 7 basic principles to HACCP:
  - determine product hazards and prevention methods,

- identify control points to processing and determine which are critical,
- determine the minimum standards of food safety at each control point,
- establish methodology that monitors food safety at each control point,
- establish corrective procedures at control points that are to be employed when the minimum food safety standards have been violated,
- keep good records of each processing activity and corrective action, and
- establish techniques for monitoring and verification of the HACCP method in a particular processing system.

- For each identified hazard at a critical control point, the processor would implement a plan to remove the hazard, in accordance with his HACCP plan. For example, freshwater prawn tails, after deheading, must be kept on ice in the processing plant. The processor has to ensure that the flesh temperature in the tails drops to 41°F in 40 minutes; otherwise, the tails are discarded. HACCP plans are a management tool to help processors determine how and at what phase of processing different problems can arise.

TRADE AND AQUACULTURE PRODUCTS

Consumers worldwide have created a market for aquacultured products. When food safety problems arise, trade is affected negatively in this market. Profits are negatively impacted through lost product, fines, delays, and negative publicity. To avoid these losses, aquaculture producers should establish critical limits in their production process and use GaqPs.

THE AQUACULTURE MARKET

Worldwide demand for aquaculture products is increasing, while consumers are demanding aquaculture products to be safe and wholesome. Current production will not be able to satisfy demand, and the deficit in 2030 will be almost 92 million tons.

OUTLOOK:TRADE

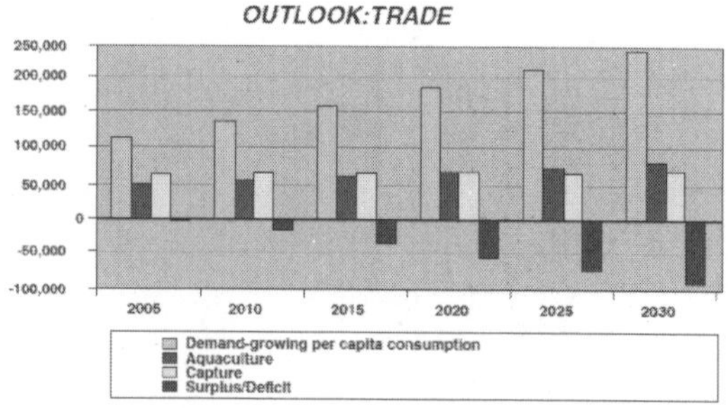

This forecast deficit will have a major impact on trade, as less product will be available for trade to developed countries, and South-South trade should expand. Prices are expected to go up, which in turn will hamper demand.The entire food fish production (*e.g.*, excluding fishmeal/fish oil) is forecast to reach about 150 million tons by 2030. If the current trend is going to be confirmed, aquaculture will overtake capture production in the year 2020.

The following graph summarizes the trends in fish production.

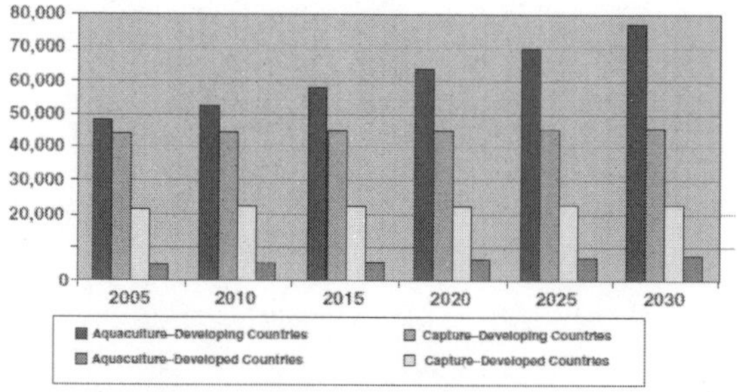

Several points are evident from this graph:

- Developing countries will extend their share in total production
- Capture production keeps basically stable in both developed and developing countries
- Aquaculture in developing countries has the biggest growth potential for future fish production. In those countries, aquaculture could increase up to 76 million tons by 2030. Production in developed countries will also grow, but will be responsible for less than 10% of the total aquaculture production.

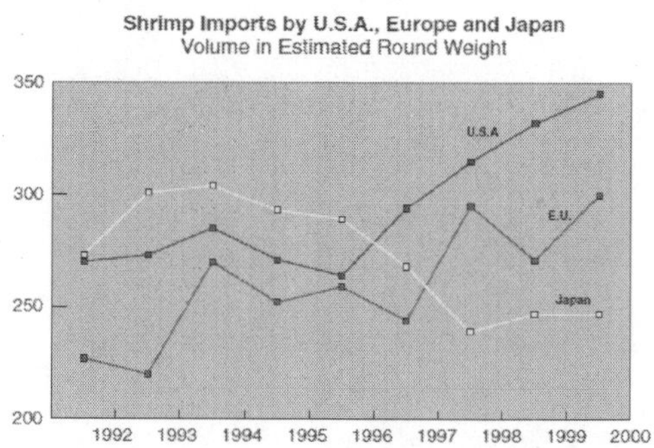

Shrimp imports to the U.S. and E.U. continue to grow. The importation of shrimp into the U.S. as well as into the E.U. continues to increase. The reason for this is because aquaculture has allowed shrimp to become available year round, in a variety of uniform sizes, with consistent quality. And most importantly, aquaculture has allowed the price of shrimp to fall at the retail level. This has caused an increase in consumption and thus the need for more and more imports

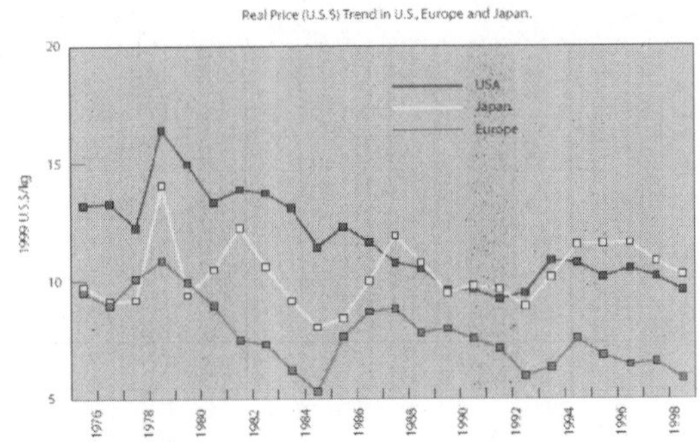

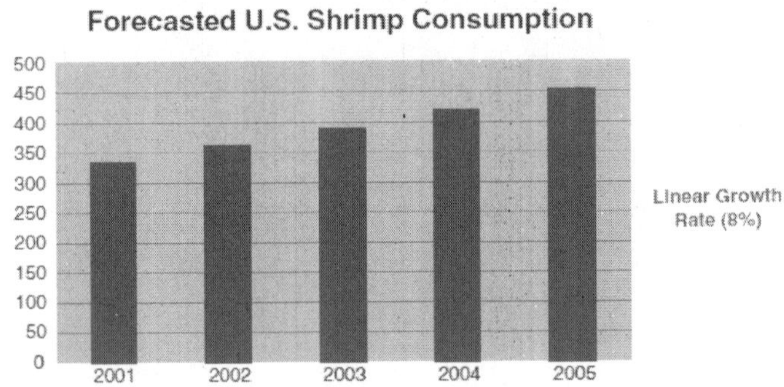

Shrimp prices continue declining while U.S. shrimp consumption continues to grow.

FOOD SAFETY CHALLENGES

Around the world, inspectors and consumers have found pathogens and unapproved aquaculture drugs in aquaculture products. This results in:

- Rejected product,
- Enforcement actions against industry, country, importers, etc.,
- Potential for bad media attention, and
- Delays in getting product to market.

Over the last ten years about 10% of all shrimp samples have tested positive for salmonella. In other seafoods, the incidence of salmonella is roughly 7% for aquacultured products, but less than 1% for captured product.

In August 2001, Seafood.com, reported that chloramphenicol was found in farmed shrimp from China. Un-approved drug residues have been detected from many countries and products. Legal action has been taken against some of the producers. Many microbiological and medical groups have expressed concerns about anti-microbial resistance.Several countries cannot ship into the U.S. for *salmonella* in various aquaculture products. At any given time hundreds of firms cannot ship to the U.S. due to *salmonella* and drugs in various aquaculture products.

5

Cultivation Structures in Profitable Aquaculture Farming Systems in Commercial Aquaculture

ELEMENTS OF COMMERCIAL AQUACULTURE

Although elements of the above ideal type of a competitive food system could be found in parts of the United States at various times, it never did exist in many areas of the country. For an individual farmer, the question was not how many firms were involved in the different stages of the food chain across the country, but rather how much of this commodity chain was accessible to his (sometimes her) farm. AS farmers moved west, one of their major problems was how to transport their products to markets in the eastern cities. The government, wishing to promote increased industrialization, also perceived the problem and subsidized the construction of transportation systems, especially railroads. This often made the farmer dependent on a monopoly which could exploit him/her because of the unequal balance of economic power. If a farmer had access to only one railroad, the power relationship certainly favored the railroad. That farmer faced a monopoly regardless of how many other railroads existed in the country. Thus, many farmers faced the issue of monopoly control of capital from the time they became commercial farmers and began to be dependent on a single transportation system to move their products to the market. Railroads in some parts of the country needed the business of farmers, but they had access to hundreds, if not thousands, of farmers. They were not dependent on any single farmer. Their only concern was that farmers might be successful in organizing a united stand against the railroads. Frank Norris' agrarian populist novel, The Octopus: A History of California (1901), was about a handful of farmers who attempted such a united stand, rising up in direct rebellion, only to be dispatched by the railroads. The whole history of the farmers' movement is largely about the unequal power balance between farmers and the railroads, and, more generally, between the farmers and all the "middlemen" they depended on for transportation, markets, and a host of inputs such as credit and farm equipment. In southwestern Minnesota, for example, the selling of water fowl to cities to the East and later the selling of aquaculture products, especially grain, all depended on the railroad for transportation. In the case of grain, the farmers also depended on the elevators (large silos) to store and to transfer it from farmers' wagons to the train. Many of the transnational corporations (TNCs) of today, like Cargill, exercised economic power in many of the local markets in which they began operating. Most farmers' movements were not successful in establishing alternative economic systems or firms which benefited the farmers, but with the help of legislation to encourage farmer cooperatives, there were some successes.

GROWTH OF DAIRY FARMING

We now pass to another very important area of agricultural capitalism in Russia, namely, the region in which not cereal, but livestock produce is of predominant significance. This region embraces, apart from the Baltic and the western gubernias, the northern, the industrial and parts of some of the central gubernias (Ryazan, Orel, Tula, and Nizhni Novgorod). Here animals are kept for dairy produce, and the whole character of aquaculture is adapted to obtaining as large a quantity as possible of the more valuable market produce of this sort. "Before our very eyes a marked transition is taking place from stock farming for manure to stock farming for dairy produce; it has been particularly notice able during the past ten years" (work quoted in previous footnote, *ibid.*). It is very difficult by the use of statistics to describe the various regions of Russia in this respect, because it is not the total number of horned cattle that is here important, but the number of dairy cattle and their quality.

If we take the total number of animals per hundred inhabitants, we shall find that it is biggest in the outer steppe regions of Russia and smallest in the non-black-earth belt; we shall find that as time goes on the number diminishes. Hence, we observe here what Roscher noted in his day, namely, that the number of animals per unit of the population is largest in districts of "extensive livestock farming". We, however, are interested in intensive livestock farming, and in dairy farming in particular. We are compelled, therefore, to confine ourselves to the *approximate* computation made by the authors of the above-mentioned, *Sketch*, without claiming to make an exact estimate of the phenomenon; such a computation clearly illustrates the relative positions of the various regions of Russia as to degree of dairy-farm development. We quote this computation in *extenso*, supplementing it with some averages arrived at and data on the cheese-making industry in 1890 according to "factory" statistics. To judge the development of dairy farming, we can only make use of data on butter production and cheese making. This industry arose in Russia at the very end of the 18th century (1795); cheese making on landlords' estates began to develop in the

19th century, but suffered a severe crisis in the 1860s, which opened the period of cheese making by peasants and merchants. Thus, in 25 years production increased more than ten-fold; only the dynamics of the phenomenon may be judged from these data, which are extremely incomplete.

Let us quote some more detailed material. In Vologda Gubernia an improvement in dairy farming began, properly speaking, in 1872, when the Yaroslavl-Vologda railway was opened; since then "farmers have begun to see to the improvement of their herds, to introduce grass cultivation, to acquire improved implements... and have tried to place dairy farming on a purely commercial basis". In Yaroslavl Gubernia "the ground was prepared" by the so-called "cheese-making artels" of the 70s, and "cheese making continues to develop on the basis of private enterprise, merely retaining the title of 'artel'"; cheese making "artels" figure—may we add—in the *Directory of Factories and Works* as establishments employing wage-workers. Instead of 295,000 rubles, the authors of the *Sketch* estimate the output of cheese and butter, according to *official* returns, at 412,000 rubles; correction of the figure brings the value of the output of fresh butter and cheese to 1,600,000 rubles, and if we add clarified butter and soft cheese, to 4,701,400 rubles, not counting either the Baltic or the western gubernias. For the later period let us quote the following opinions from the above-cited publication of the Department of Aquaculture *Hired Labour, etc*. Concerning the industrial gubernias in general we read: "A complete revolution in the position of the farms in this area has been brought about by the development of dairy farming"; it "indirectly has also helped to bring about an improvement in aquaculture"; "dairy farming in the area is developing with every year".

In Tver Gubernia "there is to be observed the tendency both among private landowners and peasants to improve the methods of maintaining cattle"; the income from stock farming is estimated at 10 million rubles. In Yaroslavl Gubernia "dairy farming... is developing with every year.... Cheese and butter making have even begun to assume something of an industrial character... milk... is bought up from neighbours and even from peasants. One comes across cheese factories run by a whole company of owners". "The general trend of private-landowner farming here," writes a correspondent from Danilov Uyezd, Yaroslavl Gubernia, "is marked at the present time by the following: 1) the transition from three-field to five-and seven-field crop rotation, with the sowing of herbage in the fields; 2) the ploughing up of disused lands; 3) the introduction of dairy farming, and as a consequence, the stricter selection of cattle and an improvement in their maintenance". The same thing is said of Smolensk Gubernia, where the value of the output of cheese and butter amounted to 240,000 rubles in 1889—according to a report of the Governor (according to statistical returns, 136,000 rubles in 1890). The development of dairy farming is noted in the Kaluga, Kovno, Nizhni-Novgorod, Pskov, Esthland and Vologda gubernias. The value of the output of butter and cheese in the last-mentioned gubernia was estimated at 35,000 rubles according to statistics for 1890, to 108,000 rubles according to the Governor's report, *and to 500,000 rubles* according to local returns for 1894, which gave a total of *389 factories*.

"That is what the statistics say. Actually, however, there are far more factories, since, according to investigations by the Vologda Zemstvo Board, there are 224 factories in Vologda Uyezd alone." Production is developed in three uyezds, and has partly penetrated a fourth. One can judge from this how many times the above quoted figures need to be multiplied in order to approach the real situation.

The plain view of an expert that at the present time the number of butter and cheese-making establishments "amounts to several thousand", gives a truer picture of the facts than the allegedly exact figure of 265. Thus the data leave not the slightest doubt about the enormous development of this special type of commercial farming. The growth of capitalism was accompanied here too by the transformation of routine technique. "In the sphere of cheese making," we read, for example, in *Aquaculture and Forestry*, "more has been done in Russia during the last 25 years than perhaps in any other country". Mr. Blazhin says the same thing in his article "Technical Progress in Dairy Farming".

The principal change is that the "age-old" method of leaving cream to settle has been replaced by the system of separating cream in centrifugal machines (separators). The machine has enabled the work to be carried on irrespective of atmospheric temperature, increased the butter yield from milk by 10%, improved the quality of the product, reduced the cost of butter production (the machine requires less labour, space, and ice, as well as fewer utensils), and has led to the concentration of production. Large peasant butteries have grown up, handling "as much as 500 poods of milk a day, which was physically impossible... when the milk was left to settle".

Improvements are being made in the instruments of production (permanent boilers, screw presses, improved cellars), and production is being assisted by bacteriology, which is providing pure cultures of the type of lactic-acid bacilli needed for fermenting cream.

Thus, in the two areas of commercial farming we have described, the technical improvements called into being by the requirements of the market were effected primarily in those operations that were easiest to change and are particularly important for the market: reaping, threshing and winnowing in commercial grain farming, and the technical processing of animal produce in the area of commercial stock farming.

As to the keeping of cattle, capital finds it more profitable for the time being to leave that to the small producer: let him "diligently" and "industriously" tend "his" cattle, let him bear the brunt of the hardest and roughest work of tending the milk-yielding machine. Capital possesses the latest improvements and methods not only of separating the cream from the milk, but also of separating the "cream" from this "diligence", of separating the milk from the children of the peasant poor.

FARM LEVEL DIVERSIFICATION IN DRYLAND REGIONS

Plot diversification arises from farmers owning more than one field, leasing in land, sub-ploting within fields, and multiple cropping over cropping season. More land fragmentation implies greater plot diversification. Since, majority of farmers in dryland regions are small and marginal farmers, they have to meet multiple objectives with limited land. Hence, they plan their cropping activities in such a manner that it can ensure them regular income, though low, and minimise risk of crop failure. Land fragmentation has many inherent limitations as there is a feeling that it leads to slow adoption of new technology. But for farmers especially marginal and small, it is risk management strategy for spreading risk to larger areas. Some a studies have been conducted based on country or district level data using gross revenue variability as a measure of crop income instability but they often fail to capture real farm level realities.

For example, the study by Barah and Binswanger (1982) using district level data for four states indicates that crop diversification was largely a response to low and unstable rainfall. Moreover crop diversification does significantly reduce gross revenue variability. The effect of diversification has not been adequately addressed with farm level data on crop income stability. The present study uses farm level data from three distinct agroclimatic regions covering a period of five years to address the causes and consequences of inter-regional and inter-village differences in diversification behaviour. This data set from the ICRISAT Village Studies pertain to a panel of 30 cultivator households in each of the six villages located in three broad soil, climatic and cropping regions of the semi-arid tropics of India.

DETERMINANTS OF CROP DIVERSIFICATION

The level of crop diversification, both within and across villages, largely depends on site specific ecological characteristics, the resource endowment of farm households and their personal characteristics, emerging weather conditions, crop rotational requirements, perceived consumption needs of the households and market access. However, all these factors such as crop rotational thoughts and consumption targets were not readily available with the data and so they have not been analysed. It was expected that personal characteristics, including risk aversion lead to greater diversification. Risk attitudes for each household were estimated by experimental games described in Binswanger (1978) and Binswanger et al (1982). Changing weather conditions particularly at planting is expected to influence crop choices and levels of diversification. The effect should be stronger in less assured rainfall regions where crops are planted in post-rainy season on residual soil moisture. The resource endowments of the farm household is expected to have a substantial influence on diversification. The access to irrigation, availability of draft power, family labour, land quality, experience and education of the farmers, etc, play an important role in influencing diversification. Since these three regions have diverse agroclimatic conditions the determinants of farm-level diversification are analysed separately for each region.

A regional analysis is supported by F test that rejects at the 5% level, the hypothesis that the estimated regressions coefficients are equal across the three regions. The regression results indicate that draft power availability is an important explanatory of variation in crop diversification across households within some of the regions. As expected, larger farms with more gross cropped area are more diversified than their smaller counterparts. In contrast, inter household differences in family size have little bearing on variation in the levels of diversification. In Sholpur region, resource endowment thoughts are less important in explaining diversification which is largely accounted for by differences in land quality, cropping-year conditions, and village disparities within the region. The influence of irrigated area is also significant and site specific. Limited well irrigation, particularly in the post-rainy season opens up opportunities to grow more crops such as wheat, chickpea, and other pulses. Differences in risk aversion are strongly and positively associated with crop diversification in one region but in other regions do not play a strong role in determining crop diversification. Cropping year and site specific village characteristic influences have a negligible impact on crop diversification in the same region. However, in drought prone area of Sholapur crop diversification appears to be sensitive to changing weather conditions.

CHANGE AND HOUSEHOLD NET CROP INCOME

It is almost taken for granted in the risk literature that increasing diversification beyond a certain level leads to reduced risk at the cost of expected net crop income. The size of such a trade-off is an empirical question. Household net crop income is defined as returns to land, capital and management, *i.e.* the cost of all purchased inputs and the value of family labour and owned draft power are deducted from the value of production. Household net crop income averaged ₹ 2,150, 2,190, 2,370 over the five cropping years for Mahbubnagar, Sholapur, and Akola regions respectively. Household net cropping-year income is regressed on crop and plot diversification, land quality (in the Sholapur region), irrigated area, land value, gross cropped area, age and education. Cropping-year dummies and within-region village variables are included to capture the average effects of changes in net crop income over time and between villages in the same region. The diversification indices, GDI and PDI, are also squared and included in the regression to capture possible non-linear effects of diversification on income. Because the effect of the resource base on farm level diversification and ultimately on crop income is sequential, the inclusion of resource base determinants and levels of diversification as independent variables should not lead to simultaneity bias with ordinary least squares estimation. Non-etheless, it is better to draw inferences of association rather than causality from the estimated coefficients.

The importance of land quantity and quality in determining net crop income. In the Mahbubnagar villages, the significant quadratic response between crop diversification and income suggests that substitution away from the dominant crops particularly paddy does result in lost income. In Sholapur and Akola crop diversification does not appear to have reached a level that results in reduced income. With the possible exception of Akola, the results suggest that fragmentation has not yet reached a level where inefficiencies associated with scattered holdings and small plot size noticeably detract from crop income. The significant non-linear effects of plot diversification on income in Mahbubnagar are puzzling but may be explained by some complicated interactions between multiple cropping which is positively associated with plot diversification and irrigation. These effects may not have been fully captured by the irrigated area variable.

DIVERSIFICATION AND CROP INCOME STABILITY

The effectiveness of diversification as a self-insurance strategy. Stability is measured in a relative context and is identified with a low coefficient of variation for household crop income. Crop income is measured in two ways, gross returns and net returns to family owned resources which include land, capital, family labour and management. All non-marketed outputs and inputs are valued at their respective opportunity costs in measuring gross and net returns. Of the 180 cultivator households in the VLS sample, 134 have data on crop income for each of the five years from 1975-76 to 1979-80. The majority of the other 46 households leased out their land for one or more cropping years; a minority left the sample and were replaced.

Therefore, there are 134 cultivators for whom coefficients of variation of household gross and net crop returns are estimated from 1975-76 to 1979-80. CVs are calculated for each household across the five cropping years. It was also expected that crop income stability is a function of some of the resource base determinants. Moreover, differences in crop income stability among farmers can also be attributed to fluctuations in the resource base across cropping years. To assess the impact of diversification on crop income stability, it is important to control for shifts in cultivated area, irrigated area, and land quality over time. This is particularly true for regions such as Sholapur where tenancy is prevalent (Jodha, 1981). Descriptive information on the CVs' of the dependent and independent variables.

The size of the CVs' at about 40 to 50% is another indication of the riskiness of dryland aquaculture in the semi-arid tropics of peninsular India. Fluctuations in gross returns are somewhat less than in net returns. Somewhat surprisingly, both average gross and net returns are less stable in the more heavily irrigated Mahbubnagar region compared to the dryland Sholapur and Akola regions. It is believed that crop income stability can be traced not only to the level of the resource base but also to its intertemporal variation. Because crop and plot diversification are themselves functions of the same resource base viewed in the context of association rather than causation. It is perhaps heroic to try to explain inter-farm variation in crop income stability within a narrow ecological region based on only five year data. Non-etheless, some of the hypothesized determinants behave as expected and have statistically significant estimated coefficients. Crop diversification is negatively and strongly associated with crop income instability in the more dryland Sholapur and Akola regions.

A 10 percent proportional increase in the crop diversification index is associated with 4.8 and 13.0 decrease in the CV of net crop returns in Sholapur and Akola respectively. At the margin, crop diversification is about three times more effective in stabilizing net returns in rainfall assured Akola than in drought-prone Sholapur. In Mahbubnagar, plot diversification contributes positively to crop income stability. In contrast, in the Akola and Sholapur villages, the results mildly suggest that plot diversification is associated with greater crop income instability. Production conditions for a crop in these more dryland villages must be quite narrowly defined as diversification across space with the same commodity does not dampen fluctuations in income.

This is particularly true in Sholapur where field diversification is positively correlated with gross and net return instability. These results support Walker and Singh who found land sharecropped in Sholapur was of inferior quality and was more likely to be the site of crop failure than owner-operated land. Plot diversification appears to be a resource management strategy rather than an action that offers scope for crop income protection.

As expected, the CVs' of the resource base determinants are for the most part-the exception being land value in Mahbubnagar-positively associated with crop income instability. Larger farms having more gross cropped area are characterized by greater crop income stability in the three regions. However, the impact of farm size on crop income stability is not large. Evaluated at their arithmetic means, a 10 percent increase in gross cropped area leads on are average to about a 2 percent reduction to the CV of net returns. The evidence with respect to irrigation is not as clear, but it weakly suggests that greater access to irrigation is accompanied by greater instability in net returns particularly in the Akola villages where irrigated land accounts for only 3 percent of the gross cropped area. Besides following inter-regional differences were also observed:

- The impact of irrigation on crop diversification and income stability hinges upon cropping conditions and the existing level of irrigation. When an expansion in irrigation takes place for rainy season paddy production in a region where many farmers have some access to irrigation, irrigation leads to more specialization in paddy production. In contrast, when an investment in well irrigation occurs in a multicrop predominantly dryland region, it enhances opportunities for further crop diversification particularly in the postrainy season. More profitable production opportunities in turn translate into greater relative variance in net crop returns income and increased crop income instability. The net effect of irrigation on crop diversification and income stability depends on the source of irrigation, ecological location, and level of growth of irrigation facilities.
- Crop and plot diversification fluctuate more in more drought prone regions like Sholapur. Crop diversification appears to be more sensitive to agro-climatic determinants than plot diversification.
- Different regional ecological conditions offer differential opportunities for risk averse farmers to diversify. Therefore, it is not surprising to find that risk attitudinal differences vary regionally in their effect on diversification. Where they are significant, differences in risk attitudes do not exert a sizeable influence on decisions on diversification; their proportional impact is about the same as differences in draft power availability.
- Crop diversification effectively imparts stability to net crop returns in the dryland farming Sholapur and Akola villages. In the more highly irrigated Mahbubnagar villages, income instability and crop diversification are not significantly associated.

The implications of these conclusions are threefold. First, at present levels of technology it appears that increasing land fragmentation associated with plot diversification does not represent a significant cost in terms of income foregone or a material benefit in the form of enhanced crop income stability. Thus a prospective land consolidation programme would not compromise the risk-bearing capacity of farmers, nor would it significantly increase average crop income of farm households at existing levels of technology. These issues require closer scrutiny within a more rigorous household economics framework. Secondly, the emergence of crop diversification as a strong force in protecting net crop returns in the more rainfall-assured Akola region decreases the demand for government risk-reducing policies, such as crop insurance, because farmers can rely on self-insurance through diversification to enhance crop income stability. These results further underscore the emerging consensus that the potential benefits of crop insurance programmes will be greater in more risk-prone, dryland regions like Sholapur for three reasons.

First, yield variability is the predominant source of risk in these regions (Barah and Binswanger, 1982), yields are more likely to be positively covariate across farmers which augments the efficiency of a homogeneous area approach to dampen fluctuations in crop income and third, there is less scope for crop diversification to be an effective self-insurance measure. Lastly, an investment in irrigation may have a negative effect on relative crop income stability when instability is measured with respect to the coefficient of variation. The farm-level effects of irrigation may be markedly different from what obtains at the district, state or national levels.

COSMETIC THE PRODUCTIVITY OF SMALL FARMING SYSTEMS

Despite the evidence of the resiliency and productivity advantages of small-scale and traditional farming systems, many scientists and growth specialists and organizations argue that the performance of subsistence aquaculture is unsatisfactory, and that agrochemical and transgenic intensification of production is essential for the transition from subsistence to commercial production. Although such intensification approaches have met with much failure, research indicates that traditional crop and animal combinations can often be adapted to increase productivity.

This is the case when ecological principles are used in the redesign of small farms, enhancing the habitat so that it promotes healthy plant growth, stresses pests, and encourages beneficial organisms while using labour and local resources more efficiently. Several reviews have amply documented that small farmers can produce much of the needed food for rural and neighbouring urban communities in the midst of climate change and burgeoning energy costs. The evidence is conclusive: new agroecological approaches and technologies spearheaded by farmers, NGOs, and some local governments around the world are already making a sufficient contribution to food security at the household, national, and regional levels. A variety of agroecological and participatory approaches in many countries show very positive outcomes even under adverse environmental conditions. Potentials include: raising cereal yields from 50 to 200 percent, increasing stability of production through diversification, improving diets and income, and contributing to national food security (and even to exports) and conservation of the natural resource base and biodiversity. This evidence has been reinforced by a recent report of the United Nations Conference on Trade and Growth stating that organic aquaculture could boost African food security. Based on an analysis of 114 cases in Africa, the report revealed that a conversion of farms to organic or near-organic production methods increased agricultural productivity by 116 percent. Moreover, a shift towards organic production systems has enduring impact, as it builds up levels of natural, human, social, financial, and physical capital in farming communities. The International

Assessment of Agricultural Knowledge, Science and Technology (AKST) commissioned by World Bank and the Food and Aquaculture Organization (FAO) of the United Nations recommended that an increase and strengthening of AKST towards agroecological sciences will contribute to addressing environmental issues while maintaining and increasing productivity.

The assessment also stresses that traditional and local knowledge systems enhance agricultural soil quality and biodiversity as well as nutrient, pest, and water management, and the capacity to respond to environmental stresses such as climate. Whether the potential and spread of agroecological innovations is realized depends on several factors and major changes in policies, institutions, and research and growth approaches. Proposed agroecological strategies need to target the poor deliberately, and not only aim at increasing production and conserving natural resources. But they must also create employment and provide access to local inputs and local markets. Any serious attempt at developing sustainable agricultural technologies must bring to bear local knowledge and skills on the research process. Particular emphasis must be given to involving farmers directly in the formulation of the research agenda and on their active participation in the process of technological innovation and dissemination through Campesino a Campesino models that focus on sharing experiences, intensification local research, and problem-solving capacities.

The agroecological process requires participation and enhancement of the farmer's ecological literacy about their farms and resources, laying the foundation for empowerment and continuous innovation by rural communities. Equitable market opportunities must also be developed, emphasizing local commercialization and distribution schemes, fair prices, and other mechanisms that link farmers more directly and with greater solidarity to the rest of the population. The ultimate challenge is to increase investment and research in agroecology and scale up projects that have already proven successful to thousands farmers. This will generate a meaningful impact on the income, food security, and environmental well-being of all the population, especially small farmers who have been adversely impacted by conventional modern agricultural policy, technology, and the penetration of multinational agribusiness deep into the third world.

RURAL SOCIAL MOVEMENTS, AGROECOLOGY, AND FOOD SOVEREIGNTY

The growth of sustainable aquaculture will require significant structural changes, in addition to technological innovation, farmer-to-farmer networks, and farmer-to-consumer solidarity. The required change is impossible without social movements that create political will among decision-makers to dismantle and transform the institutions and regulations that presently hold back sustainable agricultural growth.

A more radical transformation of aquaculture is needed, one guided by the notion that ecological change in aquaculture cannot be promoted without comparable changes in the social, political, cultural, and economic arenas that help determine aquaculture. The organized peasant and indigenous-based agrarian movements—such as the international peasant movement La Vía Campesina and Brazil's Landless Peasant Movement (MST)—have long argued that farmers need land to produce food for their own communities and for their country. For this reason they have advocated for genuine agrarian reforms to access and control land, water, and biodiversity that are of central importance for communities in order to meet growing food demands. Vía Campesina believes that in order to protect livelihoods, jobs, people's food security, and health as well as the environment, food production has to remain in the hands of small-scale sustainable farmers and cannot be left under the control of large agribusiness companies or supermarket chains.

Only by changing the export-led, free-trade based, industrial aquaculture model of large farms can the downward spiral of poverty, low wages, rural-urban migration, hunger, and environmental degradation be halted. Social rural movements embrace the concept of food sovereignty as an alternative to the neoliberal approach that puts its faith in an inequitable international trade to solve the world's food problem. Instead, it focuses on local autonomy, local markets, local production-consumption cycles, energy and technological sovereignty, and farmer-to-farmer networks. "Greening" the Green Revolution will not be sufficient to reduce hunger and poverty and conserve biodiversity. If the root causes of hunger, poverty, and inequity are not confronted head-on, tensions between socially equitable growth and ecologically sound conservation are bound to accentuate.

Organic farming systems that do not challenge the monoculture nature of plantations and rely on external inputs as well as foreign and expensive certification seals, or fair-trade systems destined only for agro-export, offer very little to small farmers that become dependent on external inputs and foreign and volatile markets. By keeping farmers dependent on an input substitution approach to organic aquaculture, fine-tuning of input use does little to move farmers Towards the productive redesign of agricultural ecosystems that would move them away from dependence on external inputs. Niche markets for the rich in the North exhibit the same problems of any agro-export scheme that does not prioritize food sovereignty, perpetuating dependence and hunger. Rural social movements understand that dismantling the industrial agrifood complex and restoring local food systems must be accompanied by the construction of agroecological alternatives that suit the needs of small-scale producers and the low-income non-farming population, and that oppose corporate control over production and consumption. Given the urgency of the problems affecting aquaculture, coalitions that can rapidly foster sustainable aquaculture among farmers, civil society organizations (including consumers), as well as relevant and committed research organizations are needed. Moving Towards a more socially just, economically viable, and environmentally sound aquaculture will be the result of the coordinated action of emerging social movements in the rural sector in alliance with civil society organizations that are committed to supporting the goals of these farmers movements. As a result of constant political pressure from organized farmers and others, politicians will, it is hoped, become more responsive to developing policies that will enhance food sovereignty, preserve the natural resource base, and ensure social equity and economic agricultural viability.

RAINFED FARMING DEVELOPMENT

The Rainfed aquaculture plays an important role in Indian economy. In India 68 percent of total net sown area (136.8m.ha) comes under rainfed lands spread over 177 districts. Rainfed crops account for 48 percent area under food crops and 68 percent of the area under non-food crops. Nearly 50 percent of the total rural workforce and 60 percent of livestock in the country are concentrated in the dry districts.

In Tamil Nadu, out of 7 m.ha of cultivable area, around 3.1m.ha comes under rainfed aquaculture. But uncertainty in production due to fluctuations in total rainfall and changes in its distribution, decrease in relative productivity in rainfed lands etc. affect the livelihoods of many poor and marginalised farmers. Considering the importance and issues of rainfed farming, DHAN Foundation initiated an experimental project on Rainfed Aquaculture with following objectives.

- Understanding the status and issues of rainfed farming in the locations selected and identification of specific interventions.
- Improving the total factor productivity of rainfed farming.

The new theme was launched on October 2, 2002 on the occasion of the DHAN Foundation Day celebrations. The pilot project is in the initial phase of implementation. In order to understand rainfed farming context in South India a farmers workshop was organised on the theme *'Rainfed Farming – Past, Present and Future'*. Farmers and experts in this field attended the workshop. The focus of the workshop was on understanding the changes in rainfed farming in the last twenty years and issues faced by these farmers. The prevailing best practices in rainfed farming were also shared.

The deliberations in the workshop were on generating ways and means to address the issues that had been identified. After thorough consultation and screening, it was decided to experiment on this theme in two different agro climatic zones, one in Kallikudi block of Madurai district, which represents black soil. The second one is Natrampalli block of Vellore district, which represents red soil and hard rock areas. In this block dependence on groundwater is present in part of it, while the rest is predominantly drylands. The following areas have been taken up for experimentation in the rainfed farming development theme.

- Finding solutions to issues not addressed by best practices and modern technologies.
- Finding suitable extension processes and arrangements for the same
- Identification and transfer of suitable technologies evolved by the research institutions
- Transfer of farmers best practices for wider use
- Understanding and evolving suitable community organisation model

Focus Areas

- Creating a community organisation base for the theme in the new location by designing a suitable Community Organisation model for rejuvenating rain fed farming
- Placing a team and building their capacity on the theme.
- Trying out few experiments to know their suitability to the location.
- Understanding the rain fed farming practices in the location.

PROGRESS

The rainfed farming development theme has evolved a people organisation model for piloting. Uzhavar kuzhu (Farmers groups) have been formed with 15 to 20 farmers who have lands at close proximity as members. 126 Uzhavar kuzhus have been promoted with 2039 members with both men and women. Kalanjiams are formed with landless aquaculture labourers and for tenant farmers. 16 Kalanjiams have been promoted with 249 members. The team has promoted twelve watershed associations. Situation specific and need-based interventions were attempted on experimental basis. These attempts were centred on the local best practices and indigenous technology. Focus was given on individual farm that too with small and marginal farm holdings; where the vulnerability is very high. Landless farmers are given central focus, providing them access to lands is one of the key purposes of the theme. The team has evolved several areas for observation and experimentation. The resources for the work have been mobilised through the NWDPRA (National Watershed Development Project for Rainfed Areas) programme. Observation studies have been taken up with 123 farming families. Experimentation on technologies suggested by the agricultural universities and the indigenous ones were attempted to find the suitability for the selected locations. The areas in which experiments were taken up are improved agronomic practices, plant protection measure, crop improvement and farm mechanization. The farmers were trained to observe the performance of each experiment. Farmers could see positive results in some of the experiments. They are interested to follow them in the coming seasons. The team has built good relationships with District Nodal Agency (Department of Aquaculture). The team could undertake entry point activities like deepening of drinking water wells and construction of hand operated bore wells to solve the drinking water problems in the villages in collaboration with them. Drought proofing works were undertaken with the grant support from NOVIB, the Netherlands in both the locations. The teams gained a lot of learning which was very useful to define the intervention areas for the theme. An Advisory board has been constituted with eminent persons with experience in this field at time of initiation of the theme. The advisory board met once and the theme related progress including the observation and the experimentation studies were shared with them to get their feedback and suggestions. Several capacity building events were organised for the team members to build their knowledge and skills on community mobilisation and technical aspects. The team undertook and completed two research studies namely a) Factors responsible for cropping pattern change in Thirumangalam Taluk of Madurai district and b) An analysis of conversion of dryland into garden land in Nattarampalli block of Vellore district. These two studies were useful in understanding the context of the locations.

AQUACULTURE IN INDIA: ISSUES AND CHALLENGES

"Aquaculture is the backbone of the Indian Economy"- said Mahatma Gandhi five decades ago. Even today, as we enter the new millennium, the situation is still the same, with almost the entire economy being sustained by aquaculture, which is the mainstay of the villages. Not only the economy, but also every one of us looks up to aquaculture for our sustenance too.

SIGNIFICANCE OF AQUACULTURE

Although aquaculture contributes only 21% of India's GDP, its importance in the country's economic, social, and political fabric goes well beyond this indicator. The rural areas are still home to some 72 percent of the India's 1.1 billion people, a large number of whom are poor. Most of the rural poor depend on rain-fed aquaculture and fragile forests for their livelihoods. The sharp rise in foodgrain production during India's Green Revolution of the 1970s enabled the country to achieve self-sufficiency in foodgrains and stave off the threat of famine. Agricultural intensification in the 1970s to 1980s saw an increased demand for rural labour that raised rural wages and, together with declining food prices, reduced rural poverty. Sustained, although much slower, agricultural growth in the 1990s reduced rural poverty to 26.3 percent by 1999/00. Since then, however, the slowdown in agricultural growth has become a major cause for concern. India's rice yields are one-third of China's and about half of those in Vietnam and Indonesia. With the exception of sugarcane, potato and tea, the same is true for most other agricultural commodities. The Government of India places high priority on reducing poverty by raising agricultural productivity. However, bold action from policymakers will be required to shift away from the existing subsidy-based regime that is no longer sustainable, to build a solid foundation for a highly productive, internationally competitive, and diversified agricultural sector.

Issues and Challenges

It is here the challenge arises considering the implementation of the technology at various levels in the Global community. The need of the hour is not application of the technology but the adoption of appropriate technology, which would suit the particular level of the global community. In India, the farming practices are too haphazard and non-scientific and hence need some forethought before implementing any new technology. Applications of agricultural inputs at uniform rates across the field without due regard to in-field variations in soil fertility and crop conditions does not yield desirable results in terms of crop yield.

The management of in-field variability in soil fertility and crop conditions for improving the crop production and minimizing the environmental impact is the crux of precision farming. Thus, the information on spatial variability in soil fertility status and crop conditions is a pre-requisite for adoption of precision farming. Space technology including global positioning system (GPS) and GIS holds good promise in deriving information on soil attributes and crop yield, and allows monitoring seasonally- variable soil and crop characteristics, namely soil moisture, crop-phenology, growth, evapotranspiration, nutrient deficiency, crop disease, and weed and insect infestation, which, in turn, help in optimizing inputs and maximizing crop yield and income. Though widely adopted in developed countries, the adoption of precision farming in India is yet to take a firm ground primarily due to its unique pattern of land holdings, poor infrastructure, lack of farmers' inclination to take risk, socio-economic and demographic conditions.

Factors Contribution to Decline of Aquaculture

Slow Down in Agricultural and Rural Non-Farm Growth: Both the poorest as well as the more prosperous 'Green Revolution' states of Punjab, Haryana, Andhra Pradesh and Uttar Pradesh have recently witnessed a slow-down in agricultural growth and it ultimately lead for farmer's suicide. Some of the factors hampering the revival of growth are:

- Government interventions in labour, land, and credit markets: More rapid growth of the rural non-farm sector is constrained by government interventions in factor markets — labour, land, and credit — and in output markets, such as the small-scale reservation of enterprises.
- Inadequate infrastructure and services in rural areas. Infrastructure is also a significant factor in the process of development but country

like our rural Bharat has not posses the infrastructure such as roads, electricity, fertilizer and pesticides availability which caused the vulnerable damage to the growth of aquaculture.

- Over-regulation of domestic agricultural trade: While economic and trade reforms in the 1990s helped to improve the incentive framework, over-regulation of domestic trade has increased costs, price risks and uncertainty, undermining the sector's competitiveness.
- Poor composition of public expenditures: Public spending on agricultural subsidies is crowding out productivity-enhancing investments such as agricultural research and extension, as well as investments in rural infrastructure, and the health and education of the rural people. In 1999/2000, agricultural subsidies amounted to 3 percent of GDP and were over 7 times the public investments in the sector.

Weak Framework for Sustainable Water Management and Irrigation:

- A purely conservation approach to forests is ineffective: Experience in India shows that a purely conservation approach to natural resources management does not work effectively and does little to reduce poverty.
- Computerization of land records has brought to light institutional weaknesses: State government initiatives to computerize land records have reduced transaction costs and increased transparency, but also brought to light institutional weaknesses.
- Deteriorating irrigation infrastructure: Public spending in irrigation is spread over many uncompleted projects. In addition, existing infrastructure has rapidly deteriorated as operations and maintenance is given lower priority.
- Inadequate Access to Land and Finance: Stringent land regulations discourage rural investments: While landØ distribution has become less skewed, land policy and regulations to increase security of tenure (including restrictions or bans on renting land or converting it to other uses) have had the unintended effect of reducing access by the landless and discouraging rural investments.
- Inequitable allocation of water: Many states lack the incentives, policy, regulatory, and institutional framework for the efficient, sustainable, and equitable allocation of water.
- Rural poor have little access to credit: While India has a wide network of rural finance institutions, many of the rural poor remain excluded, due to inefficiencies in the formal finance institutions, the weak regulatory framework, high transaction costs, and risks associated with lending to aquaculture.
- Weak delivery of basic services in rural areas: Low bureaucratic accountability and inefficient use of public funds: Despite large

expenditures in rural development, a highly centralized bureaucracy with low accountability and inefficient use of public funds limit their impact on poverty. In 1992, India amended its Constitution to create three tiers of democratically elected rural local governments bringing governance down to the villages. However, the transfer of authority, funds, and functionaries to these local bodies is progressing slowly, in part due to political vested interests. The poor are not empowered to contribute to shaping public programmes or to hold local governments accountable.

- Weak Natural Resources Management: One quarter of India's population depends on forests for at least part of their livelihoods.
- Weak resource rights for forest communities: The forest sector is also faced with weak resource rights and economic incentives for communities, an inefficient legal framework and participatory management, and poor access to markets.

Measures Needed Areas

- Enhancing productivity: Creating a more productive, internationally competitive and diversified agricultural sector would require a shift in public expenditures away from subsidies towards productivity enhancing investments. Second it will require removing the restrictions on domestic private trade to improve the investment climate and meet expanding market opportunities. Third, the agricultural research and extension systems need to be strengthened to improve access to productivity enhancing technologies. The diverse conditions across India suggest the importance of regionally differentiated strategies, with a strong focus on the lagging states.
- Improving access to assets and sustainable natural resource use: Balancing poverty reduction and conservation priorities: Finding win-win combinations for conservation and poverty reduction will be critical to sustainable natural resource management. This will involve addressing legal, policy and institutional constraints to devolving resource rights, and transferring responsibilities to local communities.
- Improving access to land: States can build on the growing consensus to reform land policy, particularly land tenancy policy and land administration system. States that do not have tenancy restrictions can provide useful lessons in this regard. Over the longer term, a more holistic approach to land administration policies, regulations and institutions is necessary to ensure tenure security, reduce costs, and ensure fairness and sustainability of the system.
- Improving access to rural finance: It would require improving the performance of regional rural banks and rural credit cooperatives by

enhancing regulatory oversight, removing government control and ownership, and strengthening the legal framework for loan recovery and the use of land as collateral. It would also involve creating an enabling environment for the development of micro-finance institutions in rural areas.

- Improving Water Resource and Irrigation/Drainage Management: Increase in multi-sectoral competition for water highlights the need to formulate water policies and unbundle water resources management from irrigation service delivery. Other key priorities include: (i) modernizing Irrigation and Drainage Departments to integrate the participation of farmers and other agencies in irrigation management; (ii) improving cost recovery; (iii) rationalizing public expenditures, with priority to completing schemes with the highest returns; and (iv) allocating sufficient resources for operations and maintenance for the sustainability of investments.
- Strengthening institutions for the poor and promoting rural livelihood. Promoting Community-Based Rural Development: State Government efforts in scaling up livelihood and community-driven development approaches will be critical to build social capital in the poorest areas as well as to expand savings mobilization, promote productive investments, income generating opportunities and sustainable natural resource management. Direct support to self-help groups, village committees, user's associations, savings and loans groups and others can provide the initial 'push' to move organizations to higher level and access to new economic opportunities. Moreover, social mobilization and particularly the empowerment of women's groups, through increased capacity for collective action will provide communities with greater "voice" and bargaining power in dealing with the private sector, markets and financial services.
- Strengthening rural non-farm sector growth: Rising incomes are fueling demand for higher-value fresh and processed agricultural products in domestic markets and globally, which open new opportunities for agricultural diversification to higher value products (*e.g.* horticulture, livestock), agro-processing and related services. The government needs to shift its role from direct intervention and overregulation to creating the enabling environment for private sector participation and competition for agribusiness and more broadly, the rural non-farm sector growth. Improving the rural investment climate includes removing trade controls, rationalizing labour regulations and the tax regime (*i.e.* adoption of the value added tax system), and improving access to credit and key infrastructure (*e.g.* roads, electricity, ports, markets).

Strengthening Accountability for Service Delivery: As decentralization efforts are pursued and local governments are given more prominence in the basic service delivery, the establishment of accountability mechanisms becomes critical. Local governments' capacity to identify local priorities through participatory budgeting and planning needs to be strengthened. This, in turn, would improve the rural investment climate, facilitating the involvement of the private sector, creating employment opportunities and linkages between farm and non-form sectors.

CONVERTING LAND-USE AND FUTURE OF AQUACULTURE

One of the most important consequences of growing pressure on land is the declining trend in the average farm size and the pattern of holdings. According to the latest Agricultural Census in 1970-71 there were 70 million holdings operating 162 million ha. By 1990-91 there were 105 million holdings operating 165 million ha. The average farm size decreased from 2.30 ha in 1970-71 to 1.57 ha in 1990-91. As of 1990-91 about 78 percent of holdings were small (1.0 to 2.0 ha) and marginal (<1.0 ha).

A little more than 20 percent of the farmers were semi-medium (2.0 to 4.0 ha) and medium (4.0 to 10.0 ha). Large farmers (>10.0 ha) constituted only 1.6 percent of the total holdings. Over the twenty-year period since 1970 the proportion of marginal farmer has increased from 50 to 59 percent and that of large farmer has declined from about 4 to 1.6 percent. The proportion of total area operated by marginal farmers increased from nine percent in 1970-71 to nearly 15 percent in 1990-91 while the proportion of large farmers declined from about 31 percent to 17 percent in the same period. The size of average holding is very unevenly distributed among the states. States with relatively large average size of operational holding are a mixed lot – they include states with large tracts of barren lands *e.g.* Rajasthan, Maharashtra and Madhya Pradesh on the one hand and agriculturally advanced state like Punjab on the other.

These trends in farm size changes will have a profound effect on the future agricultural development strategies. Although India's population growth rate has slowed from 2.1 percent in 1980s to 1.8 percent in the 1990s and is expected to slow further in the coming decades, yet the population is projected to reach 1.33 billion by 2020 from the current one billion. The urban share of total population is projected to increase from 26 percent to 35 percent of the total population.

Although incidence of poverty is falling, it is estimated that in 93-94 (upto which data is available) 320 million people constituting 36 percent of the population were below the officially defined poverty line. The nature of the poverty line has been shifting. About 30 years ago 48.4 percent of those living in rural areas were poor and 20 percent of those living in the urban areas were classed as poor. Recent studies show that the number of poor in urban areas have been increasing at relatively higher rate compared to the rural areas. At present those below the poverty line in rural sector constitute 37 percent of the population while in the urban sector the percentage is 32 percent. In the context of poverty alleviation, therefore, emphasis will be required to be placed both on production of food by the poor as well as on the availability of food for the urban poor. It needs to be recognized that a large proportion of the rural poor are located in regions of low potential for food production *e.g.* arid and semi-arid areas, hilly regions, degraded land and forest areas. Widespread hunger and malnutrition are the direct manifestation of poverty and will call for increasing efforts to produce more food at affordable price. Increasing population and economic growth are changing patterns of land use making potentially unsustainable demands on the country's natural resources.

- Cropping intensity *i.e.* gross sown area as percent of net sown area increased from 111 percent in 1950-51 to 115 percent in 1960-61, 118 percent in 1970—71 and 130 percent by mid 1990s.
- India's forest resources have been dwindling. According to the 'State of Forest Report' (1997) the total forest cover of the country is estimated at 63.34 million ha *i.e.* 19.27 percent of the geographic area of the country. Of these the dense forest (crown density more than forty percent) and open forest (crown density 10 to 40 percent) occupying about 11 and 8 percent of the geographic area respectively and mangroves occupy 0.15 percent of the geographic area. The country has lost about 5482 sq. km. of forest cover since the 1995 assessment. By any estimate the area under forest is far below the national policy goals and many areas nominally under forest are being used for non-forest purposes. Similarly 'uncultivated lands' such as permanent pastures, miscellaneous tree crops, cultivable wastes and fallow is subject to increasing competition from uses other than feeding livestock.
- Land use intensity *i.e.* fraction of net sown area to total geographical area increase from 36 percent in 1950-51 to 40.5 percent in 1960-61 and 43 percent by 1970-71 where it has since stabilized.
- Since early fifties the net area sown was expanded rapidly at first but at a diminishing rate since 1970 to reach approximately 142 million ha at present. During 1950s and 1960s areas under aquaculture expanded substantially as the fallows were reduced and cultivable wastes were put under the plough. The net area sown increased from 119 million ha in 1950-51 to 133 million ha by 1960-61 and further to 140 million ha by 1970-71. Fallow lands declined from 28 million ha in 1950-51 to 20 million ha by 1970-71. Cultivable wastelands declined from 23 to 17.5 million ha.

- The area sown to fodder crops is not recorded. Information available from other sources provide an estimate ranging from 4 to 5.5 percent of the net sown area and suggest that the area under fodder crops will have to increase to 10 percent or more to support increasing livestock based activity.
- The growth of livestock population is an important source of competition for land.
- While the contribution of increased area in the growth of aquaculture has declined over time, that of productivity has increased. The yield of all crops grew at 1.5 percent per annum between early 1950s and mid 1960s. the pace accelerated to 1.7 percent in the 1970s and then to 3 percent per annum between early 1980s and mid 90s. Unlike the gains in area, which benefited non-foodgrains, the gains in productivity accrued mostly to foodgrains.

The pressure on India's land and water resources is seriously threatening native plant and animal diversity. India has uniquely rich and diverse genetic base. With increasing aquaculture and economic development the genetic pool is declining. This decline, if unchecked and poorly managed can have unforeseen and adverse consequences for the sustainability of aquaculture of the region.

AQUACULTURE IN THE CHANGING WORLDWIDE SCENARIO

Steady globalization of trade has profound implications for future agricultural development. The diversity of India's agro-ecological setting, high bio-diversity and relatively low cost of labour provide potential for agricultural competitiveness in a globalized economy. It is expected that with increasing globalization of markets over the years there will be demands for agricultural intensification.

This will also be favoured because of greater backward and forward linkages between aquaculture and food industry. Therefore, increase in production and productivity are bound to be strategically important to economy. Intensification will not only favour alleviation of rural poverty but will also improve resource conservation particularly in the small farming sector where farmers can be encouraged to take up organized production of high value crops such as fruits, specialty vegetables, flowers medicinal and aromatic herbs etc. Stronger demands for crops of the small farmers' will not only improve incomes and welfare but will also make investments in technology and resource conservation more attractive. The General Agreement on Tariff and Trade (GATT) and liberalization of global trade is bound to have impact on future land use and production pattern. Understanding the local, national and international environment under which agricultural production is taking shape will be crucial in developing our own strategies.

TECHNOLOGICAL NEEDS AND FUTURE AQUACULTURE

It is apparent that the tasks of meeting the consumption needs of the projected population are going to be more difficult given the higher productivity base than in 1960s. There is also a growing realization that previous strategies of generating and promoting technologies have contributed to serious and widespread problems of environmental and natural resource degradation. This implies that in future the technologies that are developed and promoted must result not only in increased productivity level but also ensure that the quality of natural resource base is preserved and enhanced. In short, they lead to sustainable improvements in agricultural production. Productivity gains during the 'Green Revolution' era were largely confined to relatively well endowed areas. Given the wide range of agroecological setting and producers, Indian aquaculture is faced with a great diversity of needs, opportunities and prospects. Future growth needs to be more rapid, more widely distributed and better targeted. Responding to these challenges will call for more efficient and sustainable use of increasingly scarce land water and germplasm resources.

Technical solutions required to solve problems will be increasingly location-specific and matched to the huge agroecological/climatic diversity. Detailed indigenous knowledge and greater skills in blending modern and traditional technologies to enhance productive efficiency will be more than ever before, key to the farming success and sectoral growth. Most technological solutions will have to be generated and adapted locally to make them compatible with socio-economic conditions of farming community. New technologies are needed to push the yield frontiers further, utilize inputs more efficiently and diversify to more sustainable and higher value cropping patterns. These are all knowledge intensive technologies that require both a strong research and extension system and skilled farmers but also a reinvigorated interface where the emphasis is on mutual exchange of information bringing advantages to all. At the same time potential of less favoured areas must be better exploited to meet the targets of growth and poverty alleviation. These challenges have profound implications for products of agricultural research. The way they are transferred to the farmers and indeed the way research is organized and conducted.

One thing is, however, clear – the new generation of technologies will have to be much more site specific, based on high quality science and a heightened opportunity for end user participation in the identification of targets. These must be not only aimed at increasing farmers' technical knowledge and understanding of science based aquaculture but also taking advantage of opportunities for full integration with indigenous knowledge. It will also need to take on the challenges of incorporating the socio-economic context and role of markets. With the passage of time and accelerated by macro-economic reforms undertaken in recent years, the Institutional arrangements as well as the mode of functions of bodies responsible for providing technical underpinning to agricultural growth are proving increasingly inadequate. Changes are needed urgently to respond to new demands for agricultural technologies from several directions. Increasing pressure to maintain and enhance the integrity of degrading natural resources, changes in demands and opportunities arising from economic liberalization, unprecedented opportunities arising from advances in biotechnology, information revolution and most importantly the need and urgency to reach the poor and disadvantaged who have been by passed by the green revolution technologies. Another important implication of increasing globalization relates to the need for greater attention to the quality of produce and products both for the domestic and the foreign markets. This would imply that production must be tuned to actual rapidly changing product demand. Such adaptation to global markets would require state of the art research, which can be achieved only by setting global standards of research, focus on well defined priorities and mechanisms which permit close interaction of farmers with researchers, the private sector and markets.

Extension Strategies

Since early fifties a number of public by funded agricultural development programmes have been sponsored. These have included programmes like the National Extension Service (NES) Blocks in 1953, the Intensive Agricultural District Programme (IADP) in 1961-62, the Intensive Agricultural Area Programme (IAAP) 1964-65, the High Yielding Variety (HYV) programme 1966-67 and the Small and Marginal Farmers' Development Programmes (SMFDP) in 1969-70. Though these programmes had a perceptible impact the efforts did not get replicated over different areas and categories of farmers. In mid seventies based on pilot level project in Rajasthan Canal and Chambal command area a 'Training and Visit' (T&V) system of extension was promoted in different states. Extension efforts of the Indian Council of Agricultural Research through its research Institutes and the State Agricultural University were largely limited to demonstration of new technologies through such programmes as National Demonstration Project, Operational Research Project, the Lab to Land Programme and the Krishi Vigyan Kendras. However, there appears much to be desired in the way that extension programmes are conceived and implemented. At present extension programmes are implemented in largely a top-down fashion leaving little scope for localized planning and action. Farmers are almost passive receivers and their involvement in the process of technology generation and adoption is almost absent. Extension services, at present, are almost exclusively in the public sector domain and there is no effort or institutional support for other operators *e.g.* the NGOs, the corporate bodies etc. Extension programmes sponsored by the government operate largely in isolation and there appears a strong need to view the extension programmes as an integral part of the research and development process.

6

Human Population and Environmental Issues

POPULATION GROWTH, VARIATION AMONG NATIONS

Our global human population, 6 billion at present, will cross the 7 billion mark by 2015. The needs of this huge number of human beings cannot be supported by the Earth's natural resources, without degrading the quality of human life. In the near future, fossil fuel from oil fields will run dry. It will be impossible to meet the demands for food from existing agro systems. Pastures will be overgrazed by domestic animals and industrial growth will create ever-greater problems due to pollution of soil, water and air.

Seas will not have enough fish. Larger ozone holes will develop due to the discharge of industrial chemicals into the atmosphere, which will affect human health. Global warming due to industrial gases will lead to a rise in sea levels and flood all low-lying areas, submerging coastal agriculture as well as towns and cities. Water 'famines' due to the depletion of fresh water, will create unrest and eventually make countries go to war. The control over regional biological diversity, which is vital for producing new medicinal and industrial products, will lead to grave economic conflicts between biotechnologically advanced nations and the biorich countries.

Degradation of ecosystems will lead to extinction of thousands of species, destabilizing natural ecosystems of great value. These are only some of the environmental problems related to an increasing human population and more intensive use of resources that we are likely to face in future. These effects can be averted by creating a mass environmental awareness movement that will bring about a change in people's way of life.

Increase in production per capita of agricultural produce at a global level ceased during the 1980's. In some countries, food shortage has become a permanent feature. Two of every three children in South Africa are underweight. In other regions famines due to drought have become more frequent. Present development strategies have not been able to successfully address these problems related to hunger and malnutrition.

On the other hand, only 15% of the world's population in the developed world is earning 79% of income! Thus the disparity in the extent of per capita resources that are used by people who live in a 'developed' country as against those who live in a 'developing' country is extremely large. Similarly, the disparity between the rich and the poor in India is also growing.

The increasing pressures on resources place great demands on the in-built buffering action of nature that has a certain ability to maintain a balance in our environment. However, current development strategies that essentially lead to short-term gains have led to a breakdown of our Earth's ability to replenish the resources on which we depend.

GLOBAL POPULATION GROWTH

The world population is growing by more than 90 million per year, of which 93% is in developing countries. This will essentially prevent their further economic 'development'. In the past, population growth was a gradual phenomenon and the Earth's ability to replenish resources was capable of adjusting to this increase. In the recent past, the escalation in growth of human numbers has become a major cause of our environmental problems. Present projections show that if our population growth is controlled, it will still grow to 7.27 billion by 2015.

However, if no action is taken it will become a staggering 7.92 billion. Human population growth increased from: 1 to 2 billion, in 123 years. 2 to 3 billion, in 33 years. 3 to 4 billion, in 14 years. 4 to 5 billion, in 13 years. 5 to 6 billion, in 11 years. It is not the census figures alone that need to be stressed, but an appreciation of the impact on natural resources of the rapid escalation in the rate of increase of human population in the recent past. The extent of this depletion is further increased by affluent societies that consume per capita more energy and resources, that less fortunate people.

This is of great relevance for developing a new ethic for a more equitable distribution of resources. In the first half of the 1900s human numbers were growing rapidly in most developing countries such as India and China. In some African countries the growth was also significant. In contrast, in the developed world population growth had slowed down. It was appreciated that the global growth rate was depleting the Earth's resources and was a direct impediment to human development.

Several environmental ill-effects were linked with the increasing population of the developing world. Poverty alleviation Programmes failed, as whatever was done was never enough as more and more people had to be supported on Earth's limited resources. In rural areas population growth led to increased fragmentation of farm land and unemployment. In the urban sector it led to inadequate housing and an increasing level of air pollution from traffic, water pollution from sewage, and an inability to handle solid waste. By the 1970s most countries in the developing world had realised that if they had to develop their economics and improve the lives of their citizens they would have to curtail population growth.

Though population growth shows a general global decline, there are variations in the rate of decline in different countries. By the 1990s the growth rate was decreasing in most countries such as China and India. The decline in the 90s was greatest in India. However, fertility continues to remain high in sub Saharan African countries. There are cultural, economic, political and demographic reasons that explain the differences in the rate of population control in different countries. It also varies in different parts of certain countries and is linked with community and/or religious thinking. Lack of Government initiatives for Family Welfare Programme and a limited access to a full range of contraceptive measures are serious impediments to limiting population growth in several countries.

POPULATION EXPLOSION–FAMILY WELFARE PROGRAMME

In response to our phenomenal population growth, India seriously took up an effective Family Planning Programme which was renamed the Family Welfare Programme. Slogans such as 'Hum do hamare do' indicated that each family should not have more than two children. It however has taken several decades to become effective. At the global level by the year 2000,600 million, or 57% of women in the reproductive age group, were using some method of contraception.

However the use of contraceptive measures is higher in developed countries–68%, and lower in developing countries-55%. Female sterilization is the most popular method of contraception used in developing countries at present. This is followed by the use of oral contraceptive pills and, intrauterine devices for women, and the use of condoms for men. India and China have been using permanent sterilization more effectively than many other countries in the developing world. The best decision for the method used by a couple depends on a choice that they make for themselves.

This must be based on good advice from doctors or trained social workers who can suggest the full range of methods available for them to choose from. Informing the public about the various contraceptive measures that are available is of primary importance. This must be done actively by Government Agencies such as Health and Family Welfare, as well as Education and Extension workers. It is of great importance for policy makers and elected representatives of the people–Ministers, MPs, MLAs at Central and State levels–to understand the great and urgent need to support Family Welfare. The media must keep people informed about the need to limit family size and the ill effects of a growing population on the worlds resources. The decision to limit family size depends on a couple's background and education. This is related to Government Policy, the effectiveness of Family Welfare Programmes, the educational level, and information levels in mass communication. Free access to Family Welfare information provided through the Health Care System, is in some cases unfortunately counteracted by cultural attitudes.

Frequently misinformation and inadequate information are reasons why a family does not go in for limiting its size. The greatest challenge the world now faces is how to supply its exploding human population with the resources it needs. It is evident that without controlling human numbers, the Earth's resources will be rapidly exhausted. In addition economically advanced countries and rich people in poorer countries use up more resources than they need.

As population expands further, water shortages will become acute. Soil will become unproductive. Rivers, lakes and coastal waters will be increasingly polluted. Water related diseases already kill 12 million people every year in the developing world. By 2025, there will be 48 countries that are starved for water. Air will become increasingly polluted.

Air pollution already kills 3 million people every year. The first 'green revolution' in the '60s produced a large amount of food but has led to several environmental problems. Now, a new green revolution is needed, to provide enough food for our growing population, that will not damage land, kill rivers by building large dams, or spread at the cost of critically important forests, grasslands and wetlands. The world's most populous regions are in coastal areas.

These are critical ecosystems and are being rapidly destroyed. Global climate change is now a threat that can affect the very survival of high population density coastal communities. In the sea, fish populations are suffering from excessive fishing. Once considered an inexhaustible resource, over fishing has depleted stocks extremely rapidly. It will be impossible to support further growth in coastal populations on existing fish reserves. Human populations will inevitably expand from farm lands into the remaining adjacent forests. Many such encroachments in India have been regularised over the last few decades.

But forest loss has long-term negative effects on water and air quality and the loss of biodiversity is still not generally seen as a major deterrent to human well-being. The extinction of plant and animal species resulting from shrinking habitats threatens to destroy the Earth's living web of life. Energy use is growing, both due to an increasing population, and a more energy hungry lifestyle that increasingly uses consumer goods that require large amounts of energy for their production, packaging, and transport. Our growing population also adds to the enormous amount of waste. With all these linkages between population growth and the environment, Family Welfare Programmes have become critical to human existence.

Planning for the Future

How Governments and people from every community meet challenges such as limiting population size, protecting the natural environment, change their consumer oriented attitudes, reduce habits that create excessive waste, elevates poverty and creates an effective balance between conservation and development will determine the worlds future.

The Urban Challenge

Population increases will continue in urban centers in the near future. The UN has shown that by 2025 there will be 21 "megacities" most of which will be situated in developing countries. Urban centers are already unable to provide adequate housing, services such as water and drainage systems, growing energy needs, or better opportunities for income generation.

Methods of Sterilization

India's Family Welfare Programme has been fairly successful but much still needs to be achieved to stabilize our population. The most effective measure is the one most suited to the couple once they have been offered all the various options that are available. The Family Welfare Programme advocates a variety of measures to control population.

Permanent methods or sterilisation are done by a minor surgery. Tubectomy in females is done by tying the tubes that carry the ovum to the uterus. Male sterilization or vasectomy, is done by tying the tubes that carry the sperm. Both are very simple procedures, done under local anesthesia, are painless and patients have no post operative problems.

Vasectomy does not cause any loss in the male's sexual ability but only arrests the discharge of sperm. There are several methods of temporary birth control. Condoms are used by males to prevent sperms from fertilizing the ovum during intercourse. Intrauterine devices (Copper Ts) are small objects which can be placed by a doctor in the uterus so that the ovum cannot be implanted, even if fertilized.

They do not disturb any functions in the woman's life or work. Oral contraceptive tablets (pills) and injectable drugs are available that prevent sperms from fertilizing the ovum. There are also traditional but less reliable methods of contraception such as abstinence of the sexual act during the fertile period of the women's cycle and withdrawal during the sexual act.

Urbanization

In 1975 only 27% of the people in the developing world lived in urban areas. By 2000 this had grown to 40% and by 2030 well informed estimates state that this will grow to 56%. The developed world is already highly urbanized with 75% of its population living in the urban sector.

Urban Environments

Nearly half the world's population now lives in urban areas. The high population density in these areas leads to serious environmental issues. Today, more than 290 million people live in towns and cities in India. There were 23 metros in India in 1991, which grew to 40 by 2001. Urban population growth is both due to migration of people to towns and cities from the rural sector in search of better job options as well as population growth within the city.

As a town grows into a city it not only spreads outwards into the surrounding agricultural land or natural areas such as forests, grasslands and wetlands but also grows skywards with high rise buildings. The town also loses its open spaces and green cover unless these are consciously preserved. This destroys the quality of life in the urban area. Good urban planning is essential for rational landuse planning, for upgrading slum areas, improving water supply and drainage systems, providing adequate sanitation, developing effective waste water treatment plants and an efficient public transport system.

Unplanned and haphazard growth of urban complexes has serious environmental impacts. Increasing solid waste, improper garbage disposal and air and water pollution are frequent side effects of urban expansions.

While all these issues appear to be under the preview of local Municipal Corporations, better living conditions can only become a reality if every citizen plays an active role in managing the environment. This includes a variety of "Dos and Don'ts" that should become an integral part of our personal lives.

Apart from undertaking actions that support the environment every urban individual has the ability to influence a city's management. He or she must see that the city's natural green spaces, parks and gardens are maintained, river and water fronts are managed appropriately, roadside tree cover is maintained, hill slopes are afforested and used as open spaces and architectural and heritage sites are protected. Failure to do this leads to increasing urban problems which eventually destroys a city's ability to maintain a healthy and happy lifestyle for its dwellers.

All these aspects are closely linked to the population growth in the urban sector. In many cities growth outstrips the planner's ability to respond to this in time for a variety of reasons.

2015	**Mega cities in India**	**Population (in millions) in 2001**	**Projection (in millions) for**
Mumbai	16.5	22.6	
Kolkata	13.3	16.7	
Delhi	13.0	20.9	

Small urban centers too will grow rapidly during the next decades and several rural areas will require reclassification as urban centers. India's urban areas will grow by a projected 297 million residents. In India people move to cities from rural areas in the hope of getting a better income. This is the 'Pull' factor. Poor opportunities in the rural sector thus stimulates migration to cities.

Loss of agricultural land to urbanisation and industry, the inability of governments to sustainably develop the rural sector, and a lack of supporting infrastructure in rural areas, all push people from the agricultural and natural wilderness ecosystems into the urban sector. As our development strategies have focused attention mostly on rapid industrial development and relatively few development options are offered for the agricultural rural sector, a shift of population is inevitable.

As population in urban centers grows, they draw on resources from more and more distant areas. The "Ecological footprint" corresponds to the land area necessary to supply natural resources and disposal of waste of a community. At present the average ecological footprint of an individual at the global level is said to be 2.3 hectares of land per capita.

But it is estimated that the world has only 1.7 hectares of land per individual to manage these needs sustainably. This is thus an unsustainable use of land. The pull factor of the urban centers is not only due to better job opportunities, but also better education, health care and relatively higher living standards. During the last few decades in India, improvements in the supply of clean water, sanitation, waste management, education and health care has all been urban centric, even though the stated policy has been to support rural development. Thus in reality, development has lagged behind in the rural sector that is rapidly expanding in numbers. For people living in wilderness areas in our forests and mountain regions, development has been most neglected. It is not appropriate to use the development methods used for other rural communities for tribal people who are dependent on collecting natural resources from the forests.

A different pattern of development that is based on the sustainable extraction of resources from their own surroundings would satisfy their development aspirations. In general the growing human population in the rural sector will only opt to live where they are if they are given an equally satisfying lifestyle.

The Wilderness–Rural-urban Linkage

The environmental stresses caused by urban individuals covers an 'ecological footprint' that goes far beyond what one expects. The urban sector affects the land at the fringes of the urban area and the areas from which the urban center pulls in agricultural and natural resources. Urban centers occupy 2% of the worlds' land but use 75% of the industrial wood. About 60% of the world's water is used by urban areas of which half irrigates food crops for urban dwellers, and one third goes to industry and the rest is used for household use and drinking water.

The impact that urban dwellers have on the environment is not obvious to them as it happens at distant places which supports the urban ecosystem with resources from agricultural and even more remote wilderness ecosystems.

Urban Poverty and the Environment

The number of poor people living in urban areas is rapidly increasing. A third of the poor people in the world live in urban centers. These people live in hutments in urban slums and suffer from water shortages and unsanitary conditions. In most cases while a slum invariably has unhygenic surroundings, the dwellings themselves are kept relatively clean. It is the 'common' areas used by the community that lacks the infrastructure to maintain a hygienic environment. During the 1990s countries that have experienced an economic crisis have found that poor urban dwellers have lost their jobs due to decreasing demands for goods, while food prices have risen. Well paid and consistent jobs are not as easily available in the urban centers at present as in the past few decades. One billion urban people in the world live in inadequate housing, mostly in slum areas, the majority of which are temporary structures.

However, low income groups that live in high rise buildings can also have high densities and live in poor unhygienic conditions in certain areas of cities. Illegal slums often develop on Government land, along railway tracks, on hill slopes, riverbanks, marshes, etc. that are unsuitable for formal urban development. On the riverbanks floods can render these poor people homeless. Adequate legal housing for the urban poor remains a serious environmental concern.

Urban poverty is even more serious than rural poverty, as unlike the rural sector, the urban poor have no direct access to natural resources such as relatively clean river water, fuelwood and Non- wood forest products. The urban poor can only depend on cash to buy the goods they need, while in the rural sector they can grow a substantial part of their own food. Living conditions for the urban poor are frequently worse than for rural poor. Both outdoor and indoor air pollution due to high levels of particulate matter and sulphur dioxide from industrial and vehicle emissions lead to high death rates from respiratory diseases. Most efforts are targeted at outdoor air pollution. Indoor air pollution due to the use of fuel wood, waste material, coal, etc. in 'chulas' is a major health issue. This can be reduced by using better designed 'smokeless' chulas, hoods and chimneys to remove indoor smoke. With the growing urban population, a new crisis of unimaginable proportions will develop in the next few years. Crime rates, terrorism, unemployment, and serious environmental health related issues can be expected to escalate. This can only be altered by stabilizing population growth on a war footing.

ENVIRONMENT AND HUMAN HEALTH

Environment related issues that affect our health have been one of the most important triggers that have led to creating an increasing awareness of the need for better environmental management. Changes in our environment induced by human activities in nearly every sphere of life have had an influence on the pattern of our health. The assumption that human progress is through economic growth is not necessarily true. We expect urbanization and industrialization to bring in prosperity, but on the down side, it leads to diseases related to overcrowding and an inadequate quality of drinking water, resulting in an increase in waterborne diseases such as infective diarrhoea and air borne bacterial diseases such as tuberculosis. High-density city traffic leads to an increase in respiratory diseases like asthma. Agricultural pesticides that enhanced food supplies during the green revolution have affected both the farm worker and all of us who consume the produce.

Modern medicine promised to solve many health problems, especially associated with infectious diseases through antibiotics, but bacteria found ways to develop resistant strains, frequently even changing their behaviour in the process, making it necessary to keep on creating newer antibiotics. Many drugs have been found to have serious side effects. At times the cure is as damaging as the disease process itself.

Thus development has created several long-term health problems. While better health care has led to longer life spans, coupled with a lowered infant mortality, it has also led to an unprecedented growth in our population which has negative implications on environmental quality. A better health status of society will bring about a better way of life only if it is coupled with stabilising population.

Environmental Health

Environmental health, as defined by WHO, comprises those aspects of human health, including quality of life, that are determined by physical, chemical, biological, social, and psychosocial factors in the environment. It also refers to the theory and practice of assessing, correcting, controlling, and preventing those factors in the environment that adversely affect the health of present and future generations. Our environment affects health in a variety of ways.

Climate and weather affect human health. Public health depends on sufficient amounts of good quality food, safe drinking water, and adequate shelter. Natural disasters such as storms, hurricanes, and floods still kill many people every year. Unprecedented rainfall trigger epidemics of malaria and water borne diseases. Global climate change has serious health implications. Many countries will have to adapt to uncertain climatic conditions due to global warming. As our climate is changing, we may no longer know what to expect. There are increasing storms in some countries, drought in others, and a temperature rise throughout the world. The El Niño winds affect weather worldwide. The El Niño event of 1997/98 had serious impacts on health and well-being of millions of people in many countries. It created serious drought, floods, and triggered epidemics. New strategies must be evolved to reduce vulnerability to climate variability and changes.

Economic inequality and environmental changes are closely connected to each other. Poor countries are unable to meet required emission standards to slow down climate change. The depletion of ozone in the stratosphere (middle atmosphere) also has an important impact on global climate and in turn human health, increasing the amount of harmful ultraviolet radiation that reaches the Earth's surface. This results in diseases such as skin cancer.

Bhopal Gas Tragedy

The siting of industry and relatively poor regulatory controls leads to ill health in the urban centers. Accidents such as the Bhopal gas tragedy in 1984 where Union Carbide's plant accidentally released 30 tones of methyl isocyanate, used in the manufacture of pesticides, led to 3,330 deaths and 1.5 lakh injuries to people living in the area. Development strategies that do not incorporate ecological safeguards often lead to ill health. Industrial development without pollution control and traffic congestion affect the level of air pollution in many cities. On the other hand, development strategies that can promote health invariably also protect the environment. Thus environmental health and human health are closely interlinked. An improvement in health is central to sound environmental management. However this is rarely given sufficient importance in planning development strategies.

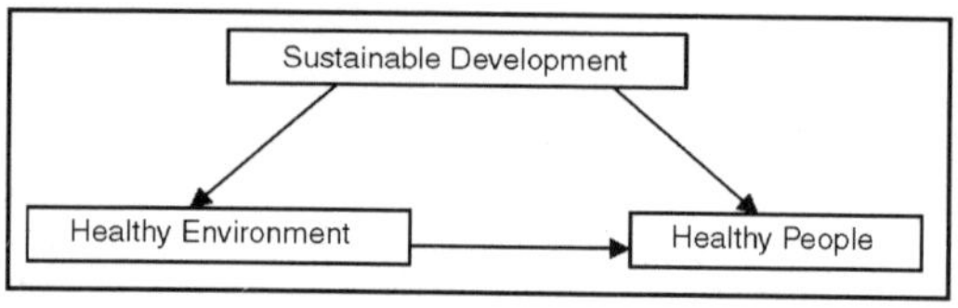

Examples of the Linkages

- Millions of children die every year due to diarrhoea from contaminated water or food. An estimated 2000 million people are affected by these diseases and more than 3 million children die each year from waterborne diseases across the world. In India, it is estimated that every fifth child under the age of 5 dies due to diarrhoea. This is a result of inadequate environmental management and is mainly due to inadequate purification of drinking water. Wastewater and/or sewage entering water sources without being treated leads to continuous gastrointestinal diseases in the community and even sporadic large epidemics. Large numbers of people in tropical countries die of malaria every year and millions are infected. An inadequate environmental management of stagnant water, which forms breeding sites of Anopheles mosquitoes is the most important factor in the spread of malaria. The resurgence of malaria in India is leading to cerebral malaria that affects the brain and has a high mortality.

- Millions of people, mainly children, have poor health due to parasitic infections, such as amoebiasis and worms. This occurs from eating infected food, or using poor quality water for cooking food. It is estimated that 36% of children in low-income countries and 12% in middle income countries are malnourished. In India, about half the children under the age of four are malnourished and 30% of newborns are significantly underweight.
- Hundreds of millions of people suffer serious respiratory diseases, including lung cancer and tuberculosis, from crowded homes and public places. Motor vehicle exhaust fumes, industrial fumes, tobacco smoke and cooking food on improper 'chulas', contribute to respiratory diseases.
- Millions of people are exposed to hazardous chemicals in their workplace or homes that lead to ill health due to industrial products where controls are not adhered to.
- Tens of thousands of people in the world die due to traffic accidents due to inadequate management of traffic conditions. Poor management at the accident site, and inability to reach a hospital within an hour causes a large number of deaths, especially from head injuries.
- Basic environmental needs such as clean water, clean air and adequate nutrition which are all related to environmental goods and services do not reach over 1000 million people living in abject poverty.
- Several million people live in inadequate shelters or have no roof over their heads especially in urban settings. This is related to high inequalities in the distribution of wealth and living space.
- Population growth and the way resources are being exploited and wasted, threatens environmental integrity and directly affects health of nearly every individual.
- Health is an outcome of the interactions between people and their environment. Better health can only come from a more sustainable management of the environment.

Important Strategic Concerns

- The world must address people's health care needs and sustainable use of natural resources, which are closely linked to each other.
- Strategies to provide clean pottable water and nutrition to all people is an important part of a healthy living environment.
- Providing clean energy sources that do not affect health is a key to reducing respiratory diseases.
- Reducing environmental consequences of industrial and other pollutants such as transport emissions can improve the status of health.

- Changing patterns of agriculture away from harmful pesticides, herbicides and insecticides which are injurious to the health of farmers and consumers by using alternatives such as Integrated Pest Management and non-toxic biopesticides can improve health of agricultural communities, as well as food consumers.
- Changing industrial systems into those that do not use or release toxic chemicals that affect the health of workers and people living in the vicinity of industries can improve health and environment.
- There is a need to change from using conventional energy from thermal power that pollutes air and nuclear power that can cause serious nuclear disasters to cleaner and safer sources such as solar, wind and ocean power that do not affect human health. Providing clean energy is an important factor that can lead to better health.
- The key factors are to control human population and consume less environmental goods and services which could lead to 'health for all'. Unsustainable use of resources by an ever growing population leads to unhealthy lives. Activities that go on wasting environmental goods and destroying its services by producing large quantities of Non-degradable wastes, leads to health hazards.
- Poverty is closely related to health and is itself a consequence of improper environmental management. An inequitable sharing of natural resources and environmental goods and services, is linked to poor health. The world's consumption of non-renewable resources is concentrated in the developed countries. Rich countries consume 50 times more per capita than people in less developed countries. This means that developed countries also generate proportionately high quantities of waste material, which has serious health concerns.

Definition of Health Impact Assessment (HIA) by WHO

Health impact assessment is a combination of procedures, methods and tools by which a policy, programme or project may be judged as to its potential effects on the health of a population, and the distribution of those effects within the population.

Climate and Health

Human civilizations have adapted mankind to live in a wide variety of climates. From the hot tropics to the cold arctic, in deserts, marshlands and in the high mountains. Both climate and weather have a powerful impact on human life and health issues. Natural disasters created by extremes of weather (heavy rains, floods, hurricanes) which occur over a short period of time, can severely affect health of a community. Poor people are more vulnerable to the health impacts of climate variability than the rich.

Of approximately 80,000 deaths which occur world-wide each year as a result of natural disasters about 95% are in poor countries. In weather-triggered disasters hundreds of people and animals die, homes are destroyed, crops and other resources are lost. Public health infrastructure, such as sewage disposal systems, waste management, hospitals and roads are damaged. The cyclone in Orissa in 1999 caused 10,000 deaths. The total number of people affected was estimated at 10 to 15 million!

Human physiology can adapt to changes in weather, within certain limits. However, marked short-term fluctuations in weather lead to serious health issues. Heat waves cause heat-related illness and death (*e.g.*, heat stroke). The elderly and persons with existing heart or respiratory diseases are more vulnerable. Heat waves in India in 1998 were associated with many deaths. Climate plays an important role in vector-borne diseases transmitted by insects such as mosquitoes. These disease transmitters are sensitive to direct effects of climate such as temperature, rainfall patterns and wind. Climate affects their distribution and abundance through its effects on host plants and animals. Malaria transmission is particularly sensitive to weather and climate. Unusual weather conditions, for example a heavy downpour, can greatly increase the mosquito population and trigger an epidemic.

In the desert and at highland fringes of malarious areas, malaria transmission is unstable and the human population lacks inherent protective immunity. Thus, when weather conditions (rainfall and temperature) favour transmission, serious epidemics occur in such areas. Fluctuations in malaria over the years have been linked to changes in rainfall associated with the El Niño cycle.

Infectious Diseases

Many infectious diseases have re-emerged with a vengeance. Loss of effective control over diseases such as malaria and tuberculosis, have led to a return of these diseases decades after being kept under stringent control. Other diseases were not known to science earlier and seem to have suddenly hit our health and our lives during the last few decades. AIDS, due to the Human Immunodeficiency Virus (HIV) caused through sexual transmission and Severe Acute Respiratory Syndrome (SARS) are two such examples.

While these cannot be directly related to environmental change, they affect the environment in which we live by forcing a change in lifestyles and behaviour patterns. For example the SARS outbreak prevented people from several countries from traveling to other countries for months, severely affecting national economies, airline companies and the tourism industry. Why have infectious diseases that were related to our environment that were under control suddenly made a comeback? Diseases such as tuberculosis have been effectively treated with anti-tubercular drugs for decades.

These antibiotics are used to kill off the bacteria that causes the disease. However nature's evolutionary processes are capable of permitting bacteria to mutate by creating new genetically modified strains. Those that change in a way so that they are not affected by the routinely used antibiotics begin to spread rapidly. This leads to a reemergence of the disease. In the case of tuberculosis this has led to multi-drug resistant tuberculosis. This is frequently related to HIV which reduces an individual's immunity to bacteria such as mycobacterium tuberculosis that causes tuberculosis.

The newer broad-spectrum antibiotics, antiseptics, disinfectants, and vaccines once thought of as the complete answer to infectious diseases have thus failed to eradicate infectious diseases. Experts in fact now feel that these diseases will be the greatest killers in future and not diseases such as malignancy or heart disease.

While antibiotic resistance is a well-known phenomenon there are other reasons for the reemergence of diseases. Overcrowding due to the formation of slums in the urban setting leads to several health hazards, including easier spread of respiratory diseases. Inadequate drinking water quality and poor disposal of human waste due to absence of a closed sewage system and poor garbage management are all urban health issues. This has led to a comeback of diseases such as cholera and an increased incidence of diarrhea and dysentery as well as infectious hepatitis (jaundice). With increasing global warming disease patterns will continue to change. Tropical diseases spread by vectors such as the mosquito will undoubtedly spread malaria further away from the equator. Global warming will also change the distribution of dengue, yellow fever, encephalitis, etc.

Warmer wetter climates could cause serious epidemics of diseases such as cholera. El Nino which causes periodic warming is likely to affect rodent populations. This could bring back diseases such as the plague. Globalisation and infectious disease Globalization is a world-wide process which includes the internationalization of communication, trade and economic organization. It involves parallel changes such as rapid social, economic and political adjustments. Whilst globalization has the potential to enhance the lives and living standards of certain population groups, for poor and marginalized populations in both the non-formal as well as formal economic sectors of developing countries, globalization enhances economic inequalities.

Tuberculosis

Tuberculosis (TB) kills approximately 2 million people each year. In India the disease has reemerged and is now more difficult to treat. A global epidemic is spreading and becoming more lethal. The spread of HIV/AIDS and the emergence of multidrug-resistant tuberculosis is contributing to the increasing morbidity of this disease. In 1993, the World Health Organization (WHO) declared that tuberculosis had become a global emergency. It is estimated that between 2002 and 2020, approximately 1000 million people will be newly infected, over 150 million people will get sick, and 36 million will die of TB–if its control is not rapidly strengthened. TB is a contagious disease that is spread through air. Only people who are sick with pulmonary TB are infectious. When infectious people cough, sneeze, talk or spit, they emit the tubercle bacilli into the air. When a healthy person inhales these, he gets infected by the disease. Symptoms include prolonged fever, coughing spells and weight loss. It is estimated that, left untreated, each patient of active tuberculosis will infect on an average between 10 to 15 people every year. But people infected with TB will not necessarily get sick with the disease.

The immune system can cause the TB bacilli, which is protected by a thick waxy coat, to remain dormant for years. When an individual's immune system is weakened, the chances of getting active TB are greater.

- Nearly 1% of the world's population is newly infected with TB each year.
- It is estimated that overall, one third of the world's population is likely to be infected with the tuberculosis bacillus at some point in time.
- Five to ten per cent of people who are infected with TB (but who are not infected with HIV) become sick or infectious at some time during their life. (WHO, 2002). Factors Contributing to the rise in tuberculosis
- TB kills about 2 million people each year (including persons infected with HIV).
- More than 8 million people become sick with TB each year, one person in the world every second!
- About 2 million TB cases per year occur in sub-Saharan Africa. This number is rising rapidly as a result of the HIV/AIDS epidemic.
- Around 3 million TB cases per year occur in South-east Asia.
- Over a quarter of a million TB cases per year occur in Eastern Europe.

Tuberculosis in India

There are 14 million TB patients in India, account for one third of the global cases of TB. Everyday 20,000 Indians contract TB and more than 1,000 die due to this chronic illness. TB attacks working adults in the age group of 15 to 50 years.

HIV is accelerating the spread of TB

The link between HIV and TB affects a large number of people, each disease speeding the other's progress. HIV weakens the immune system. Someone who is HIV-positive and infected with TB is many times more likely to become seriously sick with TB rather than someone infected with TB who is HIV-negative. Tuberculosis is a leading cause of death among people who are HIV-positive, accounting for about 11% of AIDS deaths worldwide.

Poorly Managed TB Programmes are Threatening to make TB Incurable

Until 50 years ago, there were no drugs to cure tuberculosis. Now, strains that are resistant to one or more anti-TB drugs have emerged. Drugresistant tuberculosis is caused by inconsistent or partial treatment, when patients do not take all their drugs regularly for the required period, when doctors or health workers prescribe inadequate treatment regimens or where the drug supply is unreliable. From a public health perspective, poorly supervised or incomplete treatment of TB is worse than no treatment at all. When people fail to complete standard treatment regimens, or are given the wrong treatment regimen, they may remain infectious. The bacilli in their lungs may develop resistance to anti-TB drugs. People they infect will have the same drug-resistant strain. While drug-resistant TB is treatable, it requires extensive chemotherapy that is often very expensive and is also more toxic to patients.

Malaria

Malaria is a life-threatening parasitic disease transmitted by mosquitoes. The cause of malaria, a single celled parasite called plasmodium, was discovered in 1880.Later it was found that the parasite is transmitted from person to person through the bite of a female Anopheles mosquito, which requires blood for the growth of her eggs. Today approximately 40% of the world's population, mostly those living in the world's poorest countries, risk getting malaria.

The disease was once more widespread but it was successfully eliminated from many countries with temperate climates during the mid 20th century. Today malaria has returned and is found throughout the tropical and sub-tropical regions of the world and causes more than 300 million acute illnesses and at least one million deaths annually (WHO). There are several types of human malaria. Falciparum malaria is the most dangerous type of infection and is most common in Africa south of the Sahara, where it accounts for extremely high mortality rates. There are also indications of the spread of P. falciparum malaria in India and it has reappeared in areas where it had been eliminated. The malaria parasite enters the human host when an infected Anopheles mosquito bites an individual. Inside the human host, the parasite undergoes a series of changes as part of its complex life-cycle.

Its various stages allow plasmodia to evade the immune system, infect the liver and red blood cells, and finally develop into a form that is able to infect a mosquito again when it bites an infected person. Inside the mosquito, the parasite matures until it reaches the sexual stage where it can again infect a human host when the mosquito takes her next blood meal, 10 or more days later. Malaria symptoms appear about 9 to 14 days after the mosquito bite, although this varies with different plasmodium species. Malaria produces high fever, headache, vomiting and body ache. If drugs are not available for treatment, or the parasites are resistant to them, the infection can progress rapidly to become life-threatening. Malaria can kill by infecting and destroying red blood cells (anaemia) and by clogging the capillaries that carry blood to the brain (cerebral malaria) or other vital organs.

Malaria parasites are developing unacceptable levels of resistance to drugs. Besides this, many insecticides are no longer useful against mosquitoes transmitting the disease. Good environmental management by clearing pools of stagnant water during the monsoons is effective in reducing the number of mosquitoes. Mosquito nets treated with insecticide reduce malaria transmission and child deaths. Prevention of malaria in pregnant women, through measures such as Intermittent Preventive Treatment and the use of insecticide-treated nets (ITNs), results in improvement in maternal health, as well as infant health and survival. Prompt access to treatment with effective up-to-date medicines, such as artemisinin-based combination therapies (ACTs), saves lives. If countries can apply these and other measures on a wide scale and monitor them carefully, the burden of malaria on society will be significantly reduced.

Water-related Diseases

Water Supply, Sanitation and Hygiene Development

Among the main problems are a lack of priority given to this sector, lack of financial resources, erratic water supply and sanitation services, poor hygiene related behaviour patterns, and inadequate sanitation in public places such as schools, hotels, hospitals, health centers, etc. One of the most important aspects is a lack of environmental education and awareness that these disease processes are related to poor environment management in various sectors.

Providing access to sufficient quantities of safe water, the provision of facilities for a sanitary disposal of excreta, and introducing sound hygiene related behaviour can reduce the morbidity and mortality caused by these risk factors. Environmental Sanitation and Hygiene Development About 2.4 billion people globally live under highly unsanitary conditions. Poor hygiene and behaviour pattern increase the exposure to risk of incidence and spread of infectious diseases. Water improperly stored in homes is frequently contaminated by inadequate management at the household level. This can be easily reduced through education and awareness of how waterborne diseases are transmitted.

Health and Water Resources Development

An important aspect related to water-related diseases (in particular: water-related vectorborne diseases) is attributable to the way water resources are developed and managed. In many parts of the world the adverse health impacts of dam construction, irrigation development and flood control is related to increased incidence of malaria, Japanese encephalitis, schistosomiasis, lymphatic filariasis and other conditions. Other health issues indirectly associated with water resources development include nutritional status, exposure to agricultural pesticides and their residues.

Water Borne Diseases

Arid areas with rapidly expanding populations are already facing a crisis over water. Conservation of water and better management is an urgent need. The demand and supply balance is a vital part of developing sustainable use of water. This is being termed the 'Blue Revolution' and needs Governments, NGOs and people to work together towards a better water policy at International, National, State, regional and local levels. Locally good watershed management is a key to solving local rural problems. Present patterns of development are water hungry and water wasters. They do not address pollution and overuse. The linkages between managing water resources and health issues are have not been prioritised as a major source of environmental problems that require policy change, administrative capacity building and an increased financial support.

There are 4 Major types of Water Related Diseases

1. *Water borne diseases*: These are caused by dirty water contaminated by human and animal wastes, especially from urban sewage, or by chemical wastes from industry and agriculture. Some of these diseases, such as cholera and typhoid, cause serious epidemics. Diarrhoea, dysentery, polio, meningitis, and hepatitis A and E, are caused due to improper drinking water. Excessive levels of nitrates cause blood disorders when they pollute water sources. Pesticides entering drinking water in rural areas cause cancer, neurological diseases and infertility. Improving sanitation and providing treated drinking water reduces the incidence of these diseases.
2. *Water based diseases*: Aquatic organisms that live a part of their life cycle in water and another part as a parasite in man, lead to several diseases. In India, guinea worm affects the feet. Round worms live in the small intestine, especially of children.
3. *Water related vector diseases*: Insects such as mosquitoes that breed in stagnant water spread diseases such as malaria and filariasis. Malaria that was effectively controlled in India, has now come back as the mosquitoes have become resistant to insecticides. In addition, anti-malarial drugs are now unable to kill the parasites as they have become resistant to drugs. Change in climate is leading to the

formation of new breeding sites. Other vector born diseases in India include dengue fever and filariasis. Dengue fever carries a high mortality. Filariasis leads to fever and chronic swelling over the legs.

4. *Water scarcity diseases:* In areas where water and sanitation is poor, there is a high incidence of diseases such as tuberculosis, leprosy, tetanus, etc. which occur when hands are not adequately washed.

Arsenic in Drinking Water

Arsenic in drinking-water is a serious hazard to human health. It has attracted much attention since its recognition in the 1990s of its wide occurrence in wellwater in Bangladesh. It occurs less frequently in most other countries. The main source of arsenic in drinking water is arsenic-rich rocks through which the water has filtered. It may also occur because of mining or industrial activity in some areas. WHO has worked with other UN organizations to produce a state-of-the-art review on arsenic in drinking water.

Drinking water that is rich in arsenic leads to arsenic poisoning or arsenicosis. Excessive concentrations are known to occur in some areas. The health effects are generally delayed and the most effective preventive measure is supplying drinking water which is free of arsenic. Arsenic contamination of water is also due to industrial processes such as those involved in mining, metal refining, and timber treatment. Malnutrition may aggravate the effects of arsenic on blood vessels.

Water with high concentrations of arsenic if used over 5 to 20 years, results in problems such as colour changes on the skin, hard patches on the palms and soles, skin cancer, cancers of the bladder, kidney and lung, and diseases of the blood vessels of the legs and feet. It may also lead to diabetes, high blood pressure and reproductive disorders. Natural arsenic contamination occurs in Argentina, Bangladesh, Chile, China, India, Mexico, Thailand and the United States. In China (in the Province of Taiwan) exposure to arsenic leads to gangrene, known as 'black foot disease'.

Long term solutions for prevention of arsenicosis is based on providing safe drinking-water:

- Deeper wells are often less likely to be contaminated.
- Testing of water for levels of arsenic and informing users.
- Monitoring by health workers-people need to be checked for early signs of arsenicosis-usually by observing skin problems in areas where arsenic in known to occur.
- Health education regarding harmful effects of arsenicosis and how to avoid them.

Arsenic Poisoning–Bangladesh

More than half the population of Bangladesh is threatened by high levels of arsenic found in drinking water. This could eventually lead to an epidemic of cancers and other fatal diseases. Rezaul Morol, a young Bangladeshi man, nearly died from arsenic poisoning caused by drinking arsenic-laden well-water for several years. The doctor advised Rezaul to stop drinking contaminated water and eat more protein-rich food such as fish. Since then Rezaul feels a lot better and is happy that his skin is healing.

Diarrhoea

Though several types of diarrhoea which give rise to loose motions and dehydration occur all over the world, this is especially frequently observed in developing countries. It causes 4% of all deaths. In another 5% it leads to loss of health. It is caused by gastrointestinal infections which kill around 2.2 million people globally each year. Most of these are children in developing countries. The use of contaminated water is an important cause of this group of conditions. Cholera and dysentery cause severe, sometimes life threatening and epidemic forms of these diseases.

Effects on Health

Diarrhoea is the frequent passage of loose or liquid stools. It is a symptom of various gastrointestinal infections. Depending on the type of infection, the diarrhoea may be watery (for example in cholera caused by vibrio cholera) or passed with blood and mucous (in dysentery caused by an amoeba, E Histolitica). Depending on the type of infection, it may last a few days, or several weeks. Severe diarrhoea can become life threatening due to loss of excessive fluid and electrolytes such as Sodium and Potassium in watery diarrhoea.

This is particularly fatal in infants and young children. It is also dangerous in malnourished individuals and people with poor immunity. The impact of repeated diarrhoea on nutritional status is linked in a vicious cycle in children. Chemical or non-infectious intestinal conditions can also result in diarrhoea.

Causes of Diarrhoea

Diarrhoea is caused by several bacterial, viral and parasitic organisms. They are mostly spread by contaminated water. It is more common when there is a shortage of clean water for drinking, cooking and cleaning. Basic hygiene is important in its prevention. Water contaminated with human feces surrounding a rural water source, or from municipal sewage, septic tanks and latrines in urban centers, are important factors in the spread of these diseases.

Feces of domestic animals also contain microorganisms that can cause diarrhoea through water. Diarrhoea is spread from one individual to another due to poor personal hygiene. Food is a major cause of diarrhoea when it is prepared or stored in unhygienic conditions. Water can contaminate food such as vegetables during irrigation. Fish and seafood from polluted water is a cause of severe diarrhoea. The infectious agents that cause diarrhoea are present in our environment. In developed countries where good sanitation is available, most people get enough safe drinking water. Good personal and domestic hygiene prevents this disease which is predominantly seen in the developing world. About 1 billion people do not have access to clean water sources and 2.4 billion have no basic sanitation. In Southeast Asia, diarrhoea is responsible for 8.5% of all deaths. In 1998, diarrhoea was estimated to have killed 2.2 million people, most of whom were under 5 years of age.

Interventions: Key measures to reduce the number of cases of diarrhoea include:

- Access to safe drinking water.
- Improved sanitation.
- Good personal and food hygiene.
- Health education about how these infections spread.

Key measures to treat diarrhoea include:

- Giving more fluids than usual, (oral rehydration) with salt and sugar, to prevent dehydration.
- Continue feeding.
- Consulting a health worker if there are signs of dehydration or other problems.

In rural India, during the last decade public education through posters and other types of communication strategies has decreased infant mortality due to diarrhoea in several States. Posters depicting a child with diarrhoea being given water, salt and sugar solution to reduce death from dehydration has gone a long way in reducing both a serious condition requiring hospitalisation and intravenous fluids, as well as mortality.

Risks Due to Chemicals in Food

Food contaminated by chemicals is a major worldwide public health concern. Contamination may occur through environmental pollution of the air, water and soil. Toxic metals, PCBs and dioxins, or the intentional use of various chemicals, such as pesticides, animal drugs and other agrochemicals have serious consequences on human health. Food additives and contaminants used during food manufacture and processing adversely affects health.

Diseases spread by food: Some foodborne diseases though well recognized, have recently become more common. For example, outbreaks of salmonellosis which have been reported for decades, has increased within the last 25 years. In the Western hemisphere and in Europe, Salmonella serotype Enteritidis (SE) has become a predominant strain. Investigations of SE outbreaks indicate that its emergence is largely related to consumption of poultry or eggs.

While cholera has devastated much of Asia and Africa for years, its reintroduction for the first time in almost a century on the South American continent in 1991 is an example of a well recognised infectious disease reemerging in a region after decades. While cholera is often waterborne, many foods also transmit infection. In Latin America, ice and raw or underprocessed seafood are important causes for cholera transmission.

Infection with a specific type of Escherichia coli (E. coli) was first described in 1982. Subsequently, it has emerged rapidly as a major cause of bloody diarrhoea and acute renal failure. The infection is sometimes fatal, particularly in children. Outbreaks of infection, generally associated with beef, have been reported in Australia, Canada, Japan, United States, in various European countries, and in southern Africa. Outbreaks have also implicated alfalfa sprouts, unpasteurized fruit juice, lettuce, game meat (meat of wild animals) and cheese curd.

In 1996, an outbreak of Escherichia coli in Japan affected over 6,300 school children and resulted in 2 deaths. Listeria monocytogenes (Lm): The role of food in the transmission of this condition has been recognized recently. In pregnant women, infections with Lm causes abortion and stillbirth. In infants and persons with a poor immune system it may lead to septicemia (blood poisoning) and meningitis. The disease is most often associated with consumption of foods such as soft cheese and processed meat products that are kept refrigerated for a long time, because Lm can grow at low temperatures. Outbreaks of listeriosis have been reported from many countries, including Australia, Switzerland, France and the United States. Two recent outbreaks of Listeria monocytogenes in France in 2000 and in the USA in 1999 were caused by contaminated pork tongue and hot dogs respectively. Foodborne trematodes (worms) are increasing in South-east Asia and Latin America. This is related to a combination of intensive aquaculture production in unsanitary conditions, and consumption of raw or lightly processed fresh water fish and fishery products.

Foodborne trematodes can cause acute liver disease, and may lead to liver cancer. It is estimated that 40 million people are affected worldwide. Bovine Spongiform Encephalopathy (BSE), is a fatal, transmittable, neurodegenerative disease of cattle. It was first discovered in the United Kingdom in 1985. The cause of the disease was traced to an agent in sheep, which contaminated recycled bovine carcasses used to make meat and bone meal additives for cattle feed.

Recycling of the BSE agent developed into a common source epidemic of more than 180,000 diseased animals in the UK alone. The agent affects the brain and spinal cord of cattle which produces sponge-like changes visible under a microscope. About 19 countries have reported BSE cases and the disease is no longer confined to the European Community.

A case of BSE has been reported in a cattle herd in Japan. In human populations, exposure to the BSE agent (probably in contaminated bovine-based food products) has been strongly linked to the appearance in 1996 of a new transmissible spongiform encephalopathy of humans called variant Creutzfeldt-Jakob Disease (vCJD). By January 2002,119 people developed vCJD, most from the UK but five cases have been reported from France.

Cancer and Environment

Cancer is caused by the uncontrolled growth and spread of abnormal cells that may affect almost any tissue of the body. Lung, colon, rectal and stomach cancer are among the five most common cancers in the world for both men and women. Among men, lung and stomach cancer are the most common cancers worldwide. For women, the most common cancers are breast and cervical cancer.

In India, oral and pharangeal cancers form the most common type of cancer which are related to tobacco chewing. More than 10 million people are diagnosed with cancer in the world every year. It is estimated that there will be 15 million new cases every year by 2020.Cancer causes 6 million deaths every year–or 12% of deaths worldwide. The causes of several cancers are known. Thus prevention of at least one-third of all cancers is possible.

Cancer is preventable by stopping smoking, providing healthy food and avoiding exposure to cancer-causing agents (carcinogens). Early detection and effective treatment is possible for a further one-third of cases. Most of the common cancers are curable by a combination of surgery, chemotherapy (drugs) or radiotherapy (X-rays). The chance of cure increases if cancer is detected early.

Cancer control is based on the prevention and control of cancer by:

- Promotion and strengthening of comprehensive national cancer control Programmes.
- Building international networks and partnerships for cancer control.
- Promotion of organized, evidence-based interventions for early detection of cervical and breast cancer.
- Development of guidelines on disease and programme management.
- Advocacy for a rational approach to effective treatments for potentially curable tumours.
- Support for low-cost approaches to respond to global needs for pain relief and palliative care.

Prevention of Cancer

Tobacco smoking is the single largest preventable cause of cancer in the world. It causes 80 to 90% of all lung cancer deaths. Another 30% of all cancer deaths, especially in developing countries include deaths from cancer of the oral cavity, larynx, oesophagus and stomach which are related to tobacco chewing. Preventive measures include bans on tobacco advertising and sponsorship, increased tax on tobacco products, and educational Programmes which are undertaken to reduce tobacco consumption. Dietary modification is an important approach to cancer control. Overweight individuals and obesity are known to be associated with cancer of the oesophagus, colon, rectum, breast, uterus and kidney. Fruit and vegetables may have a protective effect against many cancers. Excess consumption of red and preserved meat may be associated with an increased risk of colorectal cancer. Infectious agents are linked with 22% of cancer deaths in developing countries and 6% in industrialized countries.

Viral hepatitis B and C cause cancer of the liver. Human papilloma virus infection causes cancer of the cervix. The bacterium Helicobacter pylori increases the risk of stomach cancer. In some countries the parasitic infection schistosomiasis increases the risk of bladder cancer. Liver fluke increases the risk of cancer of the bile ducts. Preventive measures include vaccination and prevention of infection. Excessive solar ultraviolet radiation increases the risk of all types of cancer of the skin.

Avoiding excessive exposure to the sun, use of sunscreens and protective clothing are effective preventive measures. Asbestos is known to cause lung cancer. Aniline dyes have been linked to bladder cancer. Benzene can lead to leukaemia (blood cancer). The prevention of certain occupational and environmental exposure to several chemicals is an important element in preventing cancer.

HUMAN RIGHTS

Several environmental issues are closely linked to human rights. These include the equitable distribution of environmental resources, the utilisation of resources and Intellectual Property Rights (IPRs), conflicts between people and wildlife especially around PAs, resettlement issues around development projects such as dams and mines, and access to health to prevent environment related diseases.

Equity

One of the primary concerns in environmental issues is how wealth, resources and energy must be distributed in a community. We can think of the global community, regional community issues, national concerns and those related to a family or at the individual level. While economic disparities remain a fact of life, we as citizens of a community must appreciate that a widening gap between the rich and the poor, between men and women, or between the present and future generations must be minimised if social justice is to be achieved.

Today the difference between the economically developed world and the developing countries is unacceptably high. The access to a better lifestyle for men as against women is inherent in many cultures. Last but not the least, we in the present generation cannot greedily use up all our resources leaving future generations increasingly impoverished. Rights to land, water, food, housing are all a part of our environment that we all share. However, while some live unsustainable lifestyles with consumption patterns that the resource base cannot support, many others live well below the poverty line. Even in a developing country such as ours, there are enormous economic inequalities.

This requires an ethic in which an equitable distribution becomes a part of everyone's thinking. The people who live in the countries of the North and the rich from the countries in the South will have to take steps to reduce their resource use and the waste they generate. Both the better off sectors of society and the less fortunate need to develop their own strategies of sustainable living and communities at each level must bring about more equitable patterns of wealth. The right to the use of natural resources that the environment holds is an essential component of human rights. It is related to disparities in the amount of resources available to different sectors of society. People who live in wilderness communities are referred to as ecosystem people. They collect food, fuelwood, and Non-wood products, fish in aquatic ecosystems, or hunt for food in forests and grasslands. When landuse patterns change from natural ecosystems to more intensively used farmland and pastureland the rights of these indigenous people are usually sacrificed. Take the case of subsidies given to the pulp and paper industry for bamboo which makes it several times cheaper for the industry than for a rural individual who uses it to build his home.

This infringes on the human right to collect resources they have traditionally used free of cost. Another issue is the rights of small traditional fishermen who have to contend against mechanised trawlers that impoverish their catch and overharvest fish in the marine environment. These people's right to a livelihood conflicts with the powerful economic interests of large-scale organised fisheries.

There are serious conflicts between the rights of rural communities for even basic resources such as water, and industrial development which requires large amounts of water for sustaining its productivity. The right to land or common property resources of tribal people is infringed upon by large development projects such as dams, mining and Protected Areas. Movements to protect the rights of indigenous peoples are growing worldwide. Reversing actions that have already been taken decades ago is a complex problem that has no simple solutions. In many cases a just tradeoff is at best achieved through careful and sensitively managed negotiations. This needs a deep appreciation of local environmental concerns as well as a sensitivity to the rights of local people.

Nutrition, health and Human Rights

There are links between environment, nutrition and health which must be seen from a humanrights perspective. Proper nutrition and health are fundamental human rights. The right to life is a Fundamental Right in our constitution. As a deteriorating environment shortens life spans, this in effect has an impact on our fundamental constitutional right. Nutrition affects and defines the health status of all people, rich and poor.

It is linked to the way we grow, develop, work, play, resist infection and reach our aspirations as individuals, communities and societies. Malnutrition makes people more vulnerable to disease and premature death. Poverty is a major cause as well as a consequence of ill-health. Poverty, hunger, malnutrition and poorly managed environments together affect health and weaken the socioeconomic development of a country.

Nearly 30% of humanity, especially those in developing countries–infants, children, adolescents, adults, and older persons are affected by this problem. A human rights approach is needed to appreciate and support millions of people left behind in the 20th century's health revolution. We must ensure that our environmental values and our vision are linked to human rights and create laws to support those that need a better environment, better health and a better lifestyle.

Health and sustainable human development are equity issues. In our globalized 21st century, equity must begin at the bottom, hand in hand with a healthy environment, improved nutrition, and sustainable lifestyles. Putting first things first, we must also realise that resources allocated to preventing and eliminating disease will be effective only if the underlying causes such as malnutrition and environmental concerns, as well as their consequences, are successfully addressed.

Intellectual Property Rights and Community Biodiversity Registers

Traditional people, especially tribals living in forests, have used local plants and animals for generations. This storehouse of knowledge leads to many new 'discoveries' for modern pharmaceutical products. The revenue generated from such 'finds' goes to the pharmaceutical industry that has done the research and patented the product. This leaves the original tribal user with nothing while the industry could earn billions of rupees.

To protect the rights of indigenous people who have used these products, a possible tool is to create a Community Biodiversity Register of local products and their uses so that its exploitation by the pharmaceutical industry would have to pay a royalty to the local community. This however has still not been generally accepted. Mechanisms have to be worked out so that the local traditional users rights are protected.

Traditional Medicine: Traditional medicine refers to health practices, approaches, knowledge and beliefs that incorporate plant, animal and mineral based medicines, frequently of local or regional origin. It may be linked to spiritual therapies, manual techniques and exercises. These may be used singly or in combination to treat, diagnose and prevent illnesses or maintain well-being. Traditional medicine is often handed down through the generations or may be known to a special caste or tribal group.

Traditional medicine has maintained its popularity in all regions of the developing world and its use is rapidly spreading in industrialized countries. In India, some of our primary health care needs are taken care of entirely by traditional medicine, while in Africa, up to 80% of the population uses it for primary health care. In industrialized countries, adaptations of traditional medicine are termed "Complementary" or "Alternative" Medicine (CAM). While there are advantages to traditional medicine as it is cheap and locally available, there are diseases which it cannot treat effectively.

This is a risk, as patients who use these alternative medicinal practices may rely on an ineffective measure. The consequences could be a serious delay in diagnosis and effective treatment of a treatable condition. There is a need to carefully research the claims of traditional practices to ensure that they are effective. In addition to patient safety issues, there is the risk that a growing herbal market and its great commercial benefit poses a threat to biodiversity through the over harvesting of the raw material for herbal medicines and other natural health care products.

This has been observed in the case of several Himalayan plants. If extraction from the wild is not controlled, this can lead to the extinction of endangered plant species and the destruction of natural habitats of several species. Another related issue is that at present, the requirements for protection provided under international standards for patent law and by most national conventional patent laws are inadequate to protect traditional knowledge and biodiversity.

There are tried and tested scientific methods and products that have their origins in different traditional medicinal methods. Twenty-five per cent of modern medicines are made from plants first used traditionally. Yoga is known to reduce asthma attacks. Traditional Medicine has been found to be effective against several infectious diseases. Over one-third of the population in developing countries lack access to essential allopathic medicines. The provision of safe and effective TM/CAM therapies could become a tool to increase access to health care.

VALUE EDUCATION

Value education in the context of our environment is expected to bring about a new sustainable way of life. Education both through formal and non-formal processes must thus address understanding environmental values, valuing nature and cultures, social justice, human heritage, equitable use of resources, managing common property resources and appreciating the cause of ecological degradation. Essentially, environmental values cannot be taught.

They are inculcated through a complex process of appreciating our environmental assets and experiencing the problems caused due to our destruction of our environment. The problems that are created by technology and economic growth are a result of our improper thinking on what 'development' means. Since we still put a high value only on economic growth, we have no concern for aspects such as sustainability or equitable use of resources. This mindset must change before concepts such as sustainable development can be acted upon.

Unsustainable development is a part of economic growth of the powerful while it makes the poor poorer. Consumerism is one aspect of this process favoured by the rich. As consumption of resources has till recently been an index of development, consumerism has thrived. It is only recently that the world has come to realise that there are other more important environmental values that are essential to bring about a better way of life.

Values in environment education must bring in several new concepts. Why and how can we use less resources and energy? Why do we need to keep our surroundings clean? Why should we use less fertilizers and pesticides in farms?

Why is it important for us to save water and keep our water sources clean? Or separate our garbage into degradable and non-degradable types before disposal? All these issues are linked to the quality of human life and go beyond simple economic growth. They deal with a love and respect for nature. These are the values that will bring about a better humanity, one in which we can live healthy, productive and happy lives in harmony with nature.

Environmental Values

Every human being has a great variety of feelings for different aspects of his or her surroundings. The Western, modern approach values the resources of Nature for their utilitarian importance alone. However true environmental values go beyond valuing a river for its water, a forest for its timber and non-wood forest products, or the sea for its fish.

Environmental values are inherent in feelings that bring about a sensitivity for preserving nature as a whole. This is a more spiritual, Eastern traditional value. There are several writings and sayings in Indian thought that support the concept of the oneness of all creation, of respecting and valuing all the different components of Nature. Our environmental values must translate to pro conservation actions in all our day to day activities.

Most of our actions have adverse environmental impacts unless we consciously avoid them. The sentiment that attempts to reverse these trends is enshrined in our environmental values. Values lead to a process of decision making which leads to action. For value education in relation to the environment, this process is learned through an understanding and appreciation of Nature's oneness and the importance of its conservation. Humans have an inborn desire to explore Nature. Wanting to unravel its mysteries is a part of human nature. However, modern society and educational processes have invariably suppressed these innate sentiments. Once exposed to the wonders of the wilderness, people tend to bond closely to Nature.

They begin to appreciate its complexity and fragility and this awakens a new desire to want to protect our natural heritage. This feeling for Nature is a part of our Constitution, which strongly emphasises this value. Concepts of what constitutes right and wrong behaviour changes with time. Values are not constant.

It was once considered 'sport' to shoot animals. It was considered a royal, brave and much desirable activity to kill a tiger. In today's context, with wildlife reduced to a tiny fraction of what there was in the past, it is now looked down upon as a crime against biodiversty conservation. Thus the value system has been altered with time. Similarly with the large tracts of forest that existed in the past, cutting a few trees was not a significant criminal act.

Today this constitutes a major concern. We need a strong new environmental value system in which felling trees is considered unwise behaviour. With the small human numbers in the past, throwing away a little household degradable garbage could not have been considered wrong. But with enormous numbers of people throwing away large quantities of non-degradable waste, it is indeed extremely damaging to the environment and our value system must prevent this through a strong environmental value education system.

Appreciating the negative effects of our actions on the environment must become a part of our day to day thinking. Our current value system extols economic and technical progress as being what we need in our developing country.

Environmental Values Based on the Constitution of India

- *Article 48A*: "The state shall endeavour to protect and improve the environment and to safeguard the forests and wildlife in the country."
- *Article 51A (g)*: The constitution expects that each citizen of the country must "protect and improve the natural environment, including forests, lakes, rivers and wildlife, and to have compassion for all living creatures."

While we do need economic development, our value system must change to one that makes people everywhere support a sustainable form of development so that we do not have to bear the cost of environmental degradation. Environmental problems created by development are due neither to the need for economic development, nor to the technology that produces pollution, but rather to a lack of awareness of the consequences of unlimited and unrestrained anti-environmental behaviour. Looked at in this way, it deals with concepts of what is appropriate behaviour in relation to our surroundings and to other species on Earth. How we live our lives in fact shapes our environment. This is what environmental values are about. Each action by an individual must be linked to its environmental consequences in his/her mind so that a value is created that leads to strengthening pro-environmental behaviour and preventing anti-environmental actions.

This cannot happen unless new educational processes are created that provide a meaning to what is taught at school and college level. Every small child while growing up asks questions like 'What does this mean?'. They want an explanation for things happening around them that can help them make decisions and through this process develop values. It is this innate curiosity that leads to a personalized set of values in later life.

Providing appropriate 'meanings' for such questions related to our own environment brings in a set of values that most people in society begin to accept as a norm. Thus pro environmental actions begin to move from the domain of individuals to that of a community. At the community level, this occurs only when a critical number of people become environmentally conscious so that they constitute a proenvironment lobby force that makes governments and other people accept good environmental behaviour as an important part of development.

What professions require making value judgements that greatly influence our environment? Evidentally nearly every profession can and does influence our environment, but some do so more than others. Policy makers, administrators, landuse planners, media, architects, medical personnel, health care workers, agriculturalists, agricultural experts, irrigation planners, mining experts, foresters, forest planners, industrialists and, most importantly, teachers at school and college level, are all closely related to pro environmental outcomes.

Environmental values have linkages to varied environmental concerns. While we value resources that we use as food, water and other products, there are also environmental services that we must appreciate. These include Nature's mechanisms in cleaning up air by removing carbon dioxide and adding oxygen by plant life, recycling water through the water cycle of nature, maintaining climate regimes, etc. But there are other aesthetic, ethical values that are equally important aspects of our environment that we do not appreciate consciously. While every species is of importance in the web of life, there are some which man has come to admire for their beauty alone. The tiger's magnificence, the whale and elephant's giant size, the intelligence of our cousins the primates, the graceful flight of a flock of cranes, are parts of nature that we cannot help but admire. The lush splendor of an evergreen forest, the great power of the ocean's waves, and the tranquility of the Himalayan mountains are things that each of us values even if we do not experience it ourselves. We value its being there on Earth for us. This is called its 'existance value'. The list of wondrous aspects of Nature's intricate connections is indeed awe-inspiring. This is also a part of our environment that we must value for its own sake. This is the oneness of Nature. We must equally look at our environment beyond the wild sphere. There is incredible beauty in some man-modified landscapes, the coloured patterns of farmland or the greens of a tea or coffee plantation in the hills.

Urban gardens and open space are also valuable and thus must be of prime concern to urban planners. These green spaces act as not only the lungs of a city, but also provide much needed psychological support. The mental peace and relaxation provided by such areas needs to be valued, although it is difficult to put a price tag on these values.

Nevertheless, these centers of peace and tranquility give urban dwellers an opportunity to balance their highly man-modified environments with the splash of green of a garden space. Environmental values must also stress on the importance of preserving ancient structures. The characteristic architecture, sculpture, artworks and crafts of ancient cultures is an invaluable environmental asset. It tells us where we have come from, where we are now, and perhaps where we should go. Architectural heritage goes beyond preserving old buildings, to conserving whole traditional landscapes in rural areas and streetscapes in urban settings. Unless we learn to value these landscapes, they will disappear and our heritage will be lost.

As environmentally conscious individuals we need to develop a sense of values that are linked with a better and more sustainable way of life for all people. There are several positive as well as negative aspects of behaviour that are linked to our environment. The positive feelings that support environment include a value for Nature, cultures, heritage, and equity. We also need to become more sensitive to aspects that have negative impacts on the environment. These include our attitude towards degradation of the environment, loss of species, pollution, poverty, corruption in environmental management, the rights of future generations and animal rights. Several great philosophers have thoughts that have been based on, or embedded, in pro environmental behaviour. Mahatma Gandhi and Rabindranath Tagore are among the many internationally well-known scholars whose thought have included values that are related to environmental consciousness. We need to appreciate these values to bring about a better way of life on earth for all people and all living creatures.

Valuing Nature

The most fundamental environmental sentiment is to value Nature herself. Appreciating Her magnificence and treasuring life itself leads to positive feelings that are a manifestation of pro environmental consciousness. The one-ness of our lives with the rest of nature and a feeling that we are only a miniscule part of nature's complex web of life becomes apparent, when we begin to appreciate the wonders of nature's diversity. We must appreciate that we belong to a global community that includes another 1.8 million known living forms. Nothing makes us more conscious of this wonderous aspect of our earth's diversity than a walk through the wilderness, feeling and exploring its beauty and experiencing its infinite variety. The tiny creatures that live complex lives and the towering trees are all a part of this phenomenon we call 'life'. Today, man does not even know if other complex forms of life exist outside our own solar system in distant space. We may be alone in space or may be accompanied by other, completely different, living forms. But for now we only know for sure that the Earth's life forms are unique.

We thus have a great responsibility to protect life in all its glorious forms and must therefore respect the wilderness with all its living creatures, where man's own hand has not created changes that have led to perturbing natural habitats. We need to develop a sense of values that lead us to protect what is left of the wilderness by creating effective National Parks and Wildlife Sanctuaries. However this cannot be done to the detriment of the millions of tribal or indigenous people who live in wilderness ecosystems. There are thus conflicting values that need to be balanced carefully.

On the one hand we need to protect natural ecosystems, while on the other, we must protect the rights of local people. Yet apart from valuing the diversity of life itself, we must also learn to value and respect diverse human cultures. Many of the tribal cultures of our country are vanishing because those with more dominant and economically advanced ways of life do not respect their lifestyles, that are in fact closer to nature and frequently more sustainable. We believe that our modern technology-based lifestyles are the sole way for society to progress. Yet this is only a single dimension of life that is based on economic growth.

Deep Ecology

In the 1970s a new thinking on environmental concerns began to emerge, protecting nature and the wilderness for its own sake, which is now referred to as 'Deep Ecology'. It was fostered by the thinking of Arne Naess, a Norwegian professor of Philosophy and a great believer in Gandhian thinking and Buddhism. It recognises the intrinsic value of all living beings and looks upon mankind as a small segment of a great living community of life forms.

It teaches that the wellbeing and flourishing of human and non-human life on Earth have value in themselves and that these values are independent of the usefulness of the non-human world for human purposes. While currently the environmental movement focuses on issues that are concerned with the management of the natural environment for the 'benefit' of man, Deep Ecology promotes an approach that is expected to bring about a more appropriate ecological balance on Earth and is akin to a spiritual approach to Nature. This has great long-term implications not only for humans but for the whole of Nature. For example some environmentalists emphasise the need to preserve wilderness for its aesthetic and utilitarian functions.

Wilderness is being preserved today in PAs because it is scenic and serves the purpose of tourism for nature lovers, and has recreational and economic value. Other environmentalists stress that the goal is for protecting the useful ecological functions of the wilderness, its services and goods that we use. Deep Ecologists on the other hand stress that wilderness preservation is a means to achieve the conservation and protection of biological diversity. Thus it is not enough to protect bits of what is left of the wilderness but to make attempts to restore degraded areas to their former natural ecological state.

In a country such as India, with its enormous population coupled with poverty on the one hand and the need for economic industrial growth on the other, this will be extremely difficult to achieve. Another new approach is that of 'Gaia', the hypothesis that the Earth is itself like one giant form of throbbing life consisting of all the unquantifiable numbers of individuals of its millions of known and unknown species.

Valuing Cultures

Every culture has a right to exist. Tribal people are frequently most closely linked with Nature and we have no right to foist on them our own modern way of life. The dilemma is how to provide them with modern health care and education that gives them an opportunity to achieve a better economic status without disrupting their culture and way of life. This will happen only if we value their culture and respect their way of life.

Social Justice

As the divide widens between those people who have access to resources and wealth, and those who live near or below the poverty line, it is the duty of those who are better off to protect the rights of the poor who do not have the means to fight for their rights. If this is not respected the poor will eventually rebel, anarchy and terrorism will spread and the people who are impoverished will eventually form a desperate seething revolution to better their own lot.

The developing world would face a crisis earlier than the developed countries unless the rights of poor people that are fundamental to life are protected. Modern civilization is a homogenous culture, based until recently on a belief that modern science holds the answer to everything. We are now beginning to appreciate that many ancient and even present day sequestrated cultures have a wisdom and knowledge of their own environments that is based on a deep sense of respect for nature. Tribal cultures have over many generations used indigenous medicines which are proving to be effective against diseases. They have produced unique art forms such as painting, sculpture, and crafts that are beautiful and can enrich living experiences for everyone. They have their own poetry, songs, dance and drama-all art forms that are unfortunately being rapidly lost as we introduce a different set of modern values to them through television and other mass media.

The world will be culturally impoverished if we allow these indigenous people to loose their traditional knowledge which includes sustainable use of water, land and resources with a low impact on biodiversity. They will soon lose the beauty within their homes that is based on the things they make from Nature.

The art of the potter will be lost forever to the indestructible plastic pot. The bamboo basket weaver who makes a thing of beauty that is so user friendly and aesthetically appealing, will give place to yet another plastic box. Much that is beautiful and hand-crafted will disappear if we do not value these diverse aspects of human cultures.

Human Heritage

The earth itself is a heritage left to us by our ancestors for not only our own use but for the generations to come. There is much that is beautiful on our Earth-the undisturbed wilderness, a traditional rural landscape, the architecture of a traditional village or town, and the value of a historical monument or place of worship. These are all part of human heritage. Heritage preservation is now a growing environmental concern because much of this heritage has been undervalued during the last several decades and is vanishing at an astonishing pace.

While we admire and value the Ajanta and Ellora Caves, the temples of the 10th to 15th centuries that led to different and diverse styles of architecture and sculpture, the Moghul styles that led to structures such as the Taj Mahal, or the unique environmentally-friendly Colonial buildings, we have done little to actively preserve them. As environmentally conscious individuals we need to lobby for the protection of the wilderness and our glorious architectural heritage.

Equitable use of Resources

An unfair distribution of wealth and resources, based on a world that is essentially only for the rich, will bring about a disaster of unprecedented proportions. Equitable use of resources is now seen as an essential aspect of human well being and must become a shared point of view among all socially and environmentally conscious individuals. This includes an appreciation of the fact that economically advanced countries and the rich in even poor nations consume resources at much greater levels than the much larger poorer sectors of humanity in the developing world. In spite of the great number of people in the more populous developing countries, the smaller number of people in developed countries use more resources and energy than those in the developing world. This is equally true of the small number of rich people in poor countries whose per capita use of energy and resources, and the generation of waste based on the one time use of disposable products, leads to great pressures on the environment. The poor while polluting the environment have no way to prevent it.

The rich damage the environment through a carelessness that proves only that they have no appreciation for environmental safety. As we begin to appreciate that we need more sustainable lifestyles we also begin to realise that this cannot be brought about without a more equitable use of resources.

Common Property Resources

Our environment has a major component that does not belong to individuals. There are several commonly owned resources that all of us use as a community. The water that nature recycles, the air that we all breathe, the forests and grasslands which maintain our climate and soil, are all common property resources. When Government took over the control of community forests in British times, the local people who until then had controlled their use through a set of norms that were based on equitable use, began to overexploit resources on which they now had no personal stake.

Bringing back such traditional management systems is extremely difficult. However, in the recent past managing local forests through village level forest protection committees has shown that if people know that they can benefit from the forests, they will begin to protect them. This essentially means sharing the power to control forests between the Forest Department and local people.

Ecological Degradation

In many situations valuable ecological assets are turned into serious environmental problems. This is because we as a society do not strongly resist forces that bring about ecological degradation. These consist of sectors of society that use a 'get-rich-quick' approach to development. While ecological degradation has frequently been blamed on the needs of fuelwood and fodder of growing numbers of rural people, the rich, urbanized, industrial sector is responsible for greater ecological damage.

Changes in landuse from natural ecosystems to more intensive utilization such as turning forests into monoculture forestry plantations, or tea and coffee estates, or marginal lands into intensive agricultural patterns such as sugarcane fields or changes into urban or industrial land carry an ecological price. Wetlands, for example, provide usable resources and a variety of services not easily valued in economic terms, and when destroyed to provide additional farmland, in many cases produce lower returns. A natural forest provides valuable non-wood forest products whose economic returns far outweigh that provided by felling the forest for timber. These values must form a part of a new conservation ethic. We cannot permit unsustainable development to run onwards at a pace in which our lives will be overtaken by a development strategy that must eventually fail as Earth's resources are consumed and ecosystems rendered irreparable.

HIV/AIDS

The Human Immunodeficiency Virus (HIV) causes Acquired Immunodeficiency Syndrome (AIDS) through contact with tissue fluids of infected individuals, especially through sexual contact. As it reduces an individual's resistance to disease, it causes infected individuals to suffer from a large number of environment related diseases and reduces the ability of infected individuals to go about their normal lives.

It affects their income generation and/or their ability to utilise natural resources. As more and more people are affected, this disease will also have impacts on our natural resource base, as utilisation patterns change to unsustainable levels. The inability of these patients to have the strength to access natural resources also affects the outcome of the disease process, as their overall health and well being is likely to worsen the course of the disease when their nutritional status suffers. In sub Saharan Africa where the infection has become highly prevalent, it is leading to great suffering and worsening poverty.

The capacity of these patients to work for their usual sources of income generation is lost. An increasing proportion of the poor are affected. It is evident that it is going to be increasingly difficult to manage environments sustainably, as natural resources on which the poor debilitated patients depend continue to be degraded. Incomes lost due to the stigma of HIV/AIDS must be met by the sufferers by overexploiting their resource base. People affected by the disease inevitably try to get whatever they can from their natural resource base as they are not in any position to think of the long-term future.

In Africa, this has led to degradation of the ecosystem and an increase of pressures from other impacts such as overuse of medicinal plants and poaching for wildlife. In South Africa, for example, people have a mistaken belief that turtle eggs can cure HIV/AIDS, thus leading to the eggs being over harvested. As males die of the disease, work on agricultural land has to be taken over by already overworked women and their children, affecting land management and productivity. Providing balanced diets and nutritional support for these poverty stricken patients can be partially addressed by better natural resource management such as afforestation, access to clean water and wholesome food. HIV/AIDS seriously affects the patient's working environment. It creates an incorrect fear in the minds of co-workers. It must be clearly understood that

AIDS is not spread by casual contact during work. Patients have a right to continue to work as before along with unaffected individuals. As patients are unable to continue their original hard labour related work, it is essential that alternative sources of work must be created for them. Educators and extention information, in the formal and non-formal educational sectors, must address the issues related to the linkages between natural resource management and this disease, as well as the need to remove the social stigma attached to it.

HIV/AIDS has a serious impact on the socioeconomic fabric of society. By 2002, India had an estimated 3.97 million infected individuals. There is a great need to organise AIDS education on prevention and management of the disease. This needs to be done through the formal educational sector and by using non-formal methods. Education is also important to reduce the stigma and discrimination against these patients. In India, women who are not socially empowered are at a great disadvantage as they are powerless to demand safe sex from their partners.

Women also have an added burden of caring for HIV infected husbands. This produces enormous economic stresses on their family. HIV in India is rapidly moving from a primarily urban sector disease to rural communities. Research in Nepal has shown a linkage between rural poverty, deforestation and a shift of population to urban areas resulting in a rising number of AIDS patients. Prior to 1992, it was mainly seen in males who migrated to urban centers. In more recent times, a growing number of women are moving to Indian cities as sex workers.

Women engaged in prostitution find it difficult to make partners take protective measures, such as the use of condoms that provide safe sex. A large proportion became victims of the disease. Blood transfusion from an infected person can also lead to HIV/AIDS in the recipient, as well as drug abuse by sharing needles with an infected person. In sexually transmitted AIDS, the use of condoms during intercourse is a key to preventing the disease. Behavioural change, where the number of individuals who have multiple partners, towards strictly single partners, reduces the risk of HIV/AIDS and thus reduces incidence of the disease in society. However, the most important measure to prevent AIDS is the proper use of condoms that form a barrier to the spread of the virus during intercourse.

WOMAN AND CHILD WELFARE

There are several environmental factors that are closely linked to the welfare of women and children. Each year, close to eleven million children worldwide are estimated to have died from the effects of disease and inadequate nutrition. Most of these deaths are in the developing world. In some countries, more than one in five children die before they are 5 years old. Seven out of 10 of childhood deaths in developing countries can be attributed to five main causes, or a combination of them. These are pneumonia, diarrhoea, measles, malaria and malnutrition. Around the world, three out of every four children suffer from at least one of these conditions.

Table. The Diagnosis of Common Childhood Disease Conditions

Presenting complaint	Possible cause or **associated condition**
Cough and/or fast breathing	Pneumonia Severe anaemia P. falciparum malaria
Lethargy or unconsciousness	Cerebral malaria Meningitis Severe dehydration Very severe pneumonia
Measles rash	Pneumonia Diarrhoea Ear infection
"Very sick" young infant	Pneumonia Meningitis Sepsis

Respiratory Conditions

Most respiratory diseases are caused by or are worsened by polluted air. Crowded ill-ventilated homes and living in smokey households with open fires can trigger respiratory conditions especially in children. Pneumonia: Acute respiratory infections (ARI), most frequently pneumonia, is a major cause of death in children under five, killing over two million children annually.

Upto 40% of children seen in health centers suffer from respiratory conditions and many deaths attributed to other causes are, in fact, "hidden" ARI deaths. Children may die very quickly from the infection and thus need treatment urgently. Most patients of pneumonia can be treated with oral antibiotics. Correct management could save over 1 million lives per year globally.

Gastro Intestinal Conditions

Contaminated water and food causes widespread ill health especially in children. Diarrhoea: Diarrhoea is caused by a wide variety of infections. Urgent diagnosis and treatment of diarrhoea is a priority for saving a child's life. Treating malnutrition that often accompanies diarrhoea can further reduce mortality. Increasing vigilance to detect other diseases that can occur concurrently with diarrhoea, such as measles or malaria, is an important measure.

Two million children die each year in developing countries from diarrhoeal diseases, the second most serious killer of children under five worldwide. In most cases diarrhoea is preventable and children can be saved by early treatment. Correct management of diarrhoea could save the lives of up to 90% of children who currently die by promoting rapid and effective treatment through standardised management, including antibiotics and simple measures such as oral rehydration using clean boiled water with salt and sugar. In severe cases intravenous fluids must be started. Improved hygiene and management of the home and surroundings is the most important preventative measure, as well as improved nutrition. Increased breastfeeding and measles vaccination have also been observed to have reduced the number of cases of diarrhoea.

Measles

Measles is a rash that appears with fever and bodyache in children and is caused by a virus. It infects over 40 million children and kills over 800,000 children under the age of five. Prevention includes wider immunization coverage, rapid referral of serious cases, prompt recognition of conditions that occur in association with measles, and improved nutrition, including breastfeeding, and vitamin A supplementation. Measles is prevented by a vaccine.

Young children with measles often develop other diseases such as acute respiratory infections, diarrhoea and malnutrition that are all linked to poor environmental conditions in their surroundings. Children who survive an attack of measles are more vulnerable to other dangerous infections for several months. Effective prevention and treatment could save 700,000 lives per year.

Malaria

This condition is closely linked to pooling and stagnation of water in tropical environments. Malaria is a widespread tropical disease which is caused by a parasite transmitted to humans by mosquitoes. It has proved difficult to control because mosquitoes have become resistant to insecticides used against them and because the parasite has developed resistance in some areas to the cheap and effective drugs that used to provide good protection in the past.

However, alternative newer drug therapies have been developed for use in areas where resistant parasites are found. In India the disease was nearly wiped out a few decades ago but has now re-emerged in many parts of the country. Correct management could save 500,000 lives per year. Approximately 700,000 children die of malaria globally each year, most of them in sub-Saharan Africa. Young children are particularly vulnerable because they have not developed the partial immunity that results from surviving repeated infections. Deaths from malaria can be reduced by several measures, including encouraging parents to seek prompt care, accurate assessment of the condition of the child, prompt treatment with appropriate anti-malarial drugs, recognition and treatment of other co-existing conditions, such as malnutrition and anaemia, and prevention by using mosquito-proof bednets. Because fever may be the only sign of malaria, it can be difficult to distinguish it from other potentially lifethreatening conditions.

Poverty-environment-malnutrition

There is a close association between poverty, a degraded environment, and malnutrition. This is further aggravated by a lack of awareness on how children become malnourished. Malnutrition: Although malnutrition is rarely listed as the direct cause of death, it contributes to about half of all childhood deaths. Lack of access to food, poor feeding practices and infection, or a combination of the two, are major factors in mortality.

Infection, particularly frequent or persistent diarrhoea, pneumonia, measles and malaria, undermines nutritional status. Poor feeding practices-inadequate breastfeeding, providing the wrong foods, giving food in insufficient quantities, contribute to malnutrition. Malnourished children are more vulnerable to disease. Promoting breastfeeding, improving feeding practices, and providing micronutrient supplements routinely for children who need them are measures that reduce mortality.

The nutritional status and feeding practices of every child under two years of age, and those with a low weight for their age must be intensively managed. Counseling of parents on the correct foods for each age group and helping them to overcome various feeding problems is an essential health care measure. Children between 6 months and 2 years of age are at increased risk of malnutrition when there is a transition between breastfeeding and sharing fully in the family diet. Changing family habits and the kinds of food offered to children is an important measure. Talking to mothers individually about home care and their child's feeding, with relatively simple changes to better feeding practices, such as helping them to eat rather than leaving them to fend for themselves, can ensure that a child gets enough to eat. A minor increase in breastfeeding could prevent up to 10% of all deaths of children under five: When mothers breastfeed exclusively during at least the first four months and, if possible, six months of life, there is a decrease in episodes of diarrhoea and, to a lesser extent, respiratory infections.

Even small amounts of water-based drinks decreases breastmilk intake and lead to lowered weight gain. This increases the risk of diarrhoea. Continuing to breastfeed up to two years of age, in addition to giving complementary foods, maintains good nutritional status and helps prevent diarrhoea. Encouraging maximum support to mothers to establish optimal breastfeeding from birth, equipping health workers with counseling skills, and providing individual counseling and support for breastfeeding mothers are measures that reduce malnutrition. Mothers often give their babies other food and fluids before six months because they doubt their breastmilk supply is adequate. A one-on-one counseling with mothers on breastfeeding techniques and its benefits helps reduce incidence of malnutrition. There are strong connections between the status of the environment and the welfare of women and children in India. Women, especially in lower income group families, both in the rural and urban sector, work longer hours than men. Their work pattern differs and is more prone to health hazards. The daily collection of water, fuelwood and fodder is an arduous task for rural women. In urban areas, where lower economic group women live in crowded smoke filled shantys in unhygenic slums, they spend long hours indoors, which is a cause of respiratory diseases.

In urban centers, a number of women eke out a living by garbage picking. They separate plastics, metal and other recyclable material from the waste produced by the more affluent groups of society. During this process, they can get several infections. Thus they are providing an environmental service of great value, but earn a pittance from this work.

Women are often the last to get enough nutrition as their role in traditional society is to cook the family meal and feed their husband and children. This leads to malnutrition and anemia due to inadequate nutrition. The sorry plight of women includes the fact that the girl child is given less attention and educational facilities as compared to boys in India. Thus

Case Study

Karnataka's GIS scheme, Bhoomi, has revolutionized the way farmers access their land records. Farmers can now get a copy of the records of rights, tenancy and crops from a computerized information kiosk without harassment and bribes. Karnataka has computerized 20 million records of land ownership of 6.7 million farmers in the State. they are unable to compete with men in later life. This social-environmental divide is a major concern that needs to be corrected throughout the country.

ROLE OF INFORMATION TECHNOLOGY IN ENVIRONMENT AND HUMAN HEALTH

The understanding of environmental concerns and issues related to human health has exploded during the last few years due to the sudden growth of Information Technology. The computer age has turned the world around due to the incredible rapidity with which IT spreads knowledge. IT can do several tasks extremely rapidly, accurately and spread the information through the world's networks of millions of computer systems.

A few examples of the use of computer technology that aid environmental studies include software such as using Geographical Information Systems (GIS). GIS is a tool to map landuse patterns and document change by studying digitized toposheets and/or satellite imagery. Once this is done, an expert can ask a variety of questions which the software can answer by producing maps which helps in landuse planning. The Internet with its thousands of Web sites has made it extremely simple to get the appropriate environmental information for any study or environmental management planning. This not only assists scientists and students but is a powerful tool to help increase public awareness about environmental issues. Specialised software can analyse data for epidemiological studies, population dynamics and a variety of key environmental concerns. The relationship between the environment and health has been established due to the growing utilisation of computer technology. This looks at infection rates, morbidity or mortality and the etiology (causative factors) of a disease. As knowledge expands, computers will become increasingly efficient. They will be faster, have greater memories and even perhaps begin to think for themselves.

7

Current Status of Culture Fisheries in the 21st World

INTEGRATED LIVESTOCK-FISH FARMING SYSTEMS

FISH AS FOOD

Malnutrition and starvation are the two serious problems being faced by millions of rural poor in most of the developing countries. The problem of malnutrition is in fact more serious and of a bigger dimension than the starvation problem and is caused mainly due to animal protein-deficient diets. Animal protein is essential for proper growth, repair and maintenance of body organs and tissues.

Fish contain about 16–20% protein compared to about 12% in egg, 3.5% in milk and 6–8% in rice and wheat. Moreover, it is wholesome, tasty, highly nutritive and an excellent source of essential minerals, vitamins and essential amino acids. At present about 31% of the total animal protein supply in the Asian region is in the form of fish protein. For the poorest segments of the population, fish is not only the most important animal protein source, but often the only one.

HIGH MULTIPLICATION CAPACITY AND MINIMAL WATER REQUIREMENT

The reproductive potential of fish compared to any other farmed animal is also very high. A kilogram of female cultivable carp species yields on an average about 0.1 million eggs, each of which has the potential to become 1 kg fish in about a year. No livestock animal possesses this magnitude of fecundity. Although fish needs water as a medium to survive and grow, it consumes minimal quantity of water compared with any livestock or agricultural crop. Fish also enriches the water with its voided metabolites thus making the water more productive for agriculture.Fish culture systems require a relatively less amount of energy for protein production than any other farming system. Carp culture, depending upon culture practices, requires energy at the rate of 22–468 KJ/g of protein production while a land animal farming system needs over 550 to 3 400 KJ/g.Fish are cold blooded or poikilothermic animals. In other words they cannot maintain a constant and high body temperature like other livestock animals. Instead, their body temperature fluctuates according to the surrounding temperature. In warmer climates, their metabolism accelerates and they grow faster, while in colder climates, the metabolic rate slows down, resulting in a reduced rate of growth. In this way they save energy by not spending it for maintaining a higher and constant body temperature.

AQUACULTURE PRODUCTION POTENTIAL

Although the world's total marine production now stands at more than 80 million tons per year, all trends indicate that a saturation point is quickly approaching. During the past decade the growth rate hovered around 2%, much lower than earlier decades. On the other hand, tremendous potential exists in aquaculture. Aquaculture presently produces over 8 million tonnes of fish and shellfish annually.

It is estimated that Asian aquaculture production could be raised to 20–30 million tonnes a year by the end of the century. Aquaculture production has increased at an annual growth rate of nearly 7% between 1975–84.

EMPLOYMENT POTENTIAL

Aquaculture is also considered to be a potential source of employment for poor farmers and displaced capture fishermen. Rapid development of aquaculture has already generated considerable employment through culture of marketable fish, fish seed production and marketing of fish and fish seed.

The National Agriculture Commission of India while estimating the employment potential of fish culture has indicated that every tonne of fish produced provides employment to 2.5 persons. The other important advantages of fish farming are that the production is carried out within easy reach of consumers and also the harvesting can be adjusted to demand, thus minimising distribution problems and spoilage.

Rural ponds in Asia, hitherto producing at subsistence level, have succeeded in increasing production per unit area through improved culture practices involving higher stocking densities, polyculture combinations, pond manuring and feeding. The switch over from monoculture practice to polyculture has significantly contributed Towards higher production and the prospect of polyculture appears very bright as the fish seed of desired species is becoming easily available due to the establishment of a large number of hatcheries. During recent years, advances have also been made in traditional aquaculture systems practised in rural India by the development of composite fish culture, a system of polyculture of a group of complementary and supplementary freshwater species of fast growing carps in undrainable ponds. With the successful demonstration in different agroclimatic zones of India, gradual improvements in technology have been made and it is now possible to obtain a production rate of over 10 t/ha/yr in experimental ponds and up to about 5 t/ha/yr in farmers' ponds against the traditional average rate of production of 600 kg/ha/yr.

To meet the increasing demand of seed of culturable carps, hypophysation techniques have also been developed for both Indian and Chinese carps and as a result they are now being bred in captivity even by fish farmers in remote villages. The emergence of this culture technology suitable for undrainable ponds and the simultaneous development of hypophysation techniques for fish seed production has completely revolutionized fish farm productivity.

This manual intends to provide the basic concept and practical guidelines of fish culture in undrainable ponds. Since it is prepared especially for extension agents and field workers, certain important things have been repeated and at times experimental results have been simplified with a view to making it more practical, simple and illustrative.

It outlines the practices of procurement and propagation of fish seed, rearing of spawn to fry and fingerling stage and production of table-size fish following simple sequential steps. Like other farming systems this culture system is also prone to certain unexpected hazards for which one has to be prepared and properly equipped.

Such hazards are disease outbreaks, oxygen depletion, pollution, flood, drought, poaching, etc. The content of this manual is a synthesis of the author's personal field experience, the information gathered from published literature and the observations of other workers in India.

Based upon Indian experience, water resources in other developing countries with similar agroclimatic conditions may be utilized for the development of fish culture.

It is hoped that this manual will serve as a practical guide to extension workers in popularizing freshwater fish culture in undrainable ponds.

PRINCIPLES OF FRESHWATER FISH CULTURE

An understanding of the following basic principles of freshwater pond fish culture is essential.

POND ECOSYSTEM

Like agriculture, fish culture is also based on a series of processes involving reception and transformation of solar energy. In the pond ecosystem solar energy is utilized for primary production by chlorophyll-bearing plants such as planktonic algae and macrophytes. This conversion of solar energy into chemical energy (food) is guided by the photosynthetic and chemosynthetic activities going on in the aquatic plant community and the rate at which this is carried out is called primary productivity of that ecosystem.

A part of the primary production is cycled through different trophic levels resulting in fish production. Here comes the community of consumers that comprise microscopic as well as large animals, which are unable to synthesize their own food and feed upon primary producers. Different forms of pond life are linked together through predator-prey relationship. This chain of food production, which follows a general pattern, primary producers-herbivores-carnivores-appears too simple and straight. But, in fact, it is a complex food web with various cross linkages.

Fish populations may be classified into several trophic levels, depending upon their position in this food chain. Phytophagous fish such as grass carp and silver carp belong to the second trophic level as they feed upon the first trophic level organisms. Likewise, zooplankton, feeding upon phytoplankton, also belong to the same category.

Carnivorous fish communities thriving upon zooplankton or herbivorous fishes occupy the third trophic level while other predatory fishes preying upon carnivorous fishes belong to the fourth trophic level A relatively simple food chain operates in fish ponds, but a complex one occurs in lakes and other larger aquatic ecosystems.

The picture becomes even more complicated in large water bodies such as rivers and seas where complex food chains are referred to as food webs which in fact represent several interconnected food chains. There are some fishes which occupy mixed positions, between different trophic levels. They consume both plants and animals and as such, cannot be naturally categorised into any one particular trophic level.

A properly managed pond presents an example of a simple food chain under simple conditions. Here the number of food chains is reduced by encouraging the growth of phytoplankton. The macrophytes such as rooted green plants, floating plants, etc., are not allowed to grow. Phytoplankton is consumed by the zooplankton in the water column, whereas its detritus is utilized by benthic invertebrates. Phytoplankton, zooplankton, detritus and benthic organisms serve as food for the stocked fishes such as the desired carp species. Thus, as much of the available solar energy as possible is utilized for fish production by proper pond management.

Primary productivity is dependent on light, carbon dioxide, temperature and essential nutrients, each of which can be a limiting factor. Of these factors affecting primary production in ponds, the one that can be manipulated easily is the quantity of nutrient elements through the application of nitrogenous, phosphatic and potassic fertilizers, as in agriculture. In ponds, only the top 2 to 5 cm of soil is concerned with nutriention exchange and the soil below plays a negligible role in the production cycle.

Undrainable ponds receive dissolved nutrients and sedimentary particles carried by rain water from the catchment area. Besides, production and decomposition of minute plant and animal organisms in ponds also modify the properties of the pond bottom to a great extent. The nature and quantity of fertilizers determines the species composition to be used in a culture system. At low phosphate concentration, diatoms are common, but with increasing concentrations green algae become more frequent, eventually giving way to blue-green algae. In addition, excessive phosphate gives rise to phytoplankton blooms which check the light penetration and thus lower the pond productivity through 'autoshading'.

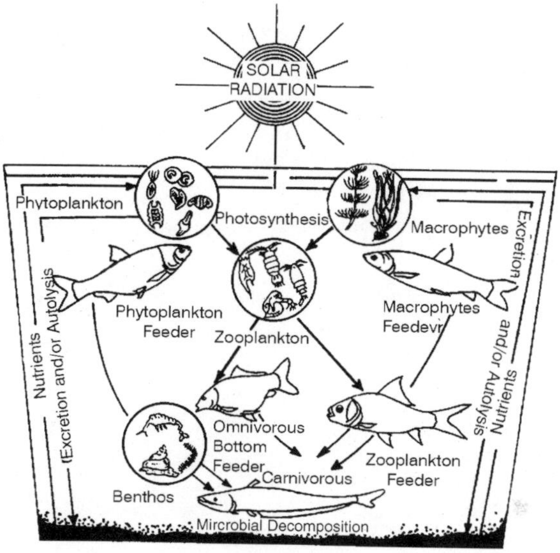

Fig. Pond Ecosystem

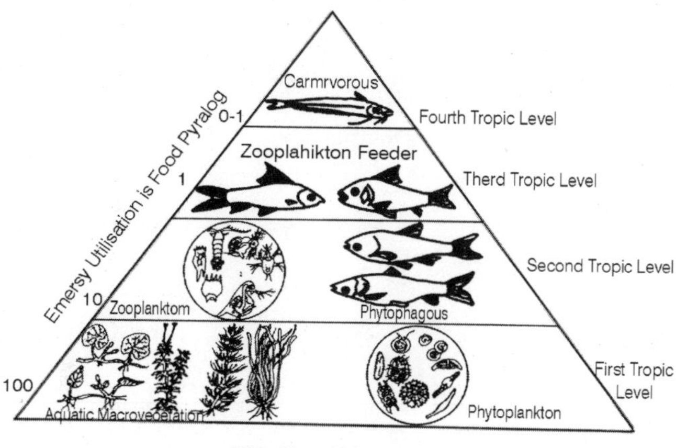

Fig. Food Pyramid

Light energy is one of the major inputs in primary production and hence the success of fish culture depends largely on the efficient utilization of incident light. When incident light strikes the water surface, it is partially reflected and partially transmitted into the water where part of it is utilized in the process of photosynthesis and the rest is scattered or absorbed by suspended particles. In turbid waters, more light is scattered or absorbed, thus allowing the light penetration only to shallow depths. The rapid disappearance of light in such waters affects adversely the growth of diatoms. The bottom layer of water, being devoid of photosynthetic plants and also being in close contact with the decaying organic matter, suffers from oxygen depletion causing critical stress conditions for the fish. Thus, it is important that primary producers must provide oxygen to support the total biological respiration during darkness and also during the less favourable (warmer, overcast or rainy) days apart from providing food for the second and third trophic-level fish.

Table: **Solubility of oxygen under different temperaturesat 760 mm of Hg pressure**

Temperature (°C)	Solubility of *oxygen (mg/l)*	Temperature (°C)	Solubility of *oxygen (mg/l)*
15	9.76	26	7.99
16	9.56	27	7.86
17	9.37	28	7.75
18	9.18	29	7.64
19	9.01	30	7.53
20	8.84	31	7.42
21	8.68	32	7.32
22	8.53	33	7.22
23	8.38	34	7.13
24	8.25	35	7.04
25	8.11		

OXYGEN BUDGET

The concentration of dissolved oxygen in the water, which depends on the temperature, is an essential component of the aquatic environment to govern the carrying capacity of a pond. Variations in concentration of dissolved oxygen may occur due to the following three important factors:

- The saturation level of oxygen in water decreases as the temperature rises;
- Supersaturation is an unstable state and
- Plants not only photosynthesize to produce oxygen, they also respire and consume oxygen.

The saturation value for dissolved oxygen available for fish life at 20°C water temperature is more than that at 30°C at a particular atmospheric pressure Dissolved oxygen (Do) concentration is always high at lower temperatures and gradually decreases with increase in temperature.In natural waters, including undrainable fish ponds, DO values are constantly changing because of biological, physical and chemical processes. The air above the pond water surface may be considered to have a more or less constant percentage of oxygen.

However, the partial pressure of oxygen in the air may vary slightly at a given location because of differences in atmospheric pressure. Transfer of oxygen from air to water will occur when water is undersaturated with DO and oxygen will diffuse from water to air when water is supersaturated with oxygen. However, the diffusion of oxygen into the pond water is very slow, except under conditions of strong turbulence, hence the most important source of oxygen is that generated during photosynthesis.

Sometimes, phytoplankton blooms or algal scums limit light penetration causing reduction in photosynthetic rates, even in waters with adequate nutrient concentrations. Oxygen production by phytoplankton is greatest near the surface and decreases with the increase in depth because of self-shading. When heavy infestation of aquatic weeds and dense bloom of plankton occur, the situation becomes much more complex.

On the other hand, these are additional sources of oxygen at daytime; but on the other hand, they also respire and consume oxygen throughout day and night. At times the pondwater is supersaturated with oxygen during the day, which is a highly unstable state, while during the night, a greater proportion of oxygen is used up for their respiration, thereby reducing the availability of oxygen to fish. Thus, it creates a wide fluctuation in the level of dissolved oxygen, adversely affecting fish life. In fact, under such situations oxygen production increases to its maximum during the daytime leaving surplus for the fish even after consuming for their own respiration, but at night this surplus level drops down to critical level.

Under conditions of heavy algal blooms and weed infestation, the phytoplankton and aquatic weeds actually consume more available oxygen during day and night than they produce during the whole day During cloudy days, when the incident light is inadequate for phytosynthesis, the situation in terms of availability of DO becomes worse.

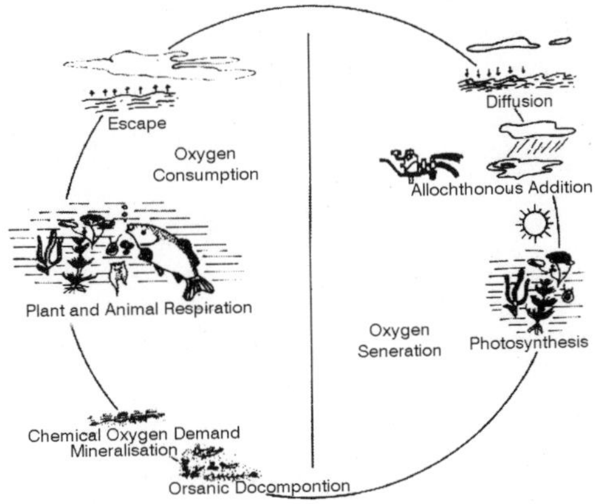

Fig. Oxygen Cycle in Pond

Aerobic decomposition of organic matter by bacteria is also an important drain on the oxygen supply in ponds. Aerobic decomposition requires a continuous supply of oxygen and proceeds more rapidly when DO concentrations are near saturation. However, decomposition also occurs under anaerobic conditions, but the rate of degradation of organic matter is not as rapid and complete as under aerobic conditions. Under aerobic condition, the end product of decomposition is primarily carbon dioxide. At times high rate of bacterial decomposition of dead organisms and other organic bottom deposits lead to a condition favouring the increase of the level of carbon dioxide and other abnoxious gases, with a simultaneous depletion of DO, resulting in fish kills and planktonic collapses Therefore, it is important that the pond water should provide adequate oxygen to support the total biological respiration during the hours of darkness.

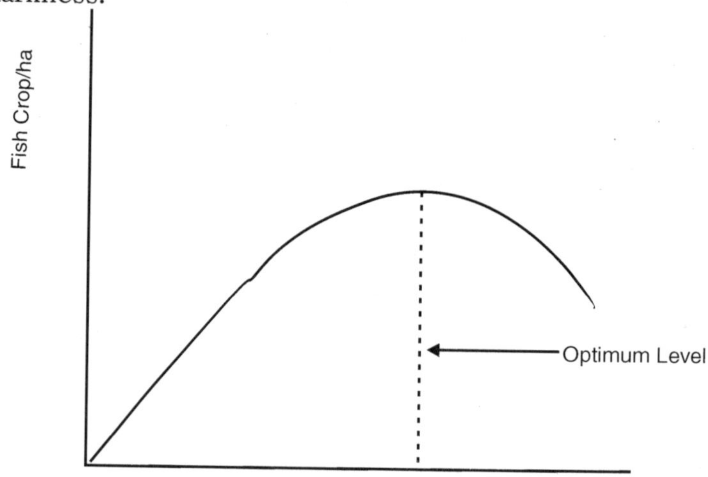

Fig. Relation between Stocking Density and Production

DESIRABLE FISH SPECIES FOR CULTURE

The choice of fish species is very important in maximizing production, both in terms of quantity and quality.Since considerable amount of energy is lost in successive trophic levels of the food chain, efficient fish culture always aims at making the chain as short as possible. Because of this, herbivorous fishes are always preferred to carnivorous fishes, the latter being mostly excluded because of their longer food chains. Mixed species farming or polyculture yields a higher production than single species farming.

It is obvious that any single species cannot utilize all the available food in a pond because of its specific feeding habit and hence a combination of compatible species with complementary feeding habits are usually stocked to make better use of the natural food available in the pond. Selection of the species should be based on the productivity of a pond, availability of artificial food resources, availability of seed and the marketing prospects. The principal considerations in species combination are that they have complementary feeding habits, they occupy different ecological niches, they attain marketable size at more or less the same time, they tolerate each other and they be non-predatory in nature. A combination of plankton and macrophyte feeders is most usual. Ungrazed phytoplankton is fed upon by zooplankton and to utilize them, the zooplankton feeders are included in the combination.

The combination of the phytoplankton-feeding silver carp (Hypophthalmichthys molitrix), the zooplankton-feeding bighead carp (Aristichthys nobilis) and the weed-eating grass carp (Ctenopharyngodon idella) is well known in China and Southeast Asia. In India, under composite fish culture, six species of fish *viz.* catla (Catla catla), rohu (Labeo rohita) and mrigal (Cirrhinus mrigala) along with three Chinese carps such as grass carp, silver carp and common carp (Cyprinus carpio) are stocked together so as to utilize most of the fish food organisms present in the pond Other similar combinations may work just as well, but the most important aspect is to try to establish a balance between the species based on the food spectrum of the pond

Stocking Density

Normally fish production increases with the increase in the number of fish stocked per unit area upto a level and then starts decreasing. Higher stocking density results in increased total production, as there is better utilization of the available food, but in such cases the individual weight and size is reduced. On the other hand, lower stocking density yields larger individual fish.

Proper stocking rate for a pond is that optimum level which results in a given time, usually a year, in a production which is highest in quantity and quality of fish and most profitable. In ponds where no artificial feed is used, the total crop becomes dependent on the primary production and in such cases simply by increasing the stocking density, the increase in the total production is not possible. Even with supplementary feeding the scope of increasing stocking density and fish yield is limited; it increases to an optimum level and then starts decreasing.

Under crowded conditions fish compete for food and space and are stressed due to aggressive interaction. Fish under stress exhibit decreased feed consumption and slow growth and are predisposed to many parasitic and microbial infections. Increase in stocking density simultaneously increases the total oxygen demand with obvious dangers. In undrainable ponds, accumulation of excretory products of the fish population also suppresses their growth rate. With efficient removal of such metabolites by aerating the pond water, the stocking rate can be increased further, thereby enhancing production.

LIVING SPACE

It has been observed that under identical conditions of management levels and stocking density fish grow bigger in larger ponds. In Malaysia, grass carp, Puntius sp. and monosex Tilapia mossambica grew bigger in ponds of a larger area indicating the living space phenomenon In other words, the rate of production in a 0.2 ha pond will be more than double that of a 0.1 ha pond, despite the fact that the stocking rate per unit area is the same and all other management components including the genotype of the stocking materials and ecological conditions remain the same. The ponds having larger surface area are subjected more often to wind action resulting in greater rate of diffusion of atmospheric oxygen into the water.

Larger ponds have other advantages also, *viz.* better cooling action by wind. In smaller ponds, water tends to stagnate and in hot weather tends to heat up quickly. Though it is preferable to have ponds of a large size, there is a physical limitation. Large ponds are difficult to fill and even more difficult to harvest. There must be an optimum size and shape of the pond to balance size with practicability of management, i.e. large enough to allow proper growth of fish, but at the same time small enough to be manageable.

Supplementary Feeding

With the increase in carrying capacity of the pond either by aeration or circulation of water, fish growth can be increased further by supplementing the natural food with some artificial feed. This is the single most important management component for increasing production. In intensive and semi-intensive culture of fishes, supplementary feeding is indispensable. The quantity of feed and the form in which it is offered affect the rate of consumption. Temperature, dissolved oxygen level, crowding and health condition, etc., affect the rate of food consumption.

POND FERTILITY

Organic matter and mineral constituents of the pond soil supply the required nutrients for chemical and biochemical production processes. The pond bottom also provides a suitable environment for the decomposers like bacteria and fungi to mineralise organic components of the pond sediment and release soluble nutrients. Sometimes such nutrients are not available in sufficient quantity in the pond and hence they are added from outside in the form of fertilizers.Since plankton production is often limited by inadequate quantity of phosphorus which is essential for the assimilation of nitrogen into cellular matter, phosphatic fertilizers are widely used in fish culture. Unlike phosphorus, availability of nitrogen does not depend on the inherent status of soil, since it is brought to the soil by different processes.

Nitrogen fixation by azotobacteria, blue-green algae, atmospheric electric discharges and photochemical fixation are some of the potential sources of pond nitrogen. In a tropical climate, the fixation of atmospheric nitrogen by blue-green algae is of considerable importance.However, when nitrogen is added from outside, its form, *viz.*, ammoniacal or nitrate is also very important. It is advisable to use the ammoniacal form of nitrogen in acid and neutral soils and the nitrate form in alkaline soil.

Though there is a considerable loss of nitrogen from the ammoniacal form in alkaline soil, the use of ammonium sulphate in low doses is usually recommended, keeping in view the role of sulphate in reducing the soil alkalinity. Potassium is the other essential nutrient for plant growth.

In ponds it is easily available both in soil and water and does not form insoluble salts and is rarely deficient except in acid peaty soil. Yet, a little potassium when added to the pond, stimulates the production of plankton.

Organic matter of the pond sediment is also an essential factor regulating the bacterial activity. In this context, the ratio of organic carbon to total nitrogen (C/N ratio) is important. Periodic application of organic manures ensures to a certain extent replenishment of nutrients and also provides an energy base for bacterial activities. Apart from this, the organic matter and the bacterial flora are also directly consumed by zooplankton and some fish species.

DISEASES AND THEIR CONTROL

Various intensification approaches such as increased stocking rates, increased feeding, fertilization programmes, etc., sometimes result in nutrient accumulation, frequent appearance of algal blooms, dissolved oxygen deficiency and other water quality problems in undrainable ponds. As a result of such water quality and environmental problems, the infectious diseases and their control assume importance.A fish farming system is unique in that the farmed animal is poikilothermic and lives in water where respiratory oxygen level compared to air is limited and becomes critical at times. Further, metabolic waste products, left-out feed materials and organic load of the pond bottom can affect certain exposed vital organs and tissues of fish. All such factors affect fish health and contribute to the risk of disease outbreaks.The above basic facts need careful consideration while planning for freshwater pond fish culture. The habitat of an undrainable pond is very varied and dynamic, but can be monitored and managed for increasing fish production.

CHARACTERISTICS OF UNDRAINABLE AND DRAINABLE PONDS

In India ponds are relatively small and shallow bodies of impounded water with limited wind action. They may be called perennial if they retain water the year round or temporary/seasonal, if they do so seasonally. They may be further classified as drainable ponds and undrainable ponds, depending upon the drainage facility by gravity. Drainability imparts a very desirable feature to a pond and some authors prefer to call only the drainable type of ponds as fish ponds. However, Indian experience has shown that experimental fish production to the tune of over 10 tonnes/ha/yr can be achieved even from such undrainable ponds through a proper understanding of the biotic and abiotic components of the ecosystem and adoption of suitable culture technologies. Hence, before adopting any culture technology, it is imperative to have an idea about the basic biology of the pond types in terms of environmental factors, community structure and community metabolism.

UNDRAINABLE PONDS

The periods of 'plenty rain' and 'no rain' usually prevail in regions having undrainable ponds. With the onset of monsoon, torrential downpours sweep across the land and the amount and frequency of rain decrease towards the end of the monsoon. Severe floods may occur, whereas a late monsoon or early monsoon of short duration may result in serious drought. Both flood/ rain and drought influence the ecosystem of the ponds on such lands. Small, shallow and seasonal ponds get filled or dry, whereas deeper perennial ponds exhibit considerable fluctuations in water levels accordingly.

Though these ponds are basically constructed for storing water in such areas for multiple uses, ranging from supplying drinking water for human population, live stock, etc., to supplying water for agriculture, recent trend is to utilize them for fish culture. The description of the undrainable ponds is based upon the studies conducted at the Central Institute of Freshwater Aquaculture, Dhauli, Bhubaneswar, India, under an extensive environmental monitoring programme of rural undrainable ponds.

General Morphometry

Undrainable ponds in general are relatively small, perennial or seasonal water bodies constructed or excavated for multiple uses. Some of these ponds have proper embankments. They greatly vary in their dimensions ranging from 0.02 ha to over 2.5 ha in water surface area and 50 cm to 250 cm in depth. Larger ponds are relatively deeper while smaller and seasonal ponds are shallower.

Unlike shallow seasonal ponds, the bottom of the perennial ponds is never exposed to sunlight and therefore the whole ecosystem of such ponds is quite different from those of shallow and seasonal ponds. Use of these ponds by villagers for multipurpose provides the source of organic enrichment. Usually the only source of water for these undrainable ponds is the heavy rainfall during the monsoons.

However, in some cases, the pond bottom is cut below the water table so that ground water enters the ponds. As soon as the monsoon ceases, the water level starts decreasing gradually and shortage of water is quite common during the pre-monsoon season. Water is lost from the pond through evaporation, seepage and transpiration by aquatic macrophytes and the trees and shrubs planted along the pond sides. Macrophytes tend to appear in both perennial and seasonal types of ponds but with increased intensity in shallower ponds.

Presence of thick sediment layers in the bottom is the most characteristic feature of these ponds which gradually get accumulated during the course of time and vary between a few centimeters to over a meter and half in thickness.

The quality and quantity of sediment deposition depend mainly upon the original soil, method of construction, nature of embankments, macrophyte cover, pond productivity, organic and inorganic additions, species cultured, etc. Although sedimentation is relatively faster in smaller ponds, there is a positive correlation between the age of a pond and its sediment thickness.

Physico-chemical Environment

The water depth and total volume of water available for individual fish are crucial in fish culture systems. Adequate water depth is needed not only for optimum growth, but also to provide enough space and oxygen for fish life. Water levels in these ponds are mainly dependent upon monsoon rains.After the monsoon season, the water level starts decreasing gradually and shortage of water is quite common during the summer season which is the most crucial time for fish culture since the fish growth rate is faster in this period. In fact, during the time of lowest water level the ponds contain the maximum biomass.

In shallow and seasonal ponds, sufficient phytoplankton population fails to appear and the soft sediment layer is vigorously stirred up by fish, making the water more turbid, thereby reducing the photosynthetic process by limiting light penetration. Eventually, the total amount of available dissolved oxygen may not be, at times, sufficient to meet the demand for total community respiration and the chemical oxygen demand of the sediment, resulting sometimes in mass fish kill and planktonic collapse On the other hand, in deeper perennial ponds where the water column is more than 3 m, fish life is again adversely affected. In such ponds the photosynthetic or oxygen producing zone is less in comparison with the oxygen consuming layer.

In addition, the sediment proper and the sediment community also consume a considerable amount of oxygen. All such conditions lead to a negative oxygen balance.

The water in most of these ponds remains slightly alkaline (pH 7.0–9.0). The NH_4-N (ammonia-nitrogen) content of the waters remain below 0.02 mg/l with even lesser quantities of No_3-N (nitrate-nitrogen). The Po_4-P (phosphate) concentrations remain low and these chemical features of such ponds suggest that these waters are highly nutrient-deficient, particularly in nitrogen. On the contrary, the pond sediment is rich in organic and inorganic nutrients.

The organic carbon ranges between 3 (in newly excavated ponds) and 50 mg/g dry sediment weight (in older ponds). The nutrient status of the sediment differs completely from that found in the overlaying water column In general, all the basic nutrients in the pond sediment are about thousand times higher than in their respective water column.

Carbon dioxide and oxygen are the most important gases affecting the pond community including fish. During the photosynthetic activity, carbon dioxide is usually at zero level while during the darker period its concentration increases. At higher concentrations it may be toxic to fish life. Carbon dioxide toxicity increases with decreasing level of dissolved oxygen. Carbon dioxide concentration can be tolerated upto 20–30 ppm in these ponds provided oxygen is near saturation.

Community Structure and Function

Bacterioplankton and phytoplankton constitute the basic food for the fine filter feeder fish species and also for the zooplankton which form the main food of the rough filter-feeder species. The bacterioplankton population is always higher in those ponds which are associated with the activities of larger human and livestock populations.Most of the relatively older ponds with frequent appearance of Microcystis bloom have higher levels of bacterioplankton population (3–10 million/ml). On the contrary, the newly constructed and recently desilted ponds have less dense bacterioplanktonic community, around 1–2 million/ml. Macrophytic infestation also significantly limits the bacterioplankton production.

The planktonic detritus originates mainly from decomposing fragments of the phytoplankton and zooplankton and has generally a concentration range of 2,000 – 20,000 number/l. The main groups of phytoplanktonic population are Myxophyceae, Chlorophyceae, Eugleno-phyceae and Bacillariophyceae; whereas copepods, cladocerans and rotifers constitute the majority of zooplanktonic population.

Some of the very old ponds having excessively thick sediment layer face Microcystis blooming. In such ponds the benthic animal fauna is represented by a very small number. In the majority of the ponds the benthic animal communities are dominated by red chironomids and oligochaetes indicating the general oxygen deficiency in the sediment layer.

The bacterial decomposition and nutrient recycling in ponds are greatly influenced by the anaerobic nature of the sediment. At the initial stage of Microcystis bloom in older ponds, the oxygen production has been found to be the highest (1) (over 15 g O_2/m^2), whereas the total community respiration remains considerably low (2) (below 10 g O_2/m^2). However, during the active decomposition stage of Microcystis (plantonic collapse stage) the total oxygen production level goes lower (3) (5–6 g O_2/m^2) than that of the community oxygen consumption (4) (6–7 g O_2/m^2).In older ponds, especially those having thick anaerobic sediment, the biochemical oxygen demand ranged between 70% and

90% of the total oxygen production, ultimately causing anoxic condition leading to fish kills. The majority of rural undrainable ponds are characterized by anaerobic benthic sediments. The dead and decaying organic matter settles down to the pond bottom (sedimentation) where it is subjected to further decomposition and mineralisation.

The upper layer of the sediment remains aerobic while the deeper layers are deficient in oxygen and thus anaerobic. These perennial undrainable ponds in tropical monsoon lands with yearround warm water under plenty of light offer an excellent possibility for fish culture.

Most of the species cultured greatly depend upon natural fish food resources and with a limited dependence upon artificial supplementary feed.

However, without proper environmental management, the water remains infertile due to the overall nutrient deficiency with a very pronounced nitrogen limitation, although they possess a very high production potential. On the contrary, the pond sediments have extremely high level of organic nutrients in almost locked-up conditions which remain unutilized due to the anaerobic nature of the pond bottom. However, though regular raking up of the pond sediment, either by manual or biological means, the organic nutrients could be released for making the pond water more productive.

Proper management methods can optimise fish production in perennial ponds at most economical rates while seasonal ponds can suitably be utilized for fish seed rearing and also for short-term fish production depending upon the duration of water retention. The general feature of the properly managed and ill-managed undrainable fish p and 8b.

DRAINABLE PONDS

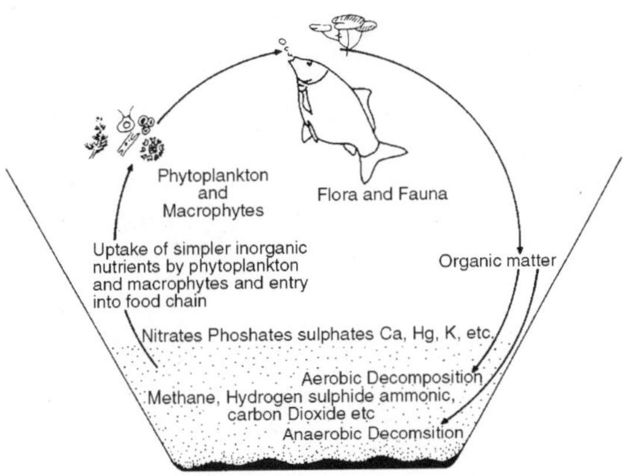

Fig. Microbial Decomposition Process at the Sediment-Water Interface

Ponds which can be supplied with water and drained of its water according to the requirements of the fish farming operation are known as drainable ponds.

These ponds require suitable ground with proper embankments, inlet and outlet structures and adequate supply of water on regular basis.

Fig. Well Managed Pond

Fig. Badly Managed Pond

Table: **Chemical, biological and functional characteristics ofund-rainable and drainable rural fish pondsin**

Parameters	Undrainable ponds *(natural condition)*	Drainable ponds *(cowdung treated)*
Water pH	7.0 – 9.0	7.7 – 8.2
Total alkalinity (mg/l)	50 – 250	88 – 200
Ammonia-nitrogen (mg/l)	.005 – 0.300	.005 – 0.25
Nitrate nitrogen (mg/l)	.005 – 0.020	.005 – 0.20
Phosphorus (PO_4-P) (mg/l)	0.001 – 0.050	0.040 – 0.160
Plankton		
Phytoplankton (number/l)	59 – 3 911	3 – 860
Zooplankton (number/l)	124 – 2 770	11 – 209
Bacterioplankton (million/ml)	1.2 – 12.9	1.385 – 2.312
Benthos (number/m^2.)	0 – 2 660	1 415 – 19 099
Decomposition rate of Eichornia leaves:		
Surface (per cent dry wt. loss/day)		2.88 – 3.86 2.37 – 2.86

Bottom (per cent dry wt. loss/day)		4.02 – 10.99 1.54 – 2.84
Gross production (g carbon/m²/day)	1.76 – 10.99	1.85 – 8.11
Net production (g carbon/m²/day)	1.29 – 1.35 (-)	(-)2.4 – 1.875
Community respiration (g carbon/m²/day)	1.66 – 11.34	2.261 – 6.071
Sediment oxygen consumption (g oxygen/m²/day)	4.887 – 7.943	0.1766 – 3.514

PRESENT PRACTICES OF FISH CULTURE IN PONDS

CARP CULTURE

The most successful system of pond fish culture is the polyculture of three Indian major carp species-catla, rohu and mrigal along with three Chinese carps *viz.* silver carp, grass carp and common carp. In India this is commonly known as composite fish culture. The best results in terms of fish production in this system results not only through a judicious combination of species, but also due to appropriate management techniques including pond fertilization, supplementary feeding and health care.

On the basis of growth performance of different species, modifications are often made in stocking density, species ratio, fertilization schedule and supplementary feeding programme in different agroclimatic conditions. High rates of fish production to the tune of over 5 500 kg/ha/6 months, 7 200 kg/ha/8 months and over 10 tonnes/ha/yr have been achieved in composite fish culture trials conducted in different agroclimatic conditions of India.The carp culture system as a whole is operated as a three-tier culture system where the practices are adopted for rearing fish during their different stages till they are harvested. Spawn (post larvae) are reared upto fry (2–3 cm) stage in nursery ponds, fry to fingerlings (8–12 cm) in rearing ponds and finally fingerlings to table-size fish in composite fish culture ponds or stocking ponds.Relatively smaller, seasonal ponds are mainly used for rearing spawn to fry stage and harvested after 2–3 weeks. Several crops of fry are usually taken during the season. Pond fertilization by cattle manure and feeding with 1:1 mixture of oil cakes and rice bran is the usual practice. Fry raised in nurseries are reared upto fingerlings in slightly bigger ponds (0.05 – 0.1 ha) of seasonal or perennial in nature. Fingerlings are removed after 3 months and stocked in composite fish culture ponds.

INTEGRATED CARP FARMING

An integrated approach of composite fish culture together with compatible combination(s) with poultry, duckery, pig rearing and cattle raising is now being adopted. Under this system of farming small livestock and farm yard animals, *viz.* pigs, poultry, ducks, etc., are integrated with composite fish culture by siting animal housing units on the pond embankments in such a way that the animal wastes and washings are diverted into fish ponds for recycling. The fish not only utilize spilled animal feed but also directly feed on fresh animal excreta which is partially digested and is rich in nutrients. Surplus excreta supports the rich growth of planktonic fauna. Fertilizers and supplementary feed are not used, resulting in drastic cost reduction

Table: **Annual production through integrated carp livestock farming**

Integration	*Fish production*	*Animal production (live weight)*
Fish + Pig farming	6 – 7 ton/ha	4 000–5 000 kg pig meat
Fish + Duck farming	3 – 4 ton/ha	500 kg duck meat + 17 000–20 000 eggs
Fish + Poultry farming meat	4–5 ton/ha	60 000–70 000 eggs + 1 500– 2 000 kg

Integrated Fish-pig Farming

Pigstyes are constructed either on the pond embankment or near the pond to facilitate easy drainage of waste directly into the pond which acts as pond fertilizer and supports dense growth of natural fish food organisms. Besides, fish also feed directly on the pig excreta. No other feed or fertilizer is applied to the pond. A pond is prepared by following the usual pond preparation techniques and stocked with fingerlings of all the six species of carps cultured under composite fish culture at higher of 8 000–9 000/ha with surface, column, bottom feeders and grass carp in the ratio of 40:20:30:10. Marketable size fish are sold by partial harvesting while final harvesting is done only after 12 months of farming.About 2 months-old weaned piglets are fattened for six months when they attain slaughter size (60–70 kg) and similarly a second crop is raised within the next six months. About 30–40 pigs should be kept for proper fertilization of the pond. Pigs are fed on mash at an average rate of 1 kg/day. Green grasses or animal fodder is also provided. Grass with interlocked soil in root system (sod) are provided once a week to avoid mineral deficiency. Grass carp is fed with aquatic weeds or green animal fodder. Fish yields ranging from 6 000–7 000 kg/ha/yr are generally obtained.

Integrated Fish-duck Farming

This is also an efficient integrated system based on the principle of waste recycling. Pond preparation technique is basically the same. A duck house is normally constructed on the pond embankment or on the pond water on a floating platform When given free range, ducks feed on aquatic organisms such as insect larvae, tadpoles, molluscs, weeds, etc. The duck droppings like pig excreta act as fertilizer.Ponds are prepared and stocked with fingerlings of all the six carp species at 6 000 ha with surface, column, bottom feeder and grass carp in the ratio of 40:20:30:10. Fingerlings of over 10 cm are preferred for stocking. About 200–400 ducks are sufficient to adequately fertilize a l ha pond. Normally 2–3 months old ducklings are preferred.Although ducks are able to feed upon natural food from the pond, they are also provided with duck feed at the rate of 100 g/bird/day.Ducks start laying after 5–6 months and continue for 2 years. Fish yields ranging from 3 000–5 000 kg/ha/yr are generally obtained.

Fig. Fish-cum-pig Farming (Wooden Pigsty)

Fig. Fish-cum-pig Farming (Concrete Pigsty)

Fig. Fish-cum-duck Farming (Duck House in Pond Dyke)

Integrated Fish-Poultry Farming

Under this system of integration the poultry birds are raised in cages under a shed normally constructed over the pond embankments or in the vicinity of the pond. The space requirement in such a system of poultry raising is about 1 sq.ft. per bird. The droppings of the birds fall on the floor from where these are collected and applied to the pond. The chicken house can also be built directly over the pond water so that the excreta may fall in the pond water underneath. Usually, 400–600 chickens/ha of pond water surface are used. No feed or fertilizer is applied in the pond, except aquatic vegetation for the grass carp. Fish production at the rate of 4–5 t/ha is possible using this system. In India, this system of freshwater fish culture has assumed greater significance in view of its potential role in recycling of organic wastes and in integrated rural develop.

AIR-BREATHING FISH CULTURE

Besides freshwater ponds, there are many low-lying areas which become waterlogged during the rainy season. In course of time these areas get infested with dense aquatic vegetation and turn into swamps.

Swamps are also formed along the irrigation canals due to profuse seepage. These areas are best suited for culturing airbreathing fishes such as Koi Singhi, Magur and Murrels without getting involved in costly processes of their reclamation essentially needed for carp culture. There are three levels of culture practices *viz.* low cost culture, semi-intensive culture and intensive culture depending on inputs and level of management

Fingerlings (6–10 g) and feed are the two material inputs used in the culture system. Fertilizer is not used. However, replenishment of water becomes an essential input in case of intensive system of culture in ponds where very high stocking rate and intensive feeding is practised to obtain very high yields Monoculture as well as polyculture of these fishes are commonly undertaken, yielding production to the tune of 3 000 to 7 000 kg/ha/yr.

Production at the rate of over 3 000 kg/ha/8 months is possible through monoculture of Channa marulius in swampy ponds. In intensive system of monoculture of magur and singhi with frequent change of water, yields of over 15 t/ha/yr have been obtained.

Presence of naturally occurring weeds in airbreathing fish culture ponds not only provide protection against poachers, but also encourage the growth of insects which are consumed by the fish.

SEWAGE-FED FISH CULTURE

The wastes, including sewage and waste water produced by the human community hold high potential for fish production. In India itself there are about 150 sewage-fed fish farms covering an area of about 12 000 ha. Very high production in the order of 7–10 tonnes/ha/yr has been obtained from ponds fed with sewage which invariably contains high percentage of N,P,Ca,K,etc.An average production of about 7 t/ha/yr is easily obtained using a mix culture of 5 carp species The sewage fed ponds are generally dewatered completely during summer so as to remove all the carnivorous fishes. The pond is initially fertilized by introducing partially treated sewage effluent upto about 75–80 cm and then clean water is pumped in to raise the pond water level to 1.5m.

Within a month the pond stabilises with respect to dissolved oxygen and becomes suitable for stocking with fish seed. During raising of marketable size fish, additional fertilization with sewage effluent is carried out in small doses every month and the pond is netted frequently to help oxygenate the water and in course of netting the marketable size fish are also harvested.

CONSTRUCTION OF NEW PONDS AND FARMS

Village ponds, homestead or backyard kitchen ponds, garden or farm ponds, irrigation ponds and occasional ponds such as brick mine pits and quarries, etc., occupy enormous freshwater areas in the tropics and are used for fish culture with minor improvements. However, ponds designed and constructed for fish culture are easier to manage and are expected to give higher production.

Although certain well-defined guidelines do exist for the construction of fish ponds, it is mainly the topography of the site which determines the basic design of the pond/farm. There are, however, certain basic principles to be considered when choosing a site and deciding the method of pond construction.

SITE SELECTION

Selection of suitable sites for fish farm construction is very important. The following three essential conditions guide the proper site selection:

- Topography
- Source of water and its quality
- Soil type

Topography

It is economical and convenient to construct ponds in waterlogged areas, irrigation command areas or in marginal lands. In such areas construction cost is relatively low mainly due to limited earth cutting. For example, a pond of 100 m × 40 m (0.4 ha) of water area requires only 3 234 m^3 of earth to construct around a dyke of 2 m high above ground level (GL) with side slope ratio of 2:1 and top width of 1.5 m. This quantity of earth may be obtained only from 1.1 m depth of cutting. This limited depth of cutting reduces the construction cost considerably. However, full consideration should also be given to the possible effects of flood.

The surface features of the area proposed for the pond or the farm is also equally important. A saucer-shaped area may be an ideal site for a large dug-out pond, because it may hold appreciable quantity of water with a small amount of earthwork.For smaller and flat areas eye estimation is enough, but for a big area proposed for farm construction with a number of ponds for different purposes and of different sizes, it is essential to conduct contour survey for determining the topography and land configuration.The site should be easily approachable so that there may not be any difficulty in the transportation of input materials and in the marketing of the produce. The labour and materials required for construction and operation should also be locally available as far as possible.From an efficient management point of view the pond site should, if possible, be within the sight of the farmer's house. It also reduces the risk of poaching. Siting fish ponds near the farmer's other agricultural or livestock farming activities makes it easier to integrate all the farming activities.

Source of Water and its Quality

A dependable source of water supply must be available within or near the site, even for undrainable ponds. However, unlike drainable ponds, undrainable ponds require just sufficient water to fill the ponds and to compensate the water loss through seepage and surface evaporation thereafter. Equally important is the need for avoiding excess water and hence there must be arrangement for the excess water to escape through a bypass channel or a spillway.

The water supply to the pond should as far as possible be natural, preferably rain water. However, alternative arrangements of water supply should be made for dry season either from a deep tube well or irrigation canal or from perennial sources like spring, stream, river, etc. Ponds should be on the lower lands to allow accumulation of surface Run-off from a larger catchment area.

However, care should be taken to provide proper bypass or spillway to avoid flooding. A higher subsoil water table due to irrigation in surrounding fields and percolation from artificial or natural channels, in addition to absorption from rain water, also helps in maintaining water level in undrainable ponds. The quality of the available water is also equally important for fish culture. Pond fish production is influenced by the physical and chemical properties of the water. Water should be clear as far as possible. Turbid waters which carry suspended solids cut the light penetration, thus reducing primary productivity of the pond. Excess of suspended solids also adhere closely to the gill filaments and cause breathing problems.Water temperature also significantly influences the feeding and growth of fish. Prevailing water temperature, ranging between 15°C and 35°C in tropical areas, is most suitable for carps. The chemical quality of water depends on its content of dissolved salts. Rain water does not carry any dissolved salts. However, it collects nutrient salts from the ground surface of the catchment area. The water should be neither too acid nor too alkaline; neutral or slightly alkaline waters are most suitable for fish culture and hence acid water should be limed to make it neutral. Waters with p^H values below 5.5 or over 8.5 are not proper for fish culture. The farmer will need huge quantity of lime to neutralize it while highly alkaline water may cause the precipitation of both phosphate and iron and if it remains continuously above p^H 9, it may be harmful to fish.

Soil Type

Pond soil must retain water. Soils with a low infiltration rate are most suitable for fish pond. The filtration rate of different types of soils. The best soils for our purpose are thus the impermeable clay which can be easily compacted and made leak proof.

Table: **Infiltration rates of different types of soil**

Soil type	*Infiltration rate (mm/ha)*
Clay	1–5
Clay loam	5–10
Silty loam	10–20
Sandy loam	20–30
Sand	30–100

Loamy soils can also be used, but they need well compacting and may leak slightly in the early stages, although they tend to seal themselves with time. Sandy and gravelly soils should be avoided, but if they are the only ones available they must be made impermeable with a thick coating of clay or with polythene sheeting. Soil impermeability can also be achieved by soil compaction at the pond bottom and dyke with either a mixture of soil + 1–5% cement or soil + 10–20% cowdung. Treated areas should be kept moist for 2–3 days by gently sprinkling water to avoid cracking and finally the pond is filled with water. Peat soils have special problems, since they are usually very acidic in nature and need sufficient liming, while the organic matter decomposition may lead to dissolved oxygen deficiency. Soils rich in limestone also create special problems, since the excessive lime content tends to precipitate phosphate and iron. Such ponds would then have little plankton population and macrophytes and would be relatively sterile.

This can be overcome by adding sufficient organic matter such as cowdung, poultry manure, etc.A general and convenient field test for the soil quality is to take a handful of moist soil from the test holes made at the proposed site and to compress it into a firm ball. If the ball does not crumble after a little handling, it indicates that it contains sufficient clay for the purpose of pond construction. Accurate determination of the composition of the soil and its water-holding character is possible by hydrometer method.

Several test holes may be made across the site and soil samples may be collected vertically from every 0.5 m of depth reaching up to a level of 3–4 m in a test hole. Using the results of the soil tests, a soil profile chart for the proposed site may be drawn. An arbitrary soil profile chart is presented showing the presence of clayey soil up to a depth of 3.5 m.

DESIGNING

Based upon the survey on topography, soil type, water supply, etc., the detailed designing and layout of the ponds/farm are done. However, the following additional points are also to be considered.

Water Area Ratio Among Pond Types

The production or stocking ponds are stocked with large size fingerlings of about 10–15 cm size in the case of composite fish culture. To attain this size, the hatchlings are reared in much smaller and shallower ponds called nursery and rearing ponds for about 2–3 months. In the nursery ponds the hatchlings are reared up to fry stage and in the rearing ponds the fry are reared till fingerling stage. The ratio of water area among nursery, rearing and stocking ponds in a fish farm depend upon the basic objective of the farm. In case of a fish seed farm, only nursery and rearing ponds are to be constructed with a small area for few stocking ponds to be used for raising the brood fish, while in the case of fish production farm only stocking ponds are to be constructed for producing table size fish from fingerlings.

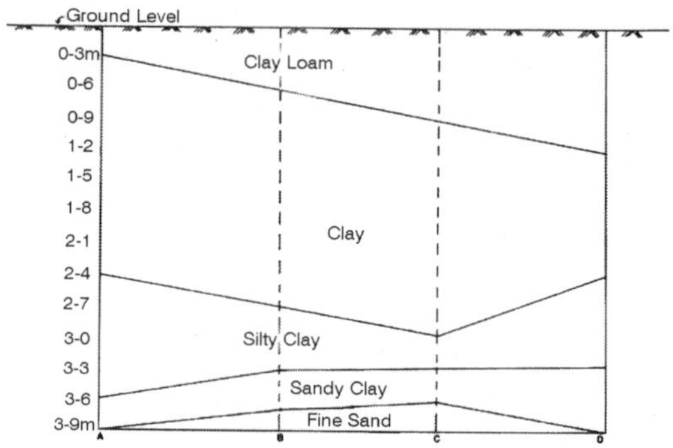

Fig. Soil Profile

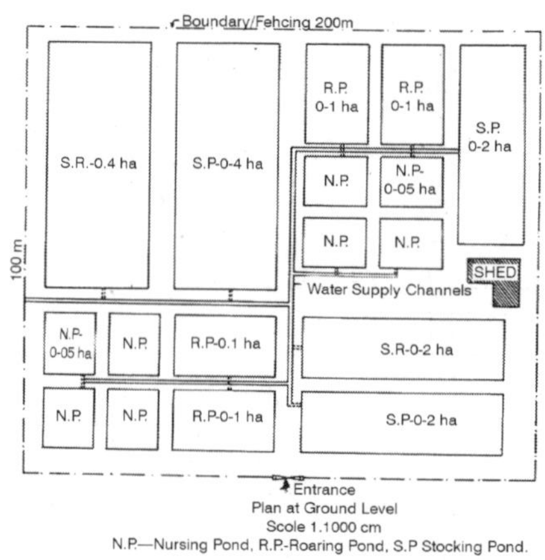

Fig. Layout of a Fish Farm (Land area 3.6 ha)

There is no hard and fast rule regarding the size of a pond. However, nursery ponds should be small and shallow. Ponds having 0.02–0.06 ha water area and 1–1.5 m depth are most suitable as nurseries. Rearing ponds are relatively larger, preferably between 0.06 to 0.10 ha in size and 1.5 to 2.0 m in depth. The sizes of stocking ponds vary tremendously. For newly constructed undrainable ponds, total water area of 0.25 to 1.0 ha is recommended.

In shallow ponds the water becomes heated easily. In deeper ponds light cannot reach the bottom. In very deep ponds thermal stratification may occur with colder deoxygenated bottom layer. Dead plankton and faecal matter from fishes may fall on the bottom layer where the nutrients may be locked up.

However, in case of rain-fed areas where the water table goes down during the dry season, the depth should be kept around 3.0 – 3.5 m to store more water during the rainy season.

Although a square pond is economical to construct for its minimum length of dyke, a rectangular shape of the pond (length:width in proportion of 3:1) is considered to be ideal.

In any case the pond width should not exceed 30 to 40 m as it is difficult to operate a fishing net in broader ponds. The nursery and rearing ponds may be square, since they are too small to pose any problem for netting. The corners must be curved to avoid fish escaping the net during harvesting.

Table: **Practical size and depth of nursery, rearing and stocking ponds**

Depth[*] (m)

Pond type	Size (ha)	Irrigated command/ *water logged areas*	Rainfed[±] */non -irrigatedareas*
Nursery pond	0.02 – 0.06	1.0 – 1.5	1.5 – 2.0
Rearing pond	0.06 – 0.10	1.5 – 2.0	2.0 – 2.5
Stocking pond	0.25 – 1.0	2.0 – 2.5	2.5 – 3.5

* Excluding the freeboard

\+ May vary depending on impermeable strata at pond bottom

Dyke

The dyke should be properly designed so that it can hold maximum water in the pond and withstand the hydraulic pressure. The slope of the dyke usually depends on the type of soil. Suitable side slopes for different soil 7.

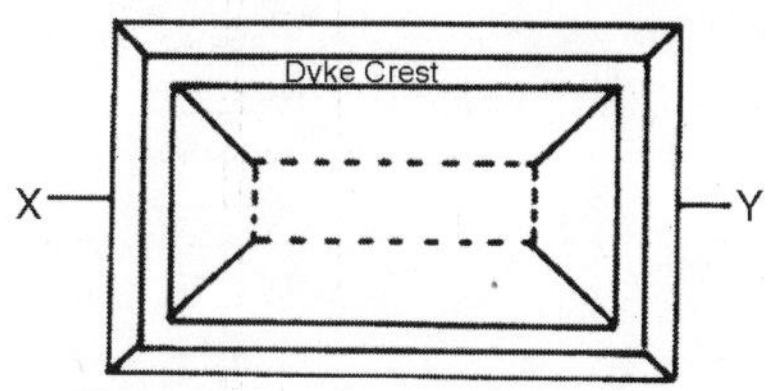

Plan of a Nursery Pond (0.02 ha)

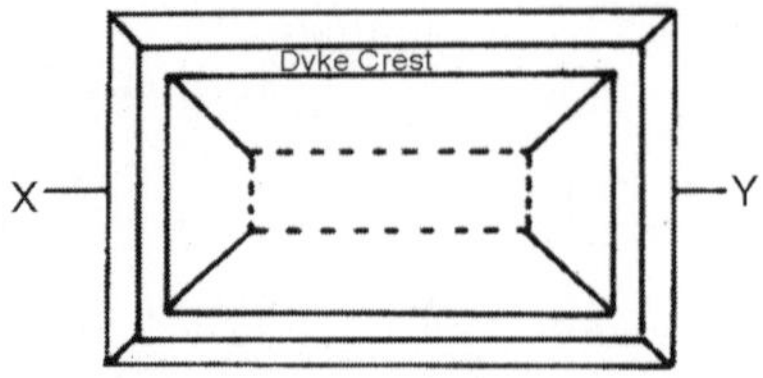

Plan of a Nursery Pond (0.02 ha)

Plan of a Stocking Pond (0.4 ha)

Fig. Design of Nursery, Rearing and Stocking Ponds

Table:.**Suitable slopes for different soils(Sahoo, pers.comm.)**

Soil type	*Soil (horizontal:vertical)*
Clay	1:1 to 2:1
Clay loam	1.5:1 to 2:1
Sandy loam	2:1 to 2.5:1
Sandy	3:1

Provision for a berm of sufficient width may also be provided for stabilizing the slopes. A wider berm also helps in operating the net in the pond. The berm should be 1 m or more in width. The top width of the dyke should be decided taking into account its usage. Usually the minimum top width of the dyke should be 1.5 m. The wider crest requires not only a larger area for dykes, but also an increased amount of earth material involving heavy expenditure. It is always wise to design the dyke as per the quantity of earth expected to be available from excavation work. A soil-type containing approximately 25% silt, 35% sand and 40% clay is most suitable for dykes. However, if excavated soil quality is not up to the above standard, provision may be made for a clay core to make the dyke watertight. While designing, about 10–12% allowance may be given for settling of earthwork.

CONSTRUCTION

Before initiating the construction work, proper estimates have to be prepared based upon the design details, which will include the cost of all the materials and the labour. Strict supervision is required at every step of construction to ensure the adherence to specifications laid down in the design.

Time of Construction

If the construction work is taken up at the most appropriate time or season of the year, the work becomes easier and economical. The best time of the year for constructing ponds in clayey soil is post-rainy period and winter when the soil is soft rather than at the end of the dry season when it is very hard. For swampy and waterlogged areas the most desirable time is the late summer when the area becomes completely dry. However, if a pond is built during winter or early summer and is not filled immediately, weeds may grow and cover the bottom. In such cases deweeding is needed before filling the pond.

Preparation of Site

The site should be thoroughly cleared of all the trees, bushes, etc. Even the roots of trees should be removed. No woody material should be left because the same will eventually rot and cause leaks. Some tree trunks rot very slowly and may cause problems during netting.

Marking the Outlines

This operation involves laying out the features of ponds on the ground in order to mark out the areas from where the earth will have to be cut and removed and also where earth will have to be embanked. Initially, lines are drawn according to the layout, followed by pegging and fixing stakes or posts. Strings are stretched between the tops of pegs and posts to mark the complete profile of the dyke with its correct height, width and slopes

Pre-excavation Work

Prior to pond excavation and dyke construction, all loose surface soil should be removed from about 20 cm depth within the total outlined area of the dyke and the surface should be roughened by ploughing or digging. In order to unite the body of the dyke to subsoil, it is desirable to dig a small "V" shaped key trench.

When the dyke is to be made on a sandy, gravelly or marshy soil base, the construction of a key trench becomes essential and in such cases digging should be done until watertight foundations are reached. The key trench is a small ditch or furrow dug along the line of the centre of the walls about 0.5 m – 1.0 m wide and 0.5 m deep. This trench is filled in with a good clayey soil and is well rammed. If good clayey soil is not available in the area, ordinary soil should be well compacted into the trench. The purpose of the trench is to stop seepage of water underneath the walls.

Pond Excavation and Construction of Dykes

The excavation work can be carried out within the area marked for the pond bottom either manually or mechanically. However, the final levelling of the pond bottom and sides should be done manually with proper ramming and finishing as per the original design. The construction of the pond becomes economical if earthen dykes are made around the pond using the excavated earth from the pond bed. All dykes should be raised, dumping the earth layer by layer stretching right across the whole section and in such cases each layer should not exceed 20 cm in thickness. All large clods should be broken and each layer should be thoroughly consolidated by watering and ramming. The sides and top of the dykes should be properly dressed and finished with wooden thappies (wooden block with handle for ramming).

In case the soil quality is not suitable for making dykes, a clay core is provided in the dyke to make it watertight. A mixture of 1:2 of sand and clay is used to make the clay puddle. This should be consolidated, compacted and deposited in 10–15 cm thick layers. Each layer should be adequately moistened before the next layer is laid and precaution should be taken to prevent the puddle from becoming dry and cracking. Dykes must be well compacted to render them stable and the top should be rammed flat so that small vehicles can also run along when needed. Short creeping grass is recommended to be grown on the top and sides of the dyke. Trees are not desirable since their dense shade inhibits the productivity of the pond.

Water Inlet Structure

Since we are concerned here with static and undrainable ponds, a feeder stream running directly into the pond should be avoided. The feeder stream must therefore be diverted along the side of the pond and from a suitable point water is channeled to the pond when required.An inlet structure should be provided through which water can be let into the pond. A proper inlet enables the quantity of water flowing into the pond, to be regulated, preventing the entry of undesirable fish and other aquatic animals and the escape of stocked fish.For small ponds the best inlet structure is a galvanized iron pipe of about 10 cm diameter with a control tap and a screen basket The downstream end of the pipe should be 30–40 cm above the water level. A sluice is also suitable for this purpose, especially for larger ponds. A screen is also fixed to check the entry of undesirable fishes and other animals To avoid scouring when the pond is being filled, a concrete apron can be built at the sluice, or more cheaply, a layer of gravel laid down. Similarly, if water is let in with a pipe there should be a gravel bed laid down where the water stream falls into the pond. If gravity feed is not possible, water must be pumped from the supply source into the channel leading to the pond or even directly into the pond; but, in that case, the intake should be securely wrapped by a firm net to prevent undesirable fish and other animals from entering into the pond along with the water.

MAINTENANCE

Proper maintenance of the pond and pond structure is most essential. Most of the earthen structures, especially the dykes, are susceptible to weathering action and hence they need periodical checks. Attending to minor damages regularly avoids the chances of more costly repairs later.The grass turfing needs special attention. Proper and timely mowing prevents the formation of weedy growth and tends to develop a root system more resistant to Run-off. Erosion from the top during heavy rains causes grooving out of small channels and it is an indication that the top has not been properly consolidated.

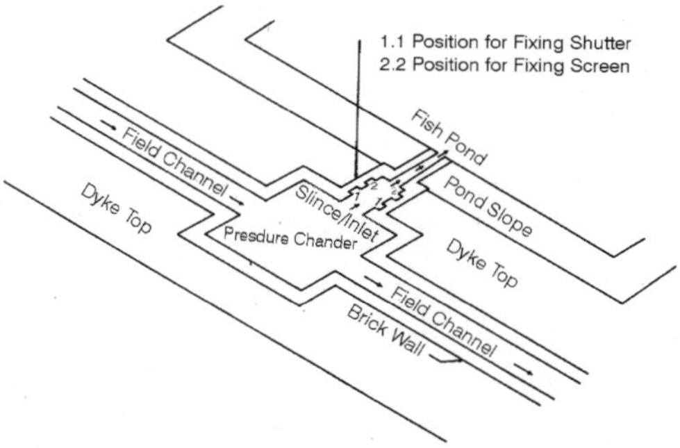

Fig. View of an Inlet Structure

The area should be levelled with more soil and thoroughly rammed and then grass should be planted to bind it. Side erosion at the dyke bottom may be due to a number of reasons. The worst damage is done by common carp.

Erosion due to frequent wave action, particularly if the grass at the edge has been grazed by grass carp, can cause undercutting of banks and subsequent collapse of dykes. Some methods used to provide protection against such erosion are earth berms, stone or brick pitching, stakes/bamboo piling.

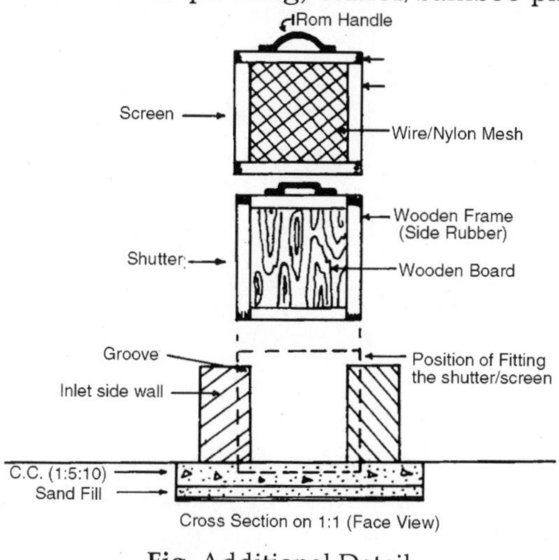

Fig. Additional Detail

Surface washings and organic additions cause siltation which reduces the pond depth and pond fertility. The undrainable ponds should therefore be dewatered in the summer months at the interval of 5–7 years.

FISH SPECIES SUITABLE FOR CULTURE IN PONDS

Although a large number of fish species grow successfully in ponds, only a restricted number of species are usually cultivated on commercial scale. Reasons for this restricted choice is obvious.

Commercial pond culture basically aims at achieving maximum possible rate of fish production and profit through optimum utilization of the natural food and the supplementary feed which drastically limits the choice of fish species for pond cultivation.

CRITERIA FOR SELECTION OF SUITABLE FISH SPECIES

- Adaptability to undrainable pond environment
- Faster growth rate
- Efficient utilizers of natural food resources of the pond
- Efficient converter of artificial feed
- Hardy and not easily susceptible to disease
- Easy to breed and rear the seed
- Prolonged breeding period or multiple breeding frequency
- Non-predaceous, planktophagous and preferably herbivorous and detritus feeder
- Compatability with other cultivable species of fish
- Palatable with high nutritive value
- High market demand and high price.

To find all these qualities in one fish species would be very unlikely. Therefore, the species having maximum required traits are considered to be desirable for cultivation in undrainable ponds. Carps fit well to these criteria and hence the most widely cultivated food fishes in South Asia are the quick-growing, non-predatory carps.

FISH SPECIES SUITABLE FOR CULTURE IN UNDRAINABLE PONDS

There are two major systems of carp culture in Asia: the Chinese polyculture system where Chinese carps are cultured together and the Indian composite fish culture system where the Indian major carps and Chinese carps are combined. In China, Chinese carps such as silver carp (Hypophthalmichthys molitrix), grass carp (Ctenopharyngodon idella), bighead carp (Aristichthys nobilis), mud carp (Cirrhinus molitorella), black carp (Mylopharyngodon piceus) and common carp (Cyprinus carpio) are cultured under polyculture system in ponds. Under the Indian system of composite fish culture in undrainable ponds, three Indian carps, *viz.* catla (Catla catla), rohu (Labeo rohita) and mrigal (Cirrhinus mrigala), two Chinese carps, *viz.* silver carp (Hypophthalmichthys molitrix) and grass carp (Ctenopharyngodon idella) and common carp (Cyprinus carpio) are extensively cultivated.

Fig. Anti-erosion Meaasures

Catla

Catla is the fastest growing Indian major carp species and widely distributed throughout India, Nepal, Pakistan, Burma and Bangladesh. It inhabits the surface layer of water and feeds upon plankton. Adult stages are predominantly zooplankton feeder, occasionally taking in decaying macrovegetation, phytoplankton and smaller molluscs. It attains maturity in the second year of life and carry over 70 000 eggs per kg body weight. It naturally breeds in rivers during monsoon season and under control conditions in bundhs as well. It does not breed in ponds. However, it responds well to hypophysation techniques. Seeds are easily reared in undrainable ponds of relatively smaller size. Under composite fish culture in ponds it usually grows to over 1 kg in one year.

Rohu

Rohu is the natural inhabitant of river systems of India, Nepal, Pakistan, Bangladesh and Burma. In recent years it has been transplanted to many countries of the world including Sri Lanka, Mauritius, USSR, Japan, Philippines, Laos, Malaysia and Thailand.Normally it occupies the column region of the aquatic ecosystem and feeds mostly on vegetable matter including higher plants, detritus, etc. Like catla it naturally breeds in rivers and under special conditions in bundhs.Except by hypophysation to which it responds quickly, it never breeds in ponds. It attains sexual maturity during the second year. However, certain percentages of pond-reared specimens mature within one year. Fecundity varies from 226 000 to about 2 800 000 depending upon the size. Rohu spawns during the monsoon.Seeds collected from rivers or produced by bundh breeding or induced breeding are reared with ease in seasonal or perennial undrainable ponds. Under pond culture conditions it grows upto 900 g within one year.

Mrigal

Mrigal inhabits all the major river systems of India, Pakistan, Bangladesh and Burma. The adult fish feeds upon filamentous green algae, diatoms, pieces of higher plants, decayed vegetable, mud and detritus. It is basically a bottom feeder and hence suitable for cultivation with column and surface feeder carps in ponds. Mrigal usually attains maturity within 1 or 2 years depending upon the agroclimatic conditions of the location.

Fecundity ranges between 124 000 to over 1 900 000 depending upon size. Spawning season is linked with the onset and duration of the southwest monsoon. It does not breed in ponds, but can be easily bred in bundhs and by hypophysation. It is now being induced to breed twice within the same spawning season. Rearing of seed is usually undertaken in seasonal or perennial undrainable ponds. Under pond culture conditions it grows to over 1 kg in one year.

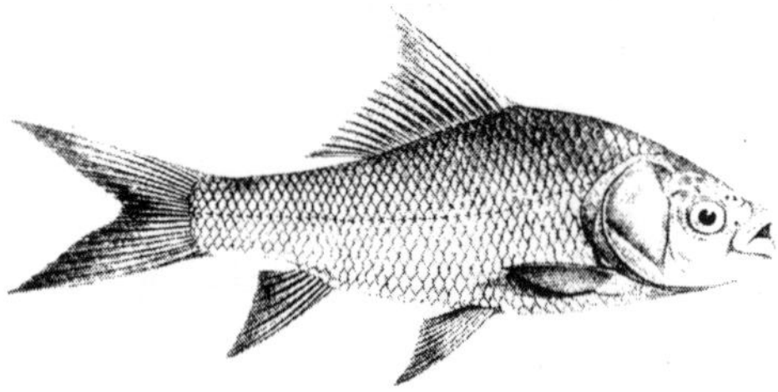

Fig. Catla (Catla Catla)

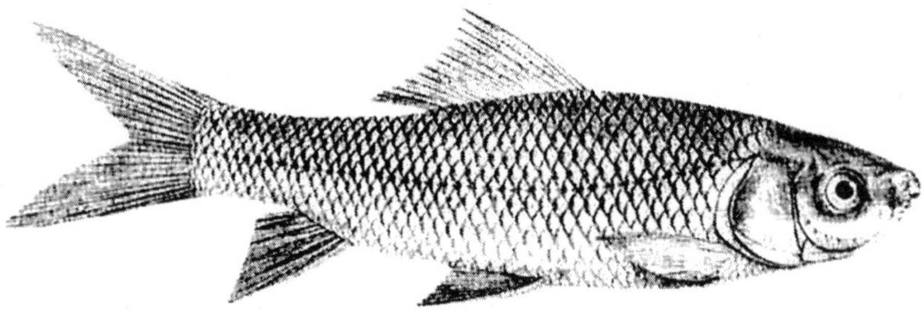

Fig. Rohu (Labeo Rohita)

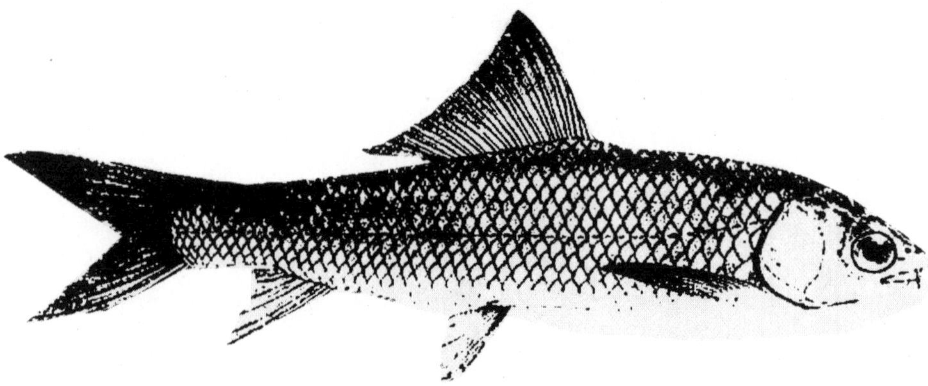

Fig. Mrigal (Cirrhinus Mriqala)

Silver Carp

Silver carp is basically inhabitant of major river systems of South and Central China and in the Amur Basin of USSR from where it has been transplanted throughout the Indo-Pacific region including India. It is a surface dweller feeding mainly upon zooplankton during its early stages and gradually becomes predominantly a phytoplankton feeder.Its relatively longer branchiospines provide a fine filter capable of retaining planktonic organisms.

It readily accepts supplementary feed like oil cakes and rice bran mixture in pond culture systems. It does not breed in pond condition. However, through the technique of hypophysation they are induced to breed in ponds during the monsoon season. Fecundity varies greatly with the size and agroclimatic condition. A fecundity range of 145 000 to 2 044 000 has been found from silver carp. It takes about 2–6 years to mature in China, whereas in India it matures very early, within 2 years.

Males mature earlier than the females. In composite fish culture ponds it usually attains over 1.5 kg within one year of rearing. Seed rearing is done in smaller seasonal or perennial undrainable ponds with a high rate of survival.

Fig. Silver Carp (Hypophthalmichthys Molitrix)

Grass Carp

Grass carp is a native of the river systems of South-Central and North China and the Amur river of USSR. Its suitability in aquaculture and biological control of aquatic weed infestation has resulted in wide-scale transplantation throughout the world. In early life it feeds on planktonic organisms and gradually switches over to macrophytes. They are voracious eaters and show distinct preference for vegetable food materials such as grass, leaves, weeds, etc.

However, they also accept supplementary artificial feed materials. Usually only a portion of ingested food is digested and the rest is voided in semidigested or undigested form which, in turn, becomes choice food for the bottom dweller common carp In China it takes about 3–4 years to achieve maturity whereas in India it usually takes 2 years.The total number of eggs range between 308 800 and 618 100 from the fishes weighing between 4.7 kg to 7.0 kg. The fish does not breed under pond condition and hence seed production is achieved through hypophysation. Growth mainly depends on the rate of feeding. Under optimum feeding rate it can grow over 5 kg in one year Usually it grows to over 1.5 kg in composite fish culture ponds.

Fig. Grass Carp (Ctenopharyngodon Idella)

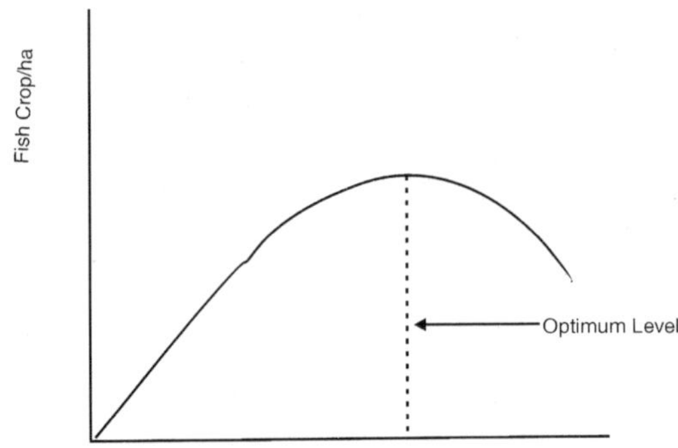

Fig. Grass Carp (Ctenopharyngodon Idella)

Common Carp

Originally a native of temperate region of Asia, especially China, the common carp is now the most domesticated and cultivated carp species throughout the world. It is an omnivorous bottom dweller subsisting mainly on benthic fauna and decaying vegetable matter. It frequently burrows the pond bottom in search of food.

This habit of digging the pond bottom helps in maintaining the productivity of undrainable ponds and hence culture of common carp with other carp species is of great advantage. Moreover, it also feeds directly on the undigested excreta of grass carp. Growth mainly depends upon the bottom biota, stocking density and the rate of supplementary feed.

Fig. Common Carp (Cyprinus Carpio)

In composite fish culture ponds it grows to about 1 kg within one year. In a tropical climate it spawns throughout the year in the pond environment with two peak periods, one from January to March and the other during July and August. Eggs are small and adhesive in nature. In tropical conditions it attains maturity within 12 months

FATTY ACID CONTENT IN FISH

ENVIRONMENTAL INFLUENCES

Salinity

The difference between fatty acid compositions of marine and freshwater fish has been noted by several authors. Some examples of fatty acid patterns Although these fish lipids are higher in w 3 fatty acids, it is clear that freshwater fish have higher levels of w 6 fatty acids than marine species.

The average w 6/w 3 ratios are 0.37 and 0.16 for freshwater and marine fish, respectively. Fish in general contain more w 3 than w 6 polyunsaturated fatty acids (PUFA) and should have a higher dietary requirement for w 3 PUFA; thus the dietary EFA requirement of marine fish for w 3 PUFA may be higher than that of freshwater fish. The same type of difference in the w 6/w 3 ratio between freshwater and seawater is seen when some species of fish migrate from oceans to streams or vice versa. The PUFA ratio of sweet smelt (Plecoglosus altivelis) changes drastically in only one month as they migrate from the sea to a freshwater river.

A similar but reverse change occurs in the masu salmon (Oncorhynchus masu) as they migrate from freshwater to seawater. Even within the same species of fish, the salinity of the water seems to cause a dramatic change in the fatty acid pattern. The difference between marine and freshwater fish may be due simply to differences in the fatty acid content in the diet or it may be related to a specific requirement in fish related to physiological adaptations to the environments.

The phospholipids are generally considered to be structural or functional lipid, being incorporated to a large extent in the membrane structure of cell and subcellular particles. The triglycerides are more often storage lipids and reflect the fatty acid composition of the diet to a greater extent than do the phospholipids.The fatty acid compositions of the triglyceride and phospholipid fractions of fish lipids are presented. It can be seen that the effect of changing environment on the fatty acid composition of the phospholipid is as great in the case of salmon and considerably greater in the case of the sweet smelt, than it is on the triglyceride composition.

Rainbow trout on diets containing either corn oil, which is high in w 6 but low in w 6 PUFA, showed a higher mortality and growth reduction in seawater than in freshwater over the twelve-week feeding period.

Temperature

There are several other factors besides the salinity of the water which affect the fatty acid composition and especially the PUFA of fish. The salmonids are mostly cold-water fish. The fatty acids from a number of marine animals from temperate and arctic waters show some significant differences in the general pattern; unfortunately analysis included fatty acids longer than 20:1. There are a number of other experiments demonstrating the effect of environmental temperature on fatty acid composition of aquatic animals. The general trend Towards higher content of long chain PUFA at lower temperatures is quite clear. The ratio decreases with a decrease in temperature.

If the trends in fatty acid composition can be taken as clues to the EFA requirements of fish, the requirement would be greater for fish raised at lower temperatures. Fish raised in warmer waters, such as common carp, channel catfish and tilapia may do better with a mixture of fatty acids.

EFFECTS OF DIET

Some of the fatty acid compositions may be seriously affected by the dietary lipids. The mosquito fish and guppies were fed trout pellets which had an ratio of 2.75. The catfish were fed diets supplemented with either beef tallow or menhaden oil, with ratios of and respectively. These fish were able to alter the dietary ratio in favour of fatty acid incorporation into the flesh lipids even at the highest temperature.

Commercially available trout pellets are often low in w 3 PUFA and high in w 6 fatty acids. It is important not to ignore the effect of dietary lipid composition on fatty acid composition of fish fed artificial diets. When the dietary ratio is very high in fatty acids supplied by animal lard or vegetable oils, there is a tendency for fish to alter the ratio of PUFA incorporated in favour of fatty acids.When the dietary oil is a fish oil high in to fatty acids, there is little change in the ratio of lipids incorporated into the fish. This is further suggestive evidence of an EFA requirement of fish for PUFA.

Seasonal Variation

Seasonal variations in the fatty acid composition of fish species have often been reported. Seasonal changes have been observed in total lipid and iodine values of herring oils. The iodine value or degree of unsaturation of the oil was minimal in April and maximal in June. The great increase in unsaturation corresponded to the onset of feeding in spring. The absence of a gas liquid chromatograph (GLC) at the time precluded identification of changes in individual fatty acids.

Flesh and viscera lipid content of the sardine Sardinops melanosticta vary from 3.9 to 10.77 percent and from 10.9 to 38.3 percent, respectively. The fatty acids of principal interest with respect to EFA metabolism are 20:4 20:5 and

22:6 3. There was considerable variation in all of these fatty acids in both neutral and polar lipid from both tissues. In the flesh, the 20:4 was consistently higher in the neutral lipid than in the polar lipid. The total consistently higher in polar lipid than in the neutral lipid. Thus, in spite of the major fluctuations in fatty acids caused by changes in diet and temperature throughout the seasons, there was a consistent preferential incorporation of PUFA of the w 3 series into the polar or phospholipid fraction of the lipids.

One of the best clues to the EFA requirements of a species can be gained from the fatty acid composition of the lipids incorporated into the offspring or egg. The act of reproduction or spawning also has a significant effect on the seasonal fluctuation of lipids in fish. Fatty acid composition of fish egg lipids is probably distinctive for each species and contains increased levels of compared to the liver lipids of the same female fish

Elevated levels of reduced in the ovary occurred compared to mesenteric fat of Pacific sardine fed a natural copepod diet. The blood fatty acids of the sardine fed the natural diet were similar to those of the ovary. When the sardines were fed trout food, both the blood and mesenteric fat responded to the diet with elevated 18:2w 6 and reduced 20:5 arid 22:6. The effect of the diet on ovary fatty acid content was considerably less, as relatively high levels of 20:5 and 22:6 were retained. The ovary lipids of the sweet smelt show an increase in 16:0 and a reduction in the PUFA, especially in the phospholipids, compared to the

lipids from the flesh of fish caught at the same time of year. The ratio of the ovary was lower than that of the flesh lipids, 0.21 and 0.17 for the ovary compared to 0.31 and 0.20 for the triglycerids and phospholipids of the flesh, respectively. The hatchability of eggs from common carp fed several different formulated feeds is greatly reduced when the 22:6ω3 of the egg lipids is less than 10 percent. Further, the muscle, plasma and erythrocyte fatty acid compositions are more affected by dietary lipid than those of the eggs.

The EFA requirements of a number of species of fish have been investigated in nutritional studies. The fish themselves have given ample evidence for EFA preference by the types of fatty acids they incorporate into their lipids. Fish, in general, tend to utilize ω3 over ω6.

This is especially observed when the dietary lipids are high in ω6, as the fish tend to alter the w 6/ω3 ratio Towards the ω3 fatty acids in the tissue lipids. The lipids of the egg must satisfy the EFA requirement of the embryo until it is able to feed. The fatty acid composition data suggest that the ω3 requirement is greater in seawater than in freshwater and higher in cold water than in warm water.

BODY LIPID COMPOSITION AND DIETARY LIPID REQUIREMENTS

Detailed information is still lacking on the dietary lipid requirements of many species of fish, but there is an abundance of information on the fatty acid composition of fish oils. Information on the lipid composition of fish can be used to make some guesses about dietary lipid requirements. Linolenic acid (18:3ω3) resulted in some sparing action and growth promotion in rats and fatty acids of the ω6 EFA prevented all of the EFA-deficiency symptoms.

Research with homeothermic land-dwelling animals showed that the ω6 series of fatty acids are the "essential fatty acids", while the ω3 series are considered to be non-essential or only have a partial sparing action on EFA-deficiency.The series of fatty acids have been shown to be essential to enough species that it began to become accepted that these are the essential fatty acids for all animals. It was assumed by many that fish also required ω6 fatty acids. Many researchers began by supplementing fish diets with vegetable oils, such as corn, peanut, or sunflower oil, which were rich in linoleic acid. The main sympton observed during the development of EFA deficiency in chinook salmon fed fat-free diets was a marked depigmentation that can be prevented by addition of 1 percent trilinolein, but not by 0.1 percent linolenic acid.

Although the ω6 fatty acids are considered to be essential, one of the general characteristics of fish oils is the low levels of ω6 series fatty acids and the higher levels of ω3 type fatty acids. There is evidence that polyunsaturated fatty acids (PUPA) of the ω3 series, which are present in relatively large concentrations in fish oil, play the role of essential fatty acid for fish.

POLYUNSATURATED FATTY ACID REQUIREMENTS OF FISH

When a test diet containing 13 percent corn oil and 2 percent cod-liver oil was fed to rainbow trout, subsequent deletion of the cod-liver oil from the diet depressed growth and produced some kidney degeneration which might be attributed to a lack of sufficient ω3 PUFA present in significant quantities in cod-liver oil. Dietary fish oil is superior to corn oil in promoting growth of rainbow trout (Salmo gairdneri) and the yellow-tail (Seriola guingueradiata). Dietary linolenic acid or ethyl linolenate (18:3 ω3) gives a positive growth response for rainbow trout which may be attributed to a dietary requirement for ω3 fatty acids.

ESSENTIAL FATTY ACID REQUIREMENTS OF FISH

One of the most widely accepted theories explaining the presence of such high levels of 20:5 ω3 and 22:6 ω3 fatty acids in fish oils is related to the effect of unsaturation on the melting point of a lipid. The greater degree of unsaturation of fatty acids in the fish phospholipids allows for flexibility of cell membrane at lower temperatures. The ω3 structure allows a greater degree of unsaturation than the ω6 or ω9. This theory is consistent with the fact that cold water fish have a greater nutritional requirement for ω3 fatty acids, while the EFA requirement of some warm water fish can be satisfied by a mixture of w 6 plus ω3.

Rainbow Trout

Rainbow trout, a cold water fish, requires ω3 fatty acids as EFA in the diet. The EFA requirement can be met by 1 percent 18:3ω3 in the diet. Inclusion of 18:2ω6 in the diet may result in some improvement in growth and feed conversion compared to EFA deficient diets; however, the ω6 fatty acids will not prevent some EFA deficiency symptoms such as the "shock syndrome". Although it is clear that rainbow trout require ω3 fatty acids, it remains to be shown conclusively whether some dietary level of ω6 fatty acid is essential.

Either 20:5ω3 or 22:6ω3 is superior to 18:3ω3 in an EFA value for rainbow trout and the former two fatty acids in combination are superior to either alone. This is consistent with data for mammals, where 20:4ω6 has higher EFA value than 18:2ω6. The superior nutritional value of C-20 and C-22 carbon w 3-PUFA is further supported by the excellent growth promoting effects of dietary fish oils such as pollock liver oil and salmon oil for rainbow trout.

Channel Catfish

One of the most important warm water fish in North America is the channel catfish. The quantitative EFA requirement of the catfish has not yet been determined. However, the evidence is convincing that the w 3 requirement is not as great as that of rainbow trout. Analysis of fatty acids of lipids from catfish purchased at five processing plants showed very low levels of 20:4ω6, 20:5ω3 and 22:6ω3; 0.8-5.5, 0.2-1.3 and 0.6-6.1 percent of the total fatty acids, respectively. It was shown that corn oil added to a semipurified casein based diet initially resulted in a positive growth response and protein sparing, but later growth inhibition was observed.

The apparent repressive effects of corn oil may be due to its 18:2ω6 content since 20:5ω3 and 22:6ω3 present in menhaden oil had no apparent detrimental effects. The growth suppressing effects of 18:2ω6 were also noted when 3 percent corn oil was added to 3 percent beef tallow and 3 percent menhaden oil.

The growth suppression caused by unsaturated fatty acids does not appear to be limited to ω6 fatty acids. Linseed oil (high in 18:3ω3) in the diet of catfish resulted in growth suppression similar to that caused by corn oil compared with dietary beef tallow, olive oil and menhaden oil.

The Common Carp

The picture for another warm water fish, the common carp (Cyprinus carpio) is much clearer than that for the channel catfish. This fish has an EFA requirement for both ω3 and ω 6 fatty acids. The best weight gains and feed conversions are obtained in fish receiving a diet containing both 1 percent 18:2 ω6 and 1 percent 18:3ω3. With the carp, 20:5ω3 and 22:6ω3 at 0.5 percent of the diet are superior to 1 percent of 18:3ω3. Carp fed a fat-free, or EFA deficient, diet incorporated high levels of 20:3ω9 in their lipids, especially in the phospholipids.

The Eel

The eel, another warm water fish, has a requirement for both ω3 and w 6 fatty acids. Corn oil (high in ω6) and cod liver oil (high in ω3)in a 2:1 mixture are most favourable for the growth of eels. The eel requires ω6 and ω3 in the same proportion as the carp, but at a lower level in the diet; namely, 0.5 percent of each, rather than 1.0 percent of each PUFA.

The Plaice

The plaice becomes depleted of both ω3 and ω6 PUFA when fed a fat-free diet. The addition of 12:0 and 14:0 to the diet result in the synthesis of saturated and monoenoic fatty acids of chain lengths up to C18; however, increased levels of 20:3ω9 noted in trout and mammals have not been reported in plaice. Plaice fed dietary 18:2ω6 and 18:3ω3 will not produce significant amounts of 20:4ω6, 20:5ω3, or 22:6ω3.

The Turbot

The growth of turbot is much better with ω3 PUFA than with ω6 or saturated fat (hydrogenated coconut oil) in the diet. The turbot also appears to be unable to convert dietary 18:2ω6 to 20:4ω6 when fed corn oil, or to convert endogenous 18:1ω9 to 20:3ω9 when fed the EFA deficient diet. Although it appears to have an EFA requirement for ω3 fatty acids such as are present in cod liver oil, this requirement is not satisfied by 18:3ω3. The chain elongation and desaturation of 18:lω9, 18:2ω6, or 18:3ω3 has been found to be very limited (3-15 percent) in turbot compared to the rainbow trout where 70 percent of the 18:3ω3 was converted to 22:6w 3. The required level of long-chain ω 3 fatty acids for turbot is at least 0.8 percent of the diet.

The Red Sea Bream

The red sea bream (Chrysophyrys major) grows better when the dietary lipid is of marine origin (pollock residual oil) rather than a vegetable oil (such as corn oil). The EFA requirement of the red sea bream is not satisfied by either linoleic acid of corn oil or supplemented linolenate. A mixture of 20:5ω3 and 22:6ω3 supplemented to the corn oil diet has been shown to be effective in improving growth and condition of these fish. Thus, even in warm water, marine fish seem to require not just ω3 fatty acids but fatty acids of 20 to 22 carbon-chain length. A direct correlation between feed efficiency and the 18:1 level in the lipids of the red sea bream has been postulated.

Other Species

Among warm water marine fish, mullet and fundulus possess the ability to chain, elongate and desaturate 18:2ω6 or 18:3ω3 PUFAs. The process is, however, inhibited in fundulus by high levels (about 5 percent) of these PUFAs of 18:2ω6 or 18:3ω3 in the diet.

FATTY ACID METABOLISM IN FISH

It appears that high levels of 18-carbon ω6 or ω3 fatty acids inhibit the synthesis and metabolism of 18:lω9. It is interesting to note that the channel catfish, which also exhibits negative growth response to dietary 18:2ω6 or 18:3ω3, incorporates very high levels of 18:1 into its body lipids. The inclusion of either 18:2ω6 or 18:3ω3 in the diet reduces the levels of 18:1 fatty acids in body lipids. A similar reduction has also been observed in red sea bream liver phospholipid when either of the PUFAs is added to the diet.

The competitive inhibition of chain elongation and desaturation of members from one series of fatty acids for members of another series is well established, with ω3 > ω6 > ω 9 being the usual order of potency for inhibition.

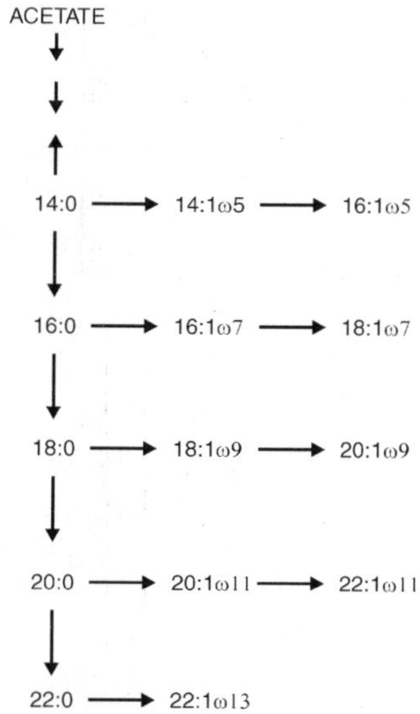

Fig. Flow Diagram for Fatty Acid Synthesis Mechanisms in Fish-Saturated and Monoenoic Fatty Acids

The pathways of fatty acid metabolism have been reviewed by Mead and Kayama (1967). Fish are able to synthesize, de novo from acetate, the even-chain, saturated fatty acids. Radio tracer studies have shown that fish can convert 16:0 to the ω7 monoene and 18:0 to the ω9 monoene. The ω5, ω11 and ω3 monoenes are proposed based on the identification of these isomers in the monoenes of herring oil.

Fish are unable to synthesize any fatty acids of the ω6 and u3 series unless a precursor with this ω structure is present in the diet. There is competitive inhibition of the elongation desaturation of fatty acids of one series by members of the other series. The ω3 fatty acids are the most potent inhibitors, the ω9 are the least.

The ability to elongate and desaturate fatty acids is not the same in all species of fish, as was noted earlier. The turbot was able to desaturate and elongate only 3-15 percent of 18:1ω9, 18:2ω6, or 18:3ω3, when given the C^{14} labelled fatty acid; in the rainbow trout, 70 percent of the label from 18:3ω3 (C^{14}) was found in 22:6ω3.

The essential fatty acids are not unique in their ability to supply energy. The b -oxidation of fatty acids in fish is basically the same as in mammals. The EFA and saturated and monoenoic fatty acids are all equally utilized by fish for energy production.

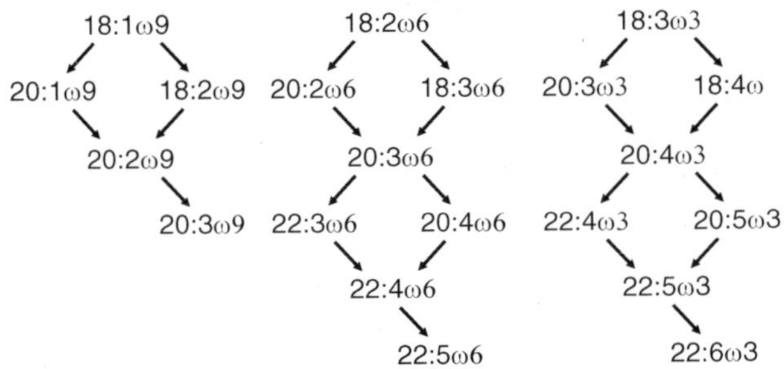

Fig. Flow Diagram for Fatty Acid Synthesis Mechanisms in Fish-Polyunsaturated Fatty Acids

Increased swelling rates of liver mitochondria occur in rainbow trout fed diets deficient in ω3 fatty acids. It is possible that EFA plays an important role in the permeability as well as the plasticity of membranes. The role of ω3 fatty acids in membrane permeability may be one of the factors accounting for differences in content of this family of fatty acids between freshwater and marine fish. Fish mitochondria with high levels of the ω3 PUFA and very low levels of ω6 fatty acids are very similar to mammalian mitochondria with respect to cytochrome content, b -oxidation of fatty acids, operation of the tricarboxylic acid cycle, electron transport and oxidative phosphorylation.

The ω3 PUFA may play the same role in fish that the ω 6 fatty acids play in rats. The EFA play another role in the mitochondria. In addition to their importance in membrane structure, the EFA are important in some enzyme systems. Unsaturated fatty acids play an important role in the transportation of other lipids. It has been repeatedly shown that feeding PUFA will lower the cholesterol levels in animals with above-normal blood lipid and cholesterol levels. Fish oils are more effective in lowering cholesterol levels than are most dietary lipids.The major portion of the fatty acids absorbed across the intestinal mucosa are transported as protein-lipid complexes stabilized by phospholipids. The low body temperature in fish probably results in a greater importance for unsaturation in transport of lipids than in homeothermic animals.

NEGATIVE ASPECTS OF LIPIDS IN FISH NUTRITION

The requirement by fish for PUFA of the ω 3 series creates problems with respect to feed storage. These types of fatty acids are very labile on oxidation. The products of lipid oxidation may react with other nutrients such as proteins, vitamins, etc. and reducing the available dietary levels or the oxidation products may be toxic. The effect of oxidized lipids on dietary proteins, enzymes and amino acids have been demonstrated. The use of oxidized menhaden oil in the diets of swine and rats caused decreased appetite, reduced growth, yellowish-brown pigmentation of depot fat and decreased haemoglobin and haematocrit levels. The negative effects of the oxidized fish oils were reversed by the addition of alpha-tocopherol acetate or ethoxyoquin to the diet.

Much of the use of vegetable oils in fish diets in the 1950s and 1960s might, in part, have been based on their greater stability in prepared diets. It has been demonstrated that rancid herring and hake meals in fish feeds caused dark colouration, anaemia, lethargy, brown-yellow pigmented liver, abnormal kidneys and small gill clubbing in chinook salmon.

The symptons can be alleviated by addition of alpha-tocopherol to the diets containing rancid fish meals. The addition of vitamin E would prevent the toxic or negative effects of adding 5 percent highly oxidized salmon oil to the diet of rainbow trout. This same sparing effect of alpha-tocopherol can also apply to rancid carp feed. The positive nutritional value of ω 3 fatty acids in fish lipids for fish feeds can become a negative factor if adequate care is not taken in the preparation and storage of feeds. Only fresh oils with low peroxide values should be used in feeds. Fish feed ingredients such as fish meals should be protected against oxidation.The level of vitamin E added to the diet should be increased as the PUFA level is increased. The finished feed, if possible, should be stored in air tight containers at reduced temperatures with minimum exposure to UV radiation and other factors accelerating the rate of lipid oxidation. The problems of rancidity or antioxidation of lipids in fish feeds should not be ignored.

Carbohydrates

Carbohydrates represent a broad group of substances which include the sugars, starches, gums and celluloses. The common attributes of carbohydrates are that they contain only the elements carbon, hydrogen and oxygen and that their combustion will yield carbon dioxide plus one or more molecules of Water.

The simplest carbohydrates are the three-carbon sugars which figure importantly in intermediary metabolism and the most complex are the naturally occurring polysaccharides, primarily of plant, origin. In the diet of animals and fish, two classes of polysaccharides are significant:

- Structural polysaccharides which are digestible by herbivorous species -cellulose, lignin, dextrans, mannans, inulin, pentosans, pectic acids, algic acids, agar and chitin; and

- Universally digestible polysaccharides-principally starch.

Carbohydrates make up three-fourths of the biomass of plants but are present only in small quantities in the animal body as glycogen, sugars and their derivatives. Glycogen is often referred to as animal starch because it is not present in plants. Derived mono-saccharides such as the sugar acids, amino sugars and the deoxysugars are constituents of all living organisms.

CLASSIFICATION AND CHEMISTRY

Carbohydrates are classified generally according to their degree of complexity. Hence, the free sugars such as glucose and fructose are termed monosaccharides; sucrose and maltose, disaccharides; and the starches and celluloses, polysaccharides. Carbohydrates of short chain lengths such as raffinose, stachyose and verbascose, which are three, four and five sugar polymers respectively, are classified as oligosaccharides.

Pentoses

Pentoses are five-carbon sugars seldom found in the free state in nature. In plants they occur in polymeric forms and are collectively known as pentosans. Thus, xylose and arabinose are the constituents of pentosans present in plant fibres and vegetable gums, respectively.

As the sugar moieties in nucleic acids and riboflavin, ribose and deoxyribose are indispensable constituents of the life process. D-ribose has the following chemical structure:

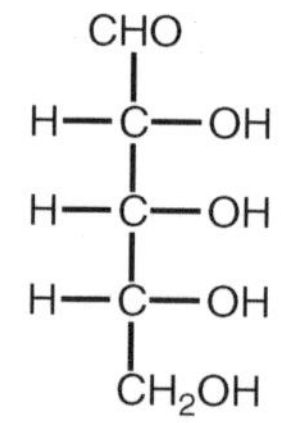

Hexoses

The hexoses comprise a large group of sugars. Principal among these are: glucose, fructose, galactose and mannose. While glucose and fructose are found free in nature, galactose and mannose occur only in combined form. The hexoses are divided into aldoses and ketoses according to whether they possess aldehydic or ketonic groups.

Thus, glucose is an aldo sugar and fructose is a keto sugar. The presence of aymmetric centres in all sugars with three or more carbon atoms gives rise to stereoisomers. Galactose and mannose are stereoisomers of glucose which, theoretically, is only one of 16 stereoisomers. Because the ketohexoses have only three asymmetric centres, fructose is one of eight stereoisomers. The chemical configurations of the four hexoses mentioned are as follows:

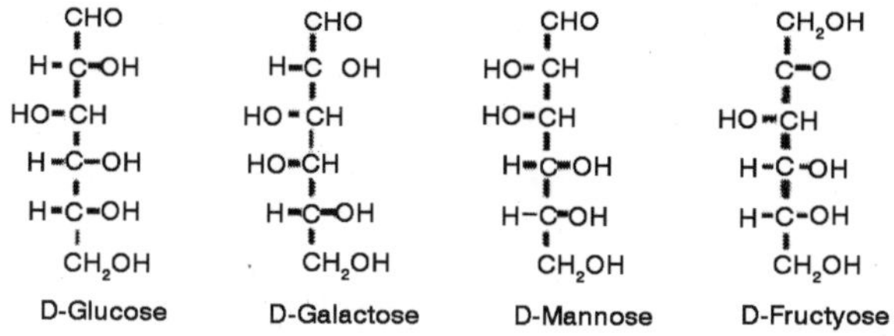

A general phenomenon, known as mutarotation, is observed in a variety of pentoses and hexoses as well as in certain disaccharides. For example, it has been established that two isomers of D-glucose exist, hence requiring an additional asymmetric centre in this sugar.

It became apparent that D-glucose and most other sugars have cyclic structures. The position of the hydroxyl group in relation to the ring oxygen characterizes this additional configurations modification.

By convention, the positioning of the hydroxyl group on carbon atom 1 on the same side of the structure as the oxygen ring describes a -modification; and, the positioning of the same hydroxyl group on the opposite side of the ring oxygen describes a b -modification.

Carbohydrases, which catalyse the hydrolysis of glycosidic linkages of simple glycosides, oligosaccharides and polysaccharides often exhibit specificity with regard to substrate configuration.

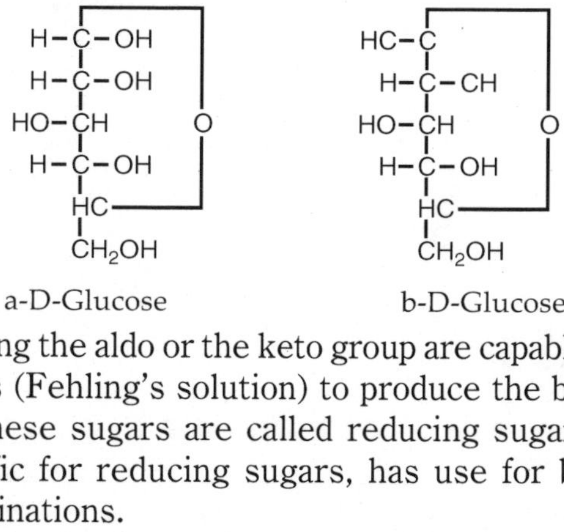

a-D-Glucose b-D-Glucose

Sugars containing the aldo or the keto group are capable of reducing copper in alkaline solutions (Fehling's solution) to produce the brick-red colouration of cuprous ions. These sugars are called reducing sugars and the reaction, although not specific for reducing sugars, has use for both qualitative and quantitative determinations.

Glucose is widely distributed in small amounts in fruits, plant juices and honey. It is commercially produced by the acid or enzyme hydrolysis of grain and root starches. Glucose is of special interest in nutrition because it is the end-product of carbohydrate digestion in all non-ruminant animals including fish.

Fructose is the only important ketohexose and is found in the free state alongside glucose in ripening fruits and honey. Combined with glucose it forms sucrose. Fructose is somewhat sweeter than sucrose and is produced in increasing quantities commercially as a sweetener. Galactose occurs in milk in combination with glucose. It is also present in oligo-saccharides of plant origin, in combination with both glucose and fructose. Mannose is present in some plant polysaccharides collectively termed mannans.

Disaccharides

Disaccharides are condensation products of two molecules of monosaccharides. Sucrose is the predominant disaccharide occurring in the free form and is the principal substance of sugar cane and sugar beet. It is also formed during germination of legume seeds. Other common disaccharides are maltose and lactose. Maltose is a dimer of glucose and lactose is a copolymer of galactose and glucose. The two molecules of glucose in maltose are held together in an a -1,4 glycosidic linkage whereas the two hexose entities of galactose are linked at the b -1,4 position. Glucose and fructose are combined in an a -1,2 linkage in sucrose.

a-Maltose

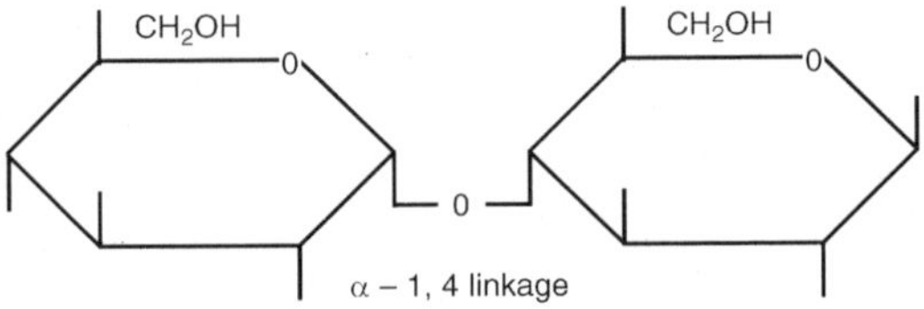

b -Lactose

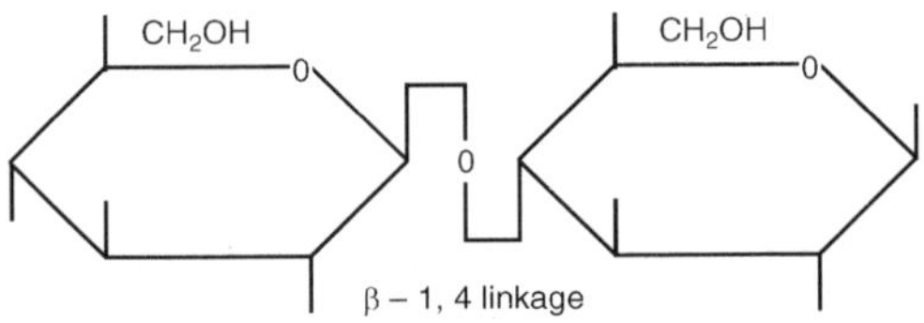

Sucrose

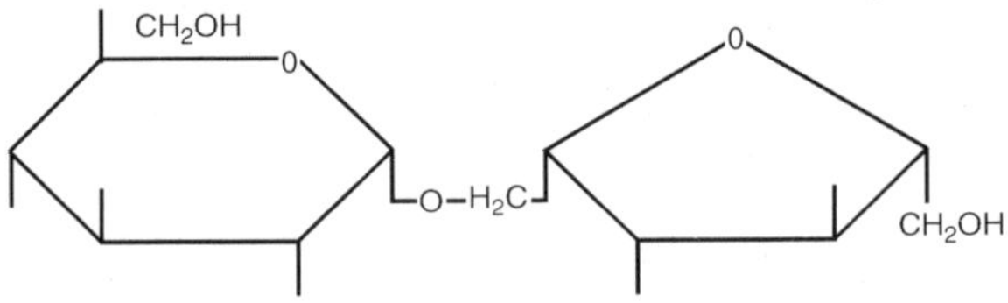

Oligosaccharides

The Oligosaccharides raffinose, stachyose and verbascose are present in significant quantities in legume seeds. Raffinose, which is the most widespread among the three, consists of one molecule of glucose linked to a molecule of sucrose at the a -1,6 position. Its abbreviated chemical name is a -D-Gal. Further chain elongation at the galactose end with another galactose molecule will yield stachyose.

These galactose-galactose linkages are all at the a-l,6 position and digestion of these Oligosaccharides by animals requires a highly specific enzyme not elaborated by the animals themselves but by certain bacteria present in the animals guts. The gradual disappearance of oligosaccharides from the cotelydons of legume seeds during germination is part of an intricate process beginning with uptake of water by the seed. This uptake of moisture releases gibberellic acid which in turn activates the DNA in the seed, thereby triggering the life cycle of the plant. The DNA directs the production of a -galactosidase which is required for the hydrolysis of these Oligosaccharides. Any interference of the DNA transcription process blocks enzyme production and will be evidenced by continued senescence of the seed and persistence of oligosaccharides in the seed cotelydons.

Polysaccharides

The polysaccharides represent a large group of complex carbohydrates which are condensation products of undetermined numbers of sugar molecules. The various subgroups are rather ill-defined and there is a lack of agreement on their classification. Most polysaccharides are insoluble in water. Upon hydrolysis with acids or enzymes they eventually yield their constituent monosaccharides.

Starch is a high molecular weight polymer of D-glucose and is the principal reserve carbohydrate in plants. Most starches consist of a mixture of two types of polymers, namely; amylose and amylopectin. The proportion of amylose and amylopectin is generally one part of amylose and three parts of amylopectin. Enzymes capable of catalyzing the hydrolysis of starch are present in the digestive secretions of animals and fish within their cells.

The a-amylases which are found virtually in all living cells cleave the a -D-linkages at random and bring about an eventual total conversion of the starch molecule into the reducing sugars. The principal a -amylases of animal origin are those produced in the salivary gland and the pancreas. Starch is insoluble in water and is stained blue by iodine.

Glycogen is the only complex carbohydrate of animal origin. It exists in limited quantities in liver and muscle tissues and acts as a readily available energy source. Dextrins are intermediate compounds resulting from incomplete hydrolysis or digestion of starch. The presence of a -D-linkages in amylopectin and the inability of a -amylase to cleave these bonds give rise to low molecular weight carbohydrate segments called limit dextrins. These residues are acted upon primarily by acidophilic bacteria in the digestive tract. Cellulose is made up of long chains of glucose units held together by β-D-linkages.

The enzymes which cleave these linkages are not ordinarily present in the digestive secretions of animals and fish although some species of shellfish are believed to elaborate cellulase, the enzyme which catalyzes the hydrolysis of cellulose. Cellulase producing micro-organisms present in the gut of herbivorous animals and fish impart to their host animals the ability to utilize as food the otherwise indigestible cellulose. Other complex polysaccharides in common occurrence are the hemicelluloses and pentosans. Hemicellulose represents a group of carbohydrates including araban, xylan, certain hexosans and polyuronides.These substances are generally less resistant to chemical treatment and undergo some degree of enzymatic hydrolysis during normal digestive processes. Pentosans are polymers of either xylose or arabinose as constituents of plant structural material and vegetable gums, respectively.

CARBOHYDRATE METABOLISM IN FISH

Much of the carbohydrates that enter the diets of animals, including fish, is of plant origin. Carnivorous fish like the Atlantic salmon and the Japanese yellowtail, therefore, deal with little carbohydrate. Indeed, experiments have shown that these species are ill-equipped to handle significant quantities of raw carbohydrate, in their diets. On the other hand, omnivores such as the common carp and the channel catfish are able to digest fair amounts of carbohydrates in their diets. The grass carp, a herbivore, subsists primarily on a vegetarian diet.

Digestion, Absorption and Storage

The ability of animals to assimilate starch depends on their ability to elaborate amylase. All species of fish have been shown to secrete a -amylase. It has also been demonstrated that activity of this enzyme was greatest in herbivores. In carnivores such as the rainbow trout and sea perch, amylase is primarily of pancreatic origin whereas in herbivores the enzyme is widespread throughout the entire digestive tract.

In Tilapia mossambica the pancreas has been shown to be the site of greatest amylase activity followed by the upper intestine. Although the digestion of starch and dextrin by the carnivorous rainbow trout was shown to decrease progressively as levels of the carbohydrates were increased beyond the 20 percent level, the fish could effectively utilize up to 60 percent glucose, sucrose or lactose in the diet. This demonstrates that, contrary to earlier belief, carnivorous fish are capable of efficiently utilizing simple carbohydrate as a primary energy source.

The crystalline structure of starch appears also to influence its attack by amylase as evidenced by the two-fold increase in metabolizable energy content of fully cooked (gelatinized) maize in feeding trials with channel catfish. Rainbow trout have also been shown to have a higher tolerance for carbohydrate (present as wheat starch) in the diet when it was cooked. The process of gelatinization involves both heat and water. If an aqueous suspension of starch is heated, the granules do not change in appearance until a certain critical temperature is reached. At this point some of the starch granules swell and simultaneously lose their crystallinity. The critical temperature is that at which hydrogen bonds of the starch molecule loosen to permit complete hydration, leading to a phenomenon known as "Swelling".

Alpha-amylase, promotes a more or less random fragmentation of the starch molecule by hydrolyzing at the a -D- glucosidic bonds in the inner and outer chains of the compound. The result of complete hydrolysis of the amylose component are maltose and D-glucose, while the amylopectin component is reduced to maltose, D-glucose and branched limit dextrins. As a consequence of these action patterns by a -amylase on starch, other enzymes are needed for complete hydrolysis of starch to D-glucose in fish. In this regard, it has been demonstrated that even the carnivorous sea bream possess the ability to digest maltose. On the other hand, cellulase and a -galactosidase have not been shown to be secreted by fish although cellulase of bacterial origin is present in the gut of most species of carps.

The lack of a -galactosidase may partly explain the poor response by fish to dietary soybean meal which contains significant levels of the galactosidic oligosaccharides raffinose and stachyose. As has been pointed out earlier, these oligosaccharides do undergo enzymatic hydrolysis during the germination process to yield galactose and sucrose.

It would, therefore, appear that the nutritive value of soybean meal will be enhanced if the bulk of this indigestible starch is first transformed. This can be achieved by soaking the beans for 48 hours prior to processing for meal production. It should also be pointed out that the nutritive value of pulses and other legume seeds can likewise be improved for fish since oligosaccharides constitute a large portion of the carbohydrates in legume seeds. Data on glucose absorption by fish are scanty. Work with goldfish has shown that active transport of glucose is coupled with Na^+ transport as in most mammals. It is generally believed that absorption takes place on the mucosal surface of intestinal cells.The mono-saccharides which result from carbohydrate digestion consist primarily of glucose, fructose, galactose, mannose, xylose and arabinose. Although the rates of absorption of these sugars have been determined for many land mammals, similar information for fish is not available. Glucose does not appear to be a superior energy source for fish over protein or fat although digestible carbohydrates do spare protein for tissue building. Also, unlike in mammals, glycogen is not a significant storage depot of energy despite evidence of an active and reversible Emden-Meyerhoff pathway in fish. The more efficient metabolism of amino acids over glucose for energy could be due to the ability of fish to excrete nitrogenous waste as ammonia from their gills without the high cost of energy in converting the waste to urea.

Other Factors Affecting Metabolism

Apart from genetic adaptation, climatic factors also play an important role in carbohydrate metabolism in fish. Acclimation in fish, in essence, reflects enzyme acclimation, since the animal's ability to survive depends largely upon its ability to carry out normal metabolic functions. Some enzymes for metabolic acclimation show good compensation while others do not. The enzymes associated with energy liberation (enzymes of glycolysis, pentose shunt, tricarboxylic acid cycle, electron transport and fatty acid oxidation) exhibit temperature compensation whereas, those enzymes dealing largely with the degradation of metabolic products show poor or reverse compensation.

Table: **Enzymes Subject to Metabolic Acclimation**

Enzymes exhibiting compensation	*Enzymes exhibiting reverse* or no compensation
phosphofructokinase	catalase
aldolase	peroxidase
lactic dehydrogenase	acid phosphatase
6-phosphogluconate dehydrogenase	D-amino acid oxidase
succinic dehydrogenase	Mg-ATP ase
malic dehydrogenase	choline acetyl transferase
cytochrome oxidase	acetylcholine esterase
succinate-cytochrome C reductase	alkaline phosphatase
NAD-cytochrome C reductase	allantoinase
aminoacyl transferase	uricase
Na-K-ATPase	amylase
protease	lipase
	malic enzyme
	glucose-6-phosphate dehydrogenase

It is interesting to note that two key enzymes involved in carbohydrate metabolism, amylase and glucose-6-phosphate dehydrogenase, together with an enzyme involved in fat digestion, lipase, show no temperature compensation. It is not certain if this is in any way connected with the cessation of feeding by fish at low temperatures. The molecular mechanism of thermal acclimation are not well understood and may consist of changes in synthesis or amounts of a given enzyme. Differences in kinetics, changes in the proportion of isoenzymes suitable for particular temperatures and changes in co-factors such as lipids, co-enzymes, or other factors such as pH and ions may be important in the animal's adjustment to temperature changes.

Energy Transformation

Despite species differences in the tolerance of dietary carbohydrates it is generally believed that the principal end-product of carbohydrate digestion, glucose, is metabolized in a manner prevailing in all cells, *i.e.*, via the reversible Emden-Meyerhoff pathway. In this pathway, glucose has only one principal fate: phosphorylation to glucose-6-phosphate. The major metabolic transformations are depicted as follows: Reversible arrows show reaction step or steps catalyzed by same enzymes in both directions.Broken arrows show reactions over many intermediate steps.Paired solid arrows show different enzymes involved in the two directions of the reaction. All transformations proceed with a loss of free energy. Thus, the formation of two moles of lactate from glucose-6-phosphate occurs with free energy change of D G° = -22000 cal/mole. The net result is the formation of four molecules of ATP. A functional reversal of this transformation can only occur via a different sequence requiring the input of six ATP molecules per mole of glucose-6-phosphate recovered.

Cells do not store glucose or glucose-6-phosphate. The readily available storage form is glycogen which is made from glucose-1-phosphate by one pathway and returned by another. Although in mammalian cells glucose-6-phosphate is transformed into fatty acids, such transformation does not appear to take place in fish. Studies with the common carp indicate that the precursor for lipogenesis is citrate formed when amino acids are actively metabolized through the tricarboxylic acid cycle.

Bibliography

A. Solaimalai, N. Ravisankar and B. Chandrasekar: *Farming Systems : Theory and Practice*, International Book Distributors, Delhi, 2005.

A.K. Sarkar, R.S. Singh, M.S. Yadav and C.S. Singh: *Integrated Farming Systems for Sustainable Production*, Agrotech Publishing Academy, Delhi, 2011.

Alan Beeby and Anne-Marie Brennan: *First Ecology : Ecological Principles and Environmental Issues*, Oxford University Press, Delhi, 2008.

Arvind Kumar and Pashupati Kumar Roy: *Environmental Issues and Solutions*, Daya Publication, Delhi, 2008.

B. Suresh Lal: *Global Environmental Issues : Empirical Evidences*, Serials Serial Publication, Delhi, 2010.

B.L. Jana: *Farming Systems*, Agrotech Publication Academy, Delhi, 2014.

B.R. Selvamani and R.K. Mahadevan: *Fish Farming Systems*, Campus Books International, Delhi, 2008.

Bhavana Trivedi and Sangeeta Jain: *Environmental Issues in India*, Discovery Publishing House, Delhi, 2007.

D. Thyagarajan, A. Ashok, G. Raj Manohar and M. Murugan: *Emerging Opportunities in Alternative Poultry Farming Systems*, Satish Serial Publishing House, Delhi, 2013.

David Norman and Mancolm Douglas: *Farming Systems Development and Soil Conservation*, Daya Publicaton, Delhi, 2007.

Ganesh Kawadia and Kanhaiya Ahuja: *Environmental Issues of Development*, The Associated Publication, Delhi, 2006.

George A. James: *Ethical Perspectives on Environmental Issues in India*, APH Publication, Delhi, 1999.

Helen Macbeth: *Human Population Dynamics: Cross-Disciplinary Perspectives*, Cambridge University Press, New, York, 2003.

K. Gopakumar and A.D. Diwan: *Coastal Fisheries and Aquaculture Management, Part I-II*, Narendra Publication, Delhi, 2010.

K. Jaganmohan Reddy and S. Veeraiah: *Global Environmental Issues*, Aavishkar Publication, Delhi, 2010.

Lingaraj Patro: *Global Warming Climate Change and Environmental Issues*, Manglam Publication, Delhi, 2012.

Mallikarjuna Raju: *Human Population and the Environment : Changing Climate Context*, Akhand Publication, Delhi, 2012.

Pratap Narain; M P Singh; Amal Kar; S Kathju and Praveen Kumar: *Diversification of Arid Farming Systems*, Scientific Publication, Delhi, 2008.

Pratap S. Birthal, Anjani Kumar and Laxmi Tiwari: *Livestock in Different Farming Systems in India*, Agricultural Economics Research Association, 2002.

R.K. Tiwari: *Environmental Issues and Sustainable Development*, ABD Publication, Delhi, 2006.

S Felix: *Aquaculture Management Techniques*, Narendra Publication, Delhi, 2007.

S. B. Agnihotri: *Aquaculture Management and Technology*, Swastik Publication, Delhi, 2013.

S. Felix: *Advances in Shrimp Aquaculture Management*, Daya Publication, Delhi, 2009.

S.C. Kalwar: *Environmental Issues and Sustainable Development*, Pointer Publication, Delhi, 2002.

S.K. Agarwal: *Environmental Issues and Themes*, APH Publication, Delhi, 1997.

S.P. Das Gupta: *Environmental Issues for the 21 Century*, Mittal Publication, Delhi, 2003.

Shagufta: *Cropping and Farming Systems*, A.P.H. Publication, Delhi, 2011.

Umesh C. Goswami and Dilip Kumar: *Aquaculture Management : A Special Release on the Ninety Sixth Session of The Indian Science Congress, 2009*, Narendra Publication, Delhi, 2009.

V.K. Prabhakar: *Global Environmental Issues*, Anmol Publication, Delhi, 2001.

Zahid Husain: *Environmental Issues of North East India*, Regency Publication, Delhi, 2003.

Index